计算机技术开发与应用丛书

Vue.js快速入门与深入实战

杨世文◎编著

清華大學出版社
北京

内 容 简 介

本书系统阐述了 Vue.js 的基本语法、体系结构、原理，以及基于 Vue.js 实现组件化编程的基本思想、第三方组件、工具，最后结合企业项目，即基于 Vue.js 实现了 ShopApp 和权限管理的部分功能。

全书分为四篇：第一篇为 Vue.js 基础（第 1～4 章），第二篇为 Vue.js 组件化编程（第 5～8 章），第三篇为 Vue.js 高级应用（第 9～13 章），第四篇为 Vue.js 实战（第 14、15 章）。书中主要内容包括 Vue.js 基本语法、自定义和使用 Vue.js 组件、组件的过渡和动画、路由、Promise 对象、axios、Vuex 状态管理和 Vue.js 实战等。

本书以浅显易懂的方式讲解 Vue.js 的相关概念、知识点，此外以配套的大量示例辅助说明了相关知识点的应用，最大程度地追求学以致用，以及即学即用的目标。最后用 ShopApp 和权限管理两个项目（模块），全面地讲解了 Vue.js 在实际项目中的真实应用。

本书可作为 Vue.js 初学者的入门书，也可作为从事跨平台移动开发的技术人员及培训机构的参考书。

图书在版编目（CIP）数据

Vue.js 快速入门与深入实战/杨世文编著. —北京：清华大学出版社，2023.1
（计算机技术开发与应用丛书）
ISBN 978-7-302-60444-0

Ⅰ. ①V… Ⅱ. ①杨… Ⅲ. ①网页制作工具－程序设计 Ⅳ. ①TP392.092.2

中国版本图书馆 CIP 数据核字（2022）第 052846 号

责任编辑：赵佳霓
封面设计：吴 刚
责任校对：时翠兰
责任印制：沈 露

出版发行：清华大学出版社
网 址：http://www.tup.com.cn，http://www.wqbook.com
地 址：北京清华大学学研大厦 A 座　　**邮 编**：100084
社 总 机：010-83470000　　**邮 购**：010-62786544
投稿与读者服务：010-62776969，c-service@tup.tsinghua.edu.cn
质量反馈：010-62772015，zhiliang@tup.tsinghua.edu.cn
课件下载：http://www.tup.com.cn，010-83470236
印 装 者：三河市科茂嘉荣印务有限公司
经 销：全国新华书店
开 本：186mm×240mm　　**印 张**：30　　**字 数**：677 千字
版 次：2023 年 1 月第 1 版　　**印 次**：2023 年 1 月第 1 次印刷
印 数：1～2000
定 价：109.80 元

产品编号：093269-01

前言

PREFACE

近年来，无论是在移动端还是在桌面端利用跨平台技术开发 App 都备受欢迎。企业为了节约时间和减少成本，基本上推崇基于框架进行开发，后端基于框架，前端也基于框架。目前比较流行的前端框架有 Angular、React 和 Vue.js(简称 Vue)。其中 Vue.js 用得越来越普遍，而且随着 Vue.js 的热度提升，还涌现了很多优秀的前端组件和框架，例如 Element UI 和 Vue-Element-Admin。为了帮助读者快速地学习、掌握和运用 Vue.js，同时本着学习交流的精神，笔者把 Vue.js 的基础知识及实践经验进行整理和总结，以供大家参考。

本书内容

第 1 章 Vue.js 简介，介绍 Vue.js 的特点、Vue.js 实现的 MVVM 模式及 Vue.js 同其他框架的对比。

第 2 章 Vue.js 快速入门，介绍 Vue.js 的安装配置，并用一个简单的样例，演示 Vue.js 的使用。

第 3 章 Vue.js 基本语法，介绍 Vue.js 中创建对象的方式、插值表达式、事件绑定和响应原理等内容。

第 4 章 compute 属性和 watch 侦听器，介绍 Vue.js 中 compute 属性和 watch 侦听器的定义和使用。

第 5 章组件化编程，介绍使用 Vue.js 自定义和使用组件，以及组件和组件之间的通信等内容。

第 6 章组件的过渡和动画，介绍组件中不同 Vue.js 元素的过渡和动画实现。

第 7 章复用和组合，介绍 Vue.js 组件中自定义和复用功能的各种方式。

第 8 章路由基础，介绍 Vue Router 的安装配置和基本使用。

第 9 章高级 Vue Router，介绍路由导航、路由元信息及获取响应数据和路由懒加载等内容。

第 10 章 Promise 对象，介绍 Promise 对象的基本思想、方法和使用。

第 11 章 axios，介绍 axios 对象的创建、各种使用方法及拦截器等。

第 12 章模板模式开发 Vue.js 应用，介绍 Node.js、webpack 和 Vue-CLI 脚手架等安装配置和使用。

第 13 章 Vuex 状态管理，介绍 Vuex 的核心概念和使用，以及在项目中的运用。

第 14 章 ShopApp 实战，介绍基于 Vue.js 实现前后端分离的 App 应用。

第15章权限管理实战，介绍基于vue-element-admin后端框架，实现前后端分离的权限管理和安全控制模块功能。

阅读建议

本书是一本基础入门加实战的书籍，既有基础知识，又有丰富示例，包括详细的操作步骤，实操性强。每个知识点都配有小例子，力求精简，还提供了完整代码，复制并执行完整代码就可以立即看到效果。这样会给读者信心，在轻松掌握基础知识的同时快速进入实战阶段。

对于没有基础的读者，建议从第1章开始学习，对于有基础的读者，可以挑选适当章节进行学习。第一篇为Vue.js基础(第1～4章)，第二篇为Vue.js组件化编程(第5～8章)，第三篇为Vue.js高级应用(第9～13章)，第四篇为Vue.js实战(第14、15章)。

致谢

感谢笔者的母亲及妻子，在笔者写作过程中承担了全部的家务(包括照顾孩子)，使笔者可以全身心地投入写作。

由于时间仓促，书中难免存在不妥之处，请读者见谅，并提宝贵意见。

杨世文

2022年10月

本书源代码

目录

CONTENTS

第一篇　Vue.js 基础

第二篇 Vue.js 组件化编程

第三篇　Vue.js 高级应用

第四篇 Vue.js 实战

第一篇　Vue.js基础

第1章 Vue.js 简介

1.1 Vue.js 概述

Vue(读音/vjuː/,类似于 view)是一套用于构建用户界面的渐进式框架。与其他大型框架不同的是,Vue.js 被设计为可以自底向上逐层应用。一方面,Vue.js 的核心库只关注视图层,不仅易于上手,还便于与第三方库或既有项目整合;另一方面,当与现代化的工具链及各种支持类库结合使用时,Vue.js 也完全能够为复杂的单页应用提供驱动。

Vue.js 的作者是尤雨溪,他也是 Vue Technology 的创始人,致力于 Vue.js 的研究开发。有这么好的框架,方便前端开发者快速开发应用前端,离不开作者的持续努力和艰辛付出,笔者对此非常感谢。我们可以访问 Vue.js 的官网(https://cn.vuejs.org/),了解 Vue.js 的最新动态。

Vue.js 有以下几个优点。

(1) 体积小,压缩后只有 33KB。

(2) 更高的运行效率:Vue.js 是基于虚拟的 DOM 技术,Vue.js 是一种可以预先通过 JavaScript 进行各种计算并把最终的 DOM 操作计算出来、进行优化的技术。

由于该 DOM 操作为预处理操作,并没有真实地操作 DOM,所以叫虚拟 DOM。

(3) 双向数据绑定:开发者不必去操作 DOM 对象,只需专注实现业务逻辑。

(4) 生态丰富,学习成本低:市场上存在大量成熟、稳定的基于 Vue.js 的 UI 框架和常用组件,开发者可以现拿现用,实现快速开发。

1.2 MVVM 模式

Vue.js 的核心是实现了 MVVM 模式,实现了数据双向绑定,减少了 DOM 操作,使开发者可以将更多精力放在数据和业务逻辑上。

MVVM 是 Model-View-ViewModel 的简写,它本质上是 MVC 的改进版。MVVM 将其中的 View 的状态和行为抽象化,并且将视图 UI 和业务逻辑分开。

(1) M：即 Model(模型)，包括数据和一些基本操作。

(2) V：即 View(视图)，指页面渲染结果。

(3) VM：即 View-Model，指模型与视图间的双向操作(无须开发者干涉)。

在 MVVM 之前，开发者从后端获取需要的数据模型，然后要通过 DOM 操作将 Model 渲染到 View 中，而后当用户操作 View 时，开发者还需要通过 DOM 获取 View 中的数据，然后同步到 Model 中。

而 MVVM 中的 VM 要做的事情就是把 DOM 操作完全封装起来，开发者不用再关心 Model 和 View 之间是如何互相影响的。

(1) 只要 Model 发生了改变，View 上自然就会表现出来。

(2) 当用户修改了 View 之后，Model 中的数据也会跟着改变。

MVVM 模式使开发者从烦琐的 DOM 操作中解放出来，把关注点放在如何操作 Model 上。如图 1.1 所示，描述了 View、Model 和 View-Model 之间的响应关系。

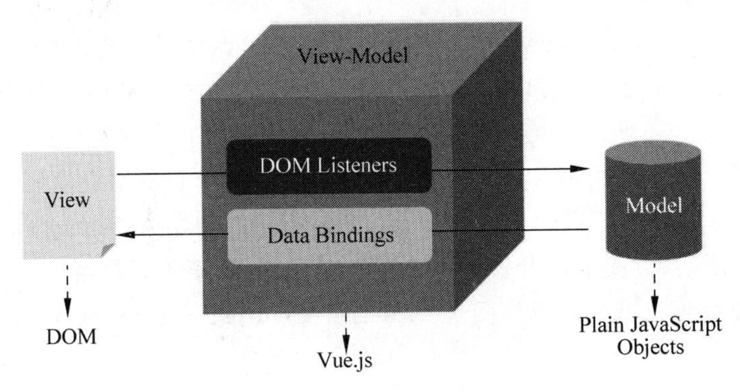

图 1.1　MVVM 响应模型

1.3　Vue.js 同其他框架的对比

Vue.js 作为一个优秀的前端框架，方便前端开发者快速开发应用的前端，在实际项目中使用得比较普遍。当然 Vue.js 也不是实际项目中唯一的前端框架，比较优秀的前端框架还有 React、AngularJS 和 Angular 等。接下来就介绍一下 Vue.js 同这 3 个框架的对比。

1.3.1　Vue.js 同 React 的对比

React 和 Vue.js 有许多相似之处，主要有下几点。

(1) 使用虚拟 DOM。

(2) 提供了响应式(Reactive)和组件化(Composable)的视图组件。

(3) 将注意力集中在核心库，而将其他功能(如路由和全局状态管理)交给相关的库。

Vue.js 同 React 的不同之处有以下几方面。

1. 运行时性能

React 和 Vue. js 的运行速度都是非常快的，所以速度并不是在它们之间做选择的决定性因素，但是在 React 应用中，当某个组件的状态发生变化时，它会以该组件为根，重新渲染整个组件的子树。如果要避免不必要的子组件重新渲染，开发者需要使用 PureComponent 或手动实现 shouldComponentUpdate()方法，而且需要非常细心。而在 Vue. js 应用中，组件的依赖是在渲染过程中自动追踪的，所以系统能精确知晓哪个组件确实需要被重新渲染，从而使开发者不再需要考虑此类优化，只需更好地专注于应用本身。

2. HTML & CSS

在 React 中，所有组件的渲染都依靠 JSX(JavaScript XML)完成。使用 JSX 的渲染函数有下面几点优势。

(1) 可以使用完整的编程语言 JavaScript 的功能来构建视图页面。例如可以使用临时变量、JS 自带的流程控制，以及直接引用当前 JS 作用域中的值等。

(2) 开发工具对 JSX 的支持相比于现有可用的其他 Vue. js 模板还是比较先进的(例如，linting、类型检查、编辑器的自动完成等)。

Vue. js 中默认推荐的是模板(Vue. js 内部也提供了渲染函数，甚至支持 JSX)。任何合乎规范的 HTML 都是合法的 Vue. js 模板，这也带来了以下几点特有的优势。

(1) 对于很多习惯了 HTML 的开发者来讲，模板比 JSX 读写起来更自然。

(2) 基于 HTML 的模板使将已有的应用逐步迁移到 Vue. js 更为容易。

(3) 使设计师和新人开发者更容易理解和参与到项目中。

(4) 可以使用其他模板预处理器(例如 Pug)来书写 Vue. js 的模板。

在 React 中通过 CSS-in-JS 方案实现 CSS 作用域，这引入了一个新的面向组件的样式范例，它和普通的 CSS 的撰写过程是有区别的。另外，虽然在构建时支持将 CSS 提取到一个单独的样式表，但 bundle 中通常还需要一个运行时程序来让这些样式生效。开发者在能够利用 JavaScript 灵活处理样式的同时，也需要权衡 bundle 的大小和运行时的开销。

Vue. js 样式设置的默认方式是在单文件组件里，用 style 的标签定义。同时还可以使用 style 标签的 scoped 属性，来指定样式的作用域。

3. 规模

从向上扩展的角度来讲，Vue. js 和 React 都提供了强大的路由来应对大型应用。React 社区在状态管理方面非常有创新精神(例如 Flux、Redux)，而这些状态管理模式使得甚至 Redux 本身也可以非常容易地被集成在 Vue. js 应用中。实际上，Vue. js 更进一步地采用了这种模式(Vuex)，更加深入集成了 Vue. js 的状态管理解决方案 Vuex，能为开发者带来更好的开发体验。

两者的另一个重要差异是，Vue. js 的路由库和状态管理库都由官方维护且与核心库同步更新。React 则选择把这些问题交给社区维护，因此创建了一个更分散的生态系统，但相对地，React 的生态系统相比 Vue. js 更加繁荣。

最后，Vue. js 提供了 CLI 脚手架，能让开发者通过交互式的脚手架引导非常容易地构建项目。开发者甚至可以使用它快速开发组件的原型。React 在这方面也提供了 create-react-app,但是现在还存在以下局限性。

(1) 它不允许在项目生成时进行任何配置,而 Vue CLI 运行于可升级的运行时依赖之上,该运行时可以通过插件进行扩展。

(2) 它只提供了一个构建单页面应用的默认选项,而 Vue.js 提供了各种用途的模板。

(3) 它不能用用户自建的预设配置构建项目,而这对企业环境下预先建立约定是特别有用的。

从向下扩展的角度来讲,React 学习曲线陡峭,在开发者开始学 React 前,需要知道 JSX 和 ES2015,因为许多示例用的是这些语法。开发者需要学习构建系统,虽然在技术上可以用 Babel 来实时编译代码,但是这种做法并不推荐用于生产环境。

就像 Vue.js 向上扩展后类似于 React 一样,Vue.js 向下扩展后就类似于 jQuery。开发者只要把如下标签放到页面中就可以运行。

```
<script src="https://cdn.jsdelivr.net/npm/vue"></script>
```

然后就可以编写 Vue.js 代码并应用到生产中,只要用 min 版 Vue.js 文件替换掉即可,不用担心其他的性能问题。由于起步阶段不需学习 JSX、ES2015 及构建系统,所以开发者只需不到一天的时间阅读指南就可以建立简单的应用程序。

4. 原生渲染

React Native 能使开发者用相同的组件模型编写有本地渲染能力的 App(iOS 和 Android)。能同时跨多平台开发,对开发者来讲是非常棒的。相应地,Vue.js 和 Weex 会进行官方合作。Weex 是阿里巴巴发起的跨平台用户界面开发框架,同时也正在 Apache 基金会进行项目孵化,Weex 允许使用 Vue.js 语法开发,不仅可以运行在浏览器端,还能被用于开发 iOS 和 Android 上的原生应用的组件,而且 Weex 还在积极发展,尽管成熟度还不能和 React Native 相抗衡,但是,Weex 的发展是由世界上最大的电子商务企业的需求在驱动,而 Vue.js 团队也会和 Weex 团队积极合作,相信会为开发者带来更好的开发体验。

1.3.2 Vue.js 同 AngularJS(Angular 1)的对比

Vue.js 的一些语法和 AngularJS 的语法很相似(例如 v-if VS ng-if),因为 AngularJS 是 Vue.js 早期开发的灵感来源,而 AngularJS 中存在的许多问题在 Vue.js 中已经得到解决。

Vue.js 同 AngularJS 的区别可以体现在如下几方面。

1. 复杂性

在 API 与设计两方面上,Vue.js 都比 AngularJS 简单得多,因此开发者可以快速地掌握它的全部特性并投入开发。

2. 灵活性和模块化

Vue.js 是一个更加灵活开放的解决方案。它允许开发者以希望的方式组织应用程序,而不是在任何时候都必须遵循 AngularJS 制定的规则,这让 Vue.js 能适用于各种项目。由开发者把握决定权是非常必要的。

Vue CLI 旨在成为 Vue.js 生态系统中标准的基础工具。它使多样化的构建工具通过妥善的默认配置无缝协作在一起，这样开发者就可以专注于应用本身，而不会在配置上花费太多时间。同时，它也提供了根据实际需求调整每个工具配置的灵活性。

3. 数据绑定

AngularJS 使用双向绑定，而 Vue.js 在不同组件间强制使用单向数据流，这使应用中的数据流更加清晰易懂。

4. 指令与组件

在 Vue.js 中指令和组件分得更清晰。指令只封装了 DOM 操作，而组件代表了一个自给自足的独立单元——有自己的视图和数据逻辑。在 AngularJS 中，每件事都由指令来做，而组件只是一种特殊的指令。

5. 运行时性能

Vue.js 有更好的性能，并且非常容易优化，因为它不使用脏检查。

在 AngularJS 中，当 watcher 越来越多时会变得越来越慢，因为作用域内的每一次变化之后，所有 watcher 都要重新计算，并且，如果一些 watcher 触发了另一个更新，则脏检查循环(Digest Cycle)可能要运行多次。AngularJS 用户常常要使用深奥的技术，以解决脏检查循环的问题。有时没有简单的办法来优化有大量 watcher 的作用域。

Vue.js 则根本没有这个问题，因为它使用基于依赖追踪的观察系统并且采用异步队列更新的方式，所有的数据变化都是独立触发的，除非它们之间有明确的依赖关系。

1.3.3 Vue.js 同 Angular(Angular 2)的对比

Angular 同 AngularJS 的名称虽然差不多，但是它们是两个完全不同的框架。Angular 具有优秀的组件系统，并且很多实现都重写了，API 也完全改变了。

Vue.js 同 Angular 的区别主要体现在以下几方面。

1. TypeScript

Angular 事实上必须用 TypeScript 来开发，因为它的文档和学习资源绝大部分是面向 TypeScript 的。TypeScript 有很多好处——静态类型检查在大规模的应用中非常有用，同时对于拥有 Java 和 C# 背景的开发者也非常容易提升开发效率。

然而，并不是所有人都想用 TypeScript——在中小型项目中，引入 TypeScript 可能并不会带来太多明显的优势。在这种情况下，用 Vue.js 会是更好的选择，因为在不用 TypeScript 的情况下使用 Angular 会很有挑战性。

最后，虽然 Vue.js 和 TypeScript 的整合可能不如 Angular 那么深入，但 Vue.js 也提供了官方的类型声明和组件装饰器，并且已有大量用户在生产环境中使用 Vue.js+TypeScript 的组合。Vue.js 也和微软的 TypeScript/VSCode 团队进行着积极的合作，其目标是为 Vue.js+TypeScript 用户提供更好的类型检查和 IDE 开发体验。

2. 体积

在体积方面，最近的 Angular 版本在使用了 AOT 和 tree-shaking 技术后最终的代码体积减小了许多，但即使如此，一个包含了 Vuex＋Vue Router 的 Vue.js 项目的大小(gzip 之后在大约为 30KB)相比上述优化后的 angular-cli 生成的默认项目的大小(大约为 65KB)还是要小得多。

3. 灵活性

Vue.js 相比于 Angular 更加灵活，Vue.js 官方提供了构建工具来协助开发者构建项目，但它并不限制开发者如何组织他们的应用代码。有人可能喜欢严格的代码组织规范，但也有开发者喜欢更灵活自由的方式。

4. 学习曲线

要学习 Vue.js，开发者只需有良好的 HTML 和 JavaScript 基础。有了这些基本的技能，开发者就可以非常快速地通过阅读指南投入开发。

Angular 的学习曲线是非常陡峭的——作为一个框架，它的 API 比 Vue.js 要大得多，开发者也因此需要理解更多的概念才能开始有效率地工作。当然，Angular 本身的复杂度是因为它的设计目标就是只针对大型的复杂应用，但不可否认的是，这也使它对于经验不甚丰富的开发者相当不友好。

第 2 章

Vue.js 快速入门

工欲善其事，必先利其器。选择一个合适的开发工具，能显著地提高学习和开发效率。目前使用率比较高的前端开发工具有以下几种。

1. Visual Studio Code

Visual Studio Code 简称 VS Code，是微软在 2015 年发布的一款针对编写现代 Web 和云应用的跨平台源代码编辑器。VS Code 功能非常强大，界面简洁明晰，操作方便快捷，设计得很人性化。软件主要改进了文档视图，完善了对 Markdown 的支持。下载网址是 https://code.visualstudio.com/。

2. HBuilder

HBuilder 是国产的一款前端开发工具，而且是免费的，对于英语不好的前端工程师是一个很大的优势，是前端开发者的首选。其优点是对其他语言的强大支持和开发 Web App 等功能，强大到“没朋友”。在语法提示、转到定义、重构、调试等方面都非常高效。下载网址是 http://www.dcloud.io/。

3. Sublime Text 3

Sublime Text 是一个轻量级的编辑器，也支持各种编程语言。Sublime Text 所有的功能都支持插件，而且快捷键十分好用，可以极大地减小开发工作的强度，使用 Sublime Text 就是要使用其快捷键和插件。Sublime Text 3 的优点是轻量级且功能强大，优雅小巧且启动速度快，有着丰富的第三方支持，能够满足各种各样的扩展。其缺点是不便于项目的管理等，而且代码提示不如 HBuilder 强大。下载网址是 http://www.sublimetext.com/。

4. WebStorm

WebStorm 是 JetBrains 公司旗下的一款 JavaScript 开发工具。官方提供插件支持，满足许多不会配置的初学者，如 ESlint、词法高亮、emmet、CSS 预处理器。新版本也添加了对 ES6 的支持，内建了服务器调试功能，目前已经被广大中国 JavaScript 开发者誉为“Web 前端开发神器”“最强大的 HTML5 编辑器”“最智能的 JavaScript IDE”等。它与 IntelliJ IDEA 同源，继承了 IntelliJ IDEA 强大的 JavaScript 部分的功能。下载网址是 https://www.jetbrains.com/webstorm/。

上面罗列的是目前比较常用的前端开发工具，当然还有很多其他的优秀开发工具，项目团队可以根据实际情况，选择最合适的开发工具。本书中所有的学习案例，使用 Visual Studio Code。接下来介绍如何安装及配置 Visual Studio Code 软件，并且使用它基于 Vue.js 快速开发猜数字游戏案例，体验 VS Code 工具的基本使用和基于 Vue.js 开发前端的基本流程。

2.1 安装配置开发环境

安装配置 VS Code 比较简单，主要步骤是下载并安装 VS Code 软件，再安装高效率的插件。为了方便测试，这里除了介绍 VS Code 的安装配置外，还会介绍在 Chrome 浏览器中如何安装 vue-devtools 插件，以便开发调试 Vue.js 项目。

1. 安装配置 VS Code

1）下载 VS Code 安装软件

用浏览器打开 https://code.visualstudio.com/页面，单击 Download 按钮，进入下载页面，如图 2.1 所示，选择适合自己计算机环境的版本下载。本书选择的是 Windows 64 的 User Installer 版本，文件名称是 VSCodeUserSetup-x64-1.33.1.exe。

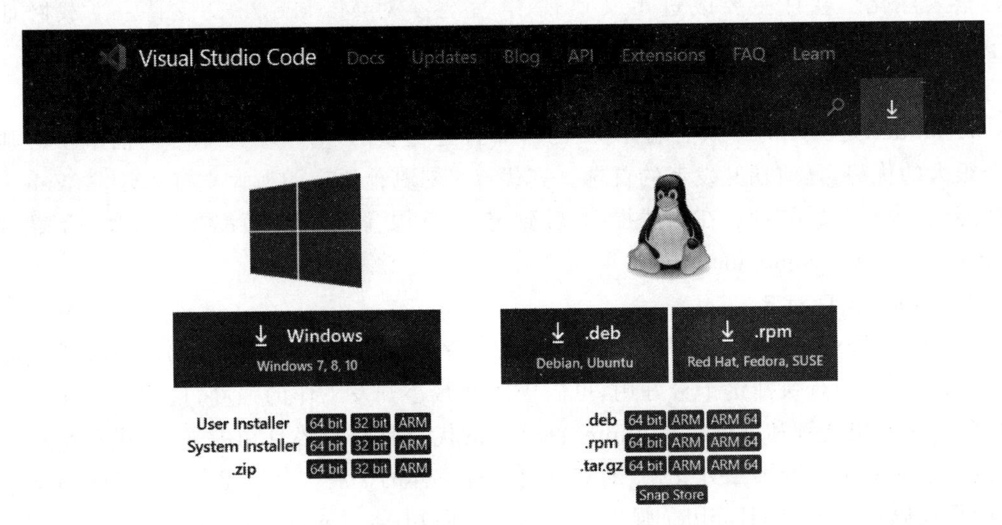

图 2.1 VS Code 下载页面

2）安装 VS Code

双击已下载的安装文件（VSCodeUserSetup-x64-1.33.1.exe），直接运行安装。安装过程比较简单，这里就不再介绍。

3）安装 Live Server 插件

Live Server 插件能方便开发者实时浏览页面的显示效果，接下来介绍如何在 VS Code 中安装 Live Server 插件。

打开 VS Code 开发工具，如图 2.2 所示。

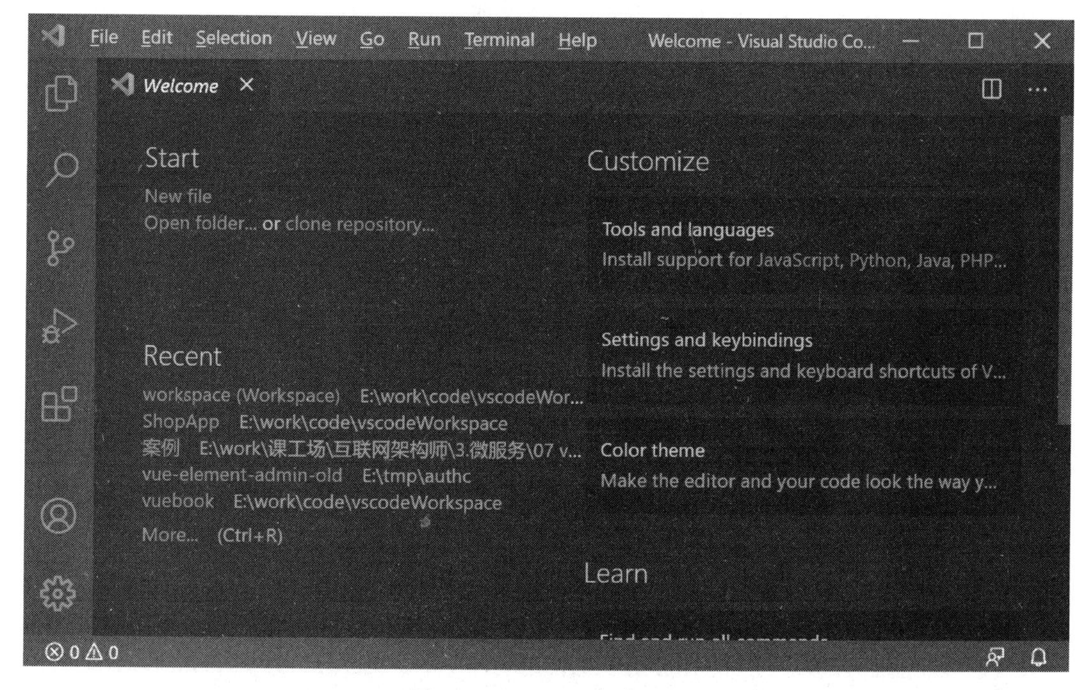

图 2.2 VS Code 启动界面

单击图 2.2 左边的 Extensions 图标，或按快捷键 Ctrl＋Shift＋X，打开 VS Code 的 Extensions 窗口，在输入框中输入 Live Server，选中 Live Server 安装选项，如图 2.3 所示，再单击 Install 按钮，完成 Live Server 插件的安装。如果没有异常，并且在 Extensions 中的 Live Server 选项中没有 Install 按钮，则表明基本安装成功。

4）安装 Vetur 插件

Vetur 是一个可以运行在 VS Code 中的插件，支持.vue 文件的语法高亮显示，支持 template 模板，还支持大多数主流的前端开发脚本和插件，例如 Sass 和 TypeScript。接下来在 VS Code 中安装 Vetur 插件。

同安装 Live Server 插件一样，打开 VS Code 的 Extensions 窗口，在输入框中输入 Vetur，选中 Vetur 安装选项，如图 2.4 所示，单击 Install 按钮，完成 Vetur 插件的安装。如果没有异常，并且在 Extensions 中的 Vetur 选项中没有 Install 按钮，则表明基本安装成功。

2. 安装配置 vue-devtools 插件

vue-devtools 是一款运行在 Chrome 浏览器中的插件，能方便开发者开发和调试 Vue.js 应用。本书后面的案例会结合这个插件进行调试说明，同时在开发过程中，它也是一个很好的调试工具。接下来介绍 vue-devtools 插件的安装。

1）下载 vue-devtools 插件源码

在浏览器中打开 https://github.com/vuejs/vue-devtools，选择 master 分支，单击

Code 按钮，然后在下拉菜单中单击 Download ZIP 选项，下载 vue-devtools 源码。本书下载的文件是 vue-devtools-dev. zip，下载界面如图 2.5 所示。

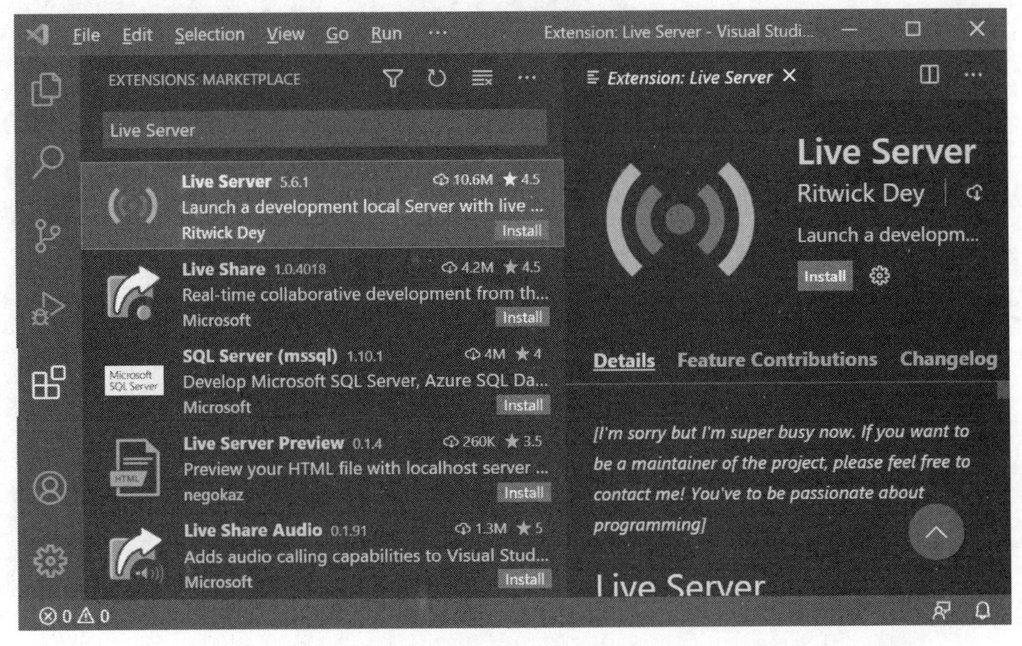

图 2.3 Live Server 安装界面

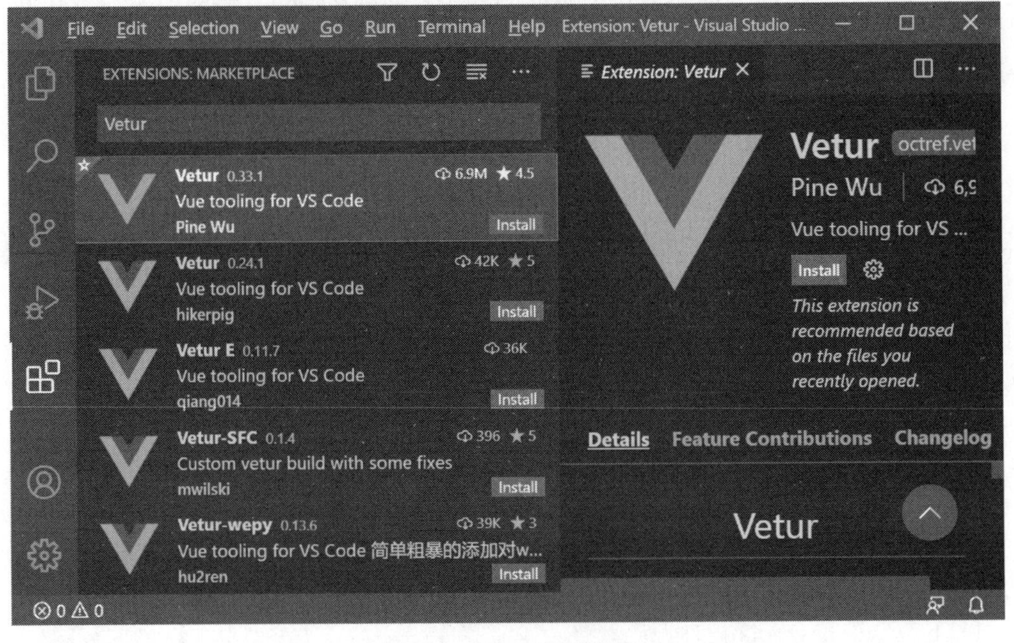

图 2.4 Vetur 安装界面

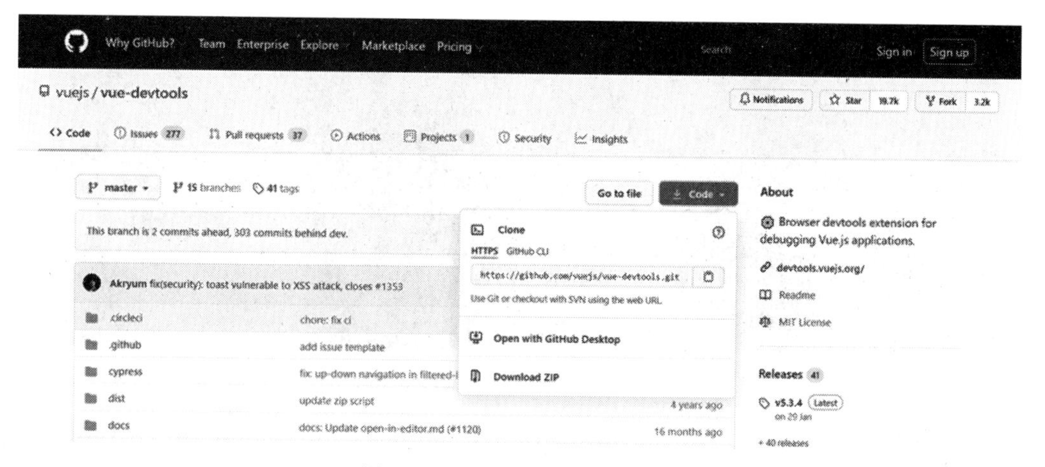

图 2.5　vue-devtools 下载界面

2）编译 vue-devtools 源码

第一步，解压 vue-devtools 源码压缩文件，打开一个 cmd 窗口，切换到 vue-devtools-master 解压目录。输入 npm install 命令，安装 vue-devtools 所需要的依赖，如图 2.6 所示。

图 2.6　安装依赖

第二步，用文本编辑器打开解压目录 vue-devtools-master 下的 shells/Chrome/manifest.json 文件，将代码"persistent": false 改成"persistent": true，代码如下：

```
//第 2 章/manifest.json
{
  …
  "background": {
    "scripts": [
      "build/background.js"
    ],
    "persistent": true
  },
  …
}
```

第三步，在解压目录下，运行 npm run build 命令，构建 vue-devtools 插件，结果如图 2.7 所示。

```
C:\Windows\System32\cmd.exe
 DONE  Compiled successfully in 6590ms                                    上午10:57:27

Hash: 3bb5fba6d331221996f3
Version: webpack 3.12.0
Time: 6590ms
                  Asset       Size  Chunks                    Chunk Names
            devtools.js     864 kB       0  [emitted]  [big]  devtools
             backend.js    32.1 kB       1  [emitted]         backend
                hook.js    2.34 kB       2  [emitted]         hook
            detector.js    3.61 kB       3  [emitted]         detector
               proxy.js  878 bytes       4  [emitted]         proxy
devtools-background.js    1.93 kB       5  [emitted]         devtools-background
          background.js    2.21 kB       6  [emitted]         background
```

图 2.7　构建 vue-devtools

第四步，打开 Chrome 浏览器，选择菜单"更多程序"→"扩展程序"，打开扩展程序界面，如图 2.8 所示。

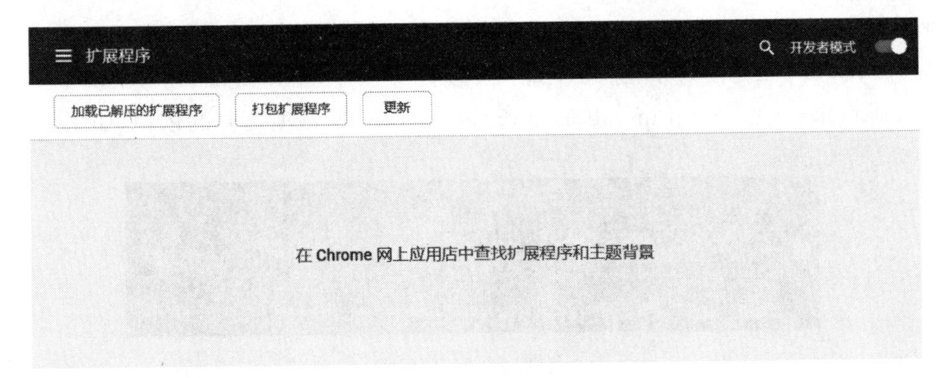

图 2.8　扩展程序

单击"加载已解压的扩展程序"按钮，选择 vue-devtools-master/shells/Chrome 目录，将 vue-devtools 插件安装到 Chrome 浏览器，安装结果如图 2.9 所示。

图 2.9　vue-devtools 安装结果

2.2 实现猜数字游戏

安装好了开发环境，接下来实现一个简单的猜数字游戏，验证开发环境是否搭建完整，同时快速体验 Vue.js 在前端页面中的使用。

2.2.1 功能说明

猜数字游戏的操作比较简单，在打开一个页面时会随机生成一个 1～10 的整数，并且显示页面，如图 2.10 所示。玩家根据提示，在输入框中输入自己所猜的数字，单击“猜一猜”按钮，程序就会比较玩家输入的数字同页面随机生成的数字是否一样。如果玩家输入的数字比随机数字小，提示“你猜的数字太小，请往大的猜”；如果玩家输入的数字比随机数字大，则提示“你输入的数字太大，请往小的猜”。玩家按照提示，在页面上重新输入一个数字，再单击“猜一猜”按钮，程序继续比较和提示，直到玩家输入的数字等于随机生成的数字，此时页面提示“恭喜你猜中了，一共猜了 *N* 次”。

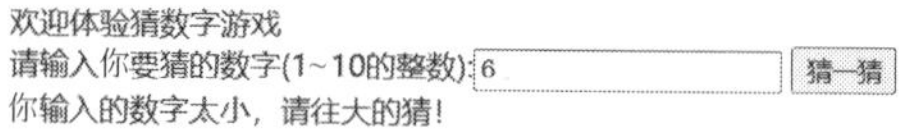

欢迎体验猜数字游戏
请输入你要猜的数字(1~10的整数):6 猜一猜
你输入的数字太小，请往大的猜!

图 2.10 猜数字游戏界面

2.2.2 实现猜数字游戏

了解了猜数字游戏的大体功能后，接下来开始按步骤实现猜数字游戏。

1. 创建工作目录

在工作计算机上，创建一个空目录作为工作目录。本书是在 E:/work/code/vscodeworkspace 目录下，创建 vuebook 目录作为所有案例的工作目录。本书的后续章节，如果没有特殊说明，则工作目录所指的都是这个目录。

2. 在工作目录下规划基本子目录

为了方便管理，在工作目录下分别创建 html 目录和 static/js 目录。html 目录用来存放本书案例中的所有 html 文件，static/js 目录用来存放本书案例用到的所有 JS 文件。

3. 用 VS Code 打开工作目录

打开 VS Code 编辑器，单击 VS Code 编辑器上的 File 菜单，选中前面创建的 vuebook 文件夹，如图 2.11 所示。

4. 添加 Vue.js 文件

在 static/js 目录下，创建 Vue.js 文件。打开浏览器，打开 https://cdn.jsdelivr.net/npm/vue/dist/vue.js 链接，将页面显示的内容全部复制到 Vue.js 文件中。

5. 创建猜数字游戏页面

在 html 目录下，创建猜数字游戏.html 文件，代码如下：

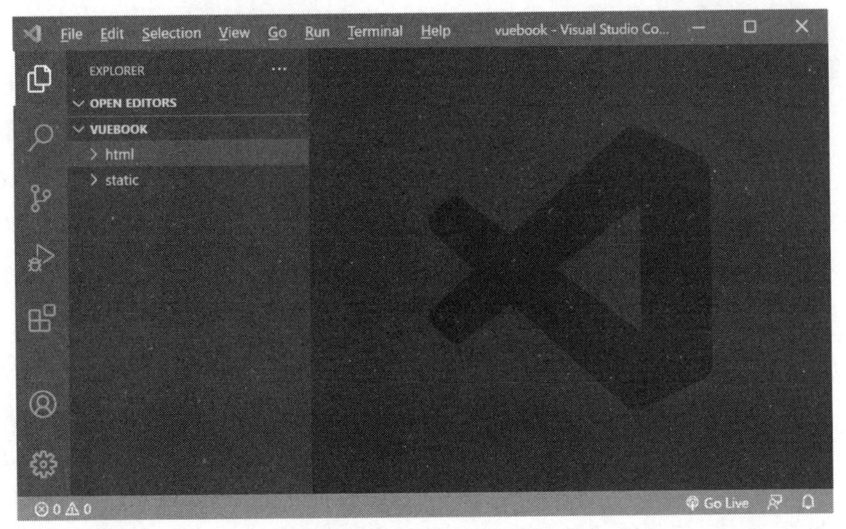

图 2.11 案例工作目录

```
//第 2 章/猜数字游戏.html
<!DOCTYPE html>
<html lang = "en">
<head>
    <meta charset = "UTF-8">
    <meta http-equiv = "X-UA-Compatible" content = "IE = edge">
    <meta name = "viewport" content = "width = device-width, initial-scale = 1.0">
    <title>猜数字游戏</title>
    <!-- 引入 Vue.js -->
    <script type = "text/JavaScript" src = "../static/js/Vue.js"></script>
</head>
<body>
  <div id = "app">
    欢迎体验猜数字游戏<br/>
    请输入你要猜的数字(1～10 的整数):<input type = 'text' v-model = "guessNumber"/>
    <input type = "button" value = "猜一猜" @click = "guess">
    <div>
        {{message}}
    </div>
  </div>

  <script type = 'text/JavaScript'>
    //创建 Vue.js 对象
    const vm = new Vue({
       el:"#app",                          //绑定 app div
       data:{                              //定义 data 属性
          randomNumber:0,
          guessNumber:'',
```

```
            times:0,
            message: ''
        },
        created(){                          //定义创建函数,在创建 Vue.js 对象的时候自动执行
            let random = parseInt((Math.random) * 10000)
            this.randomNumber = random % 10 + 1
        },
        methods:{                           //定义事件方法
            guess(){
                this.times++
                this.guessNumber = parseInt(this.guessNumber)

                if( this.guessNumber < this.randomNumber ){
                    this.message = '你输入的数字太小,请往大的猜!'
                }else if( this.guessNumber > this.randomNumber ){
                    this.message = '你输入的数字太大,请往小的猜!'
                }else{
                    this.message = "恭喜你猜中了,一共猜了" + this.times + "次"
                }

            }
        }
    })
  </script>
</body>
</html>
```

6. **运行游戏案例**

右击 html/猜数字游戏.html 文件,在弹出的菜单中选中 Open with Live Server 选项,启动 Live Server,如图 2.12 所示,VS Code 会自动启动浏览器,进入猜数字游戏界面。

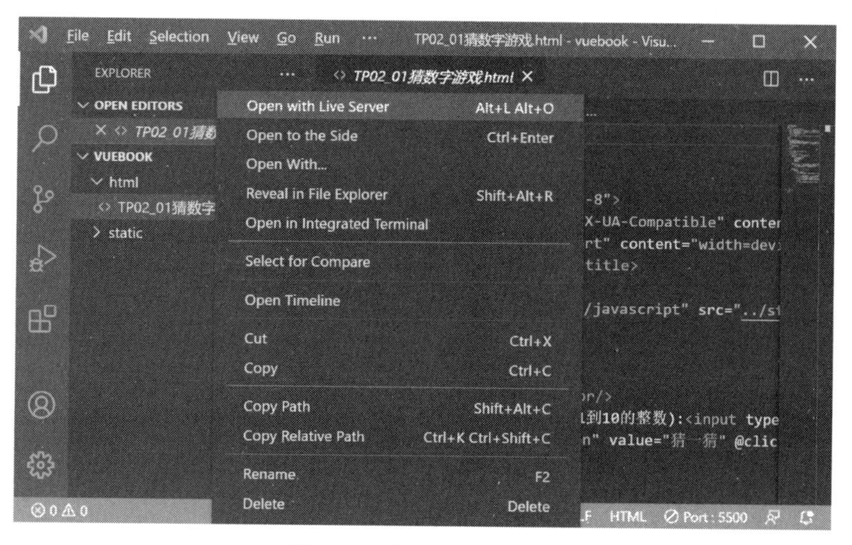

图 2.12 启动 Live Server

7. 基于 vue-devtools 插件，查看 Vue.js 对象

在 Chrome 浏览器中，按 F12 键，单击 Vue 选项，再单击视图中的组件，这样就可以查看该组件对象的 data 数据了，如图 2.13 所示。

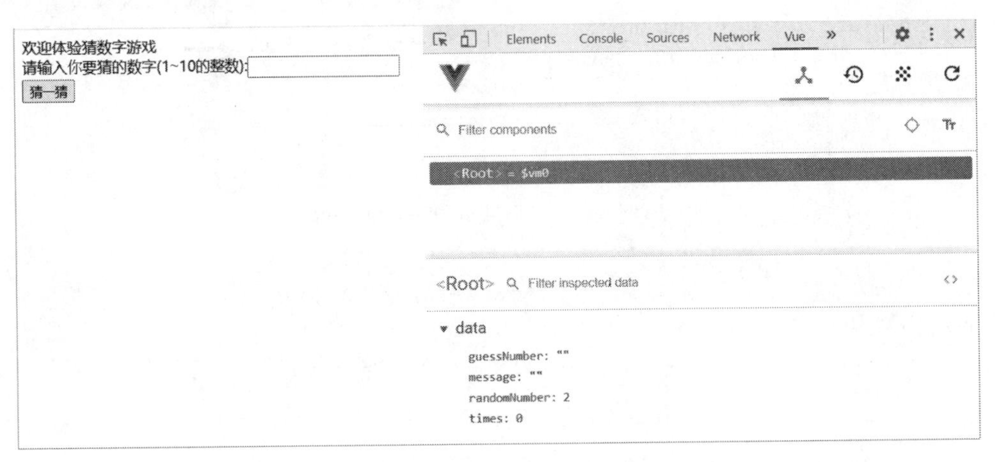

图 2.13 vue-devtools 调试

第3章 Vue.js 基本语法

万丈高楼平地起。本章主要从如下几方面，介绍 Vue.js 的基本语法。

1. Vue.js 对象

介绍如何创建 Vue.js 对象，如何在 Vue.js 对象中定义数据属性和方法，以及 Vue.js 对象的生命周期。

2. 插值表达式

介绍如何在视图上绑定 Vue.js 对象的数据属性，包括简单文本的绑定、html 值的绑定、元素属性值的绑定、样式值的绑定和 JavaScript 表达式的值绑定等。

3. 表单输入绑定

介绍如何给复杂的表单元素绑定动态的数据。

4. 事件处理

介绍如何给视图元素的事件绑定事件监听，事件触发后将执行绑定的函数。

5. 指令

介绍 Vue.js 提供的常用指令，实现页面中事件的绑定、元素属性值的绑定和显示内容的逻辑控制指令等。

6. Vue.js 响应原理

介绍 Vue.js 框架内部的响应原理及对对象、数组的检测响应，还有异步更新问题的处理。

3.1 Vue.js 对象

虽然从严格意义上讲，Vue.js 没有完全遵循 MVVM 模型，但是 Vue.js 的整体设计还是受到了 MVVM 模型的启发，在 Vue.js 中有个非常重要的概念——Vue.js 实例。在后面的章节中，会经常使用 VM(ViewModel 的缩写)表示 Vue.js 实例。

每个 Vue.js 应用都是通过 Vue.js 函数创建一个新的 Vue.js 实例开始的，而且每个 Vue.js 应用都是由通过 Vue.js 函数创建的一个根 Vue.js 实例，以及多个可选的、嵌套的、可复用的组件树组成，如一个 Todo 应用的组件树结构如图 3.1 所示。

创建 Vue.js 实例的语法代码如下：

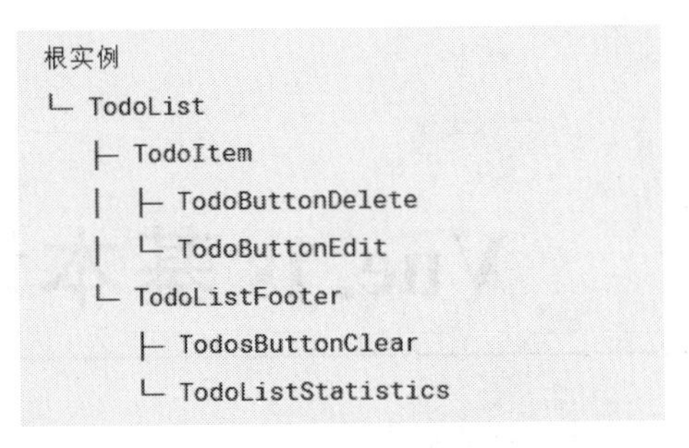

图 3.1　Todo 应用组件树

```
var vm = new Vue({
  //选项
})
```

注意语法中的 Vue.js，它是定义在 Vue.js 中的函数，所以在使用上面的语法创建 Vue.js 实例之前，需要引入 Vue.js，代码如下：

```
<script type="text/JavaScript" src="../static/js/Vue.js"></script>
```

3.1.1　Vue.js 实例的数据属性

创建 Vue.js 实例的时候，可以在创建的 JSON 对象中，使用可选属性 data 定义 Vue.js 实例的 property 属性，这些 property 属性将会被加入 Vue.js 的响应式系统中。当这些 property 属性的值发生改变的时候，视图将会产生“响应”，即匹配后更新为新值，代码如下：

```
//第 3 章/创建 Vue.js 实例和定义 data 属性.html

//数据对象
var data = { a: 1 }

//该对象被加入一个 Vue.js 实例中
var vm = new Vue({
  data: data
})

//获得这个实例上的 property
//返回源数据中对应的字段
vm.a == data.a //=> true

//设置 property 也会影响原始数据
vm.a = 2
data.a //=> 2
```

```
//反之亦然
data.a = 3
vm.a // => 3
```

当这些数据改变时，视图会进行重新渲染。值得注意的是，只有当实例被创建时就已经存在于 data 中的 property 才是响应式的。也就是说，如果添加一个新的 property，例如 vm. b='hi'，则对 b 的改动将不会触发任何视图的更新。如果开发人员知道在晚些时候需要一个或多个 property 属性，则需要在 data 属性中定义这些 property 属性，给它们赋予空值，后面需要的时候再操作或使用这些 property，代码如下：

```
data: {
 newTodoText: '',
 visitCount: 0,
 hideCompletedTodos: false,
 todos: [],
 error: null
}
```

这里唯一的例外是使用 Object. freeze()函数会阻止修改现有的 property 属性的"响应"，也意味着响应系统无法再追踪变化。因为调用了 Object. freeze()函数，改变 property 属性值后，不会渲染到视图上，代码如下：

```
//第 3 章/freeze 方法.html

var data = {
  foo: 'bar'
}
//阻止 property 属性的响应
Object.freeze(data)

const vm = new Vue({
 el: '#app',
 data: data
})
<div id="app">
 <p>{{ foo }}</p>
 <!-- 这里的 `foo` 不会更新! -->
 <button v-on:click="foo = 'baz'">Change it</button>
</div>
```

除了数据 property 属性，Vue. js 实例还暴露了一些有用的 property 属性。开发人员可以在这些 property 属性名前面添加 $前缀，以便与开发人员定义的 property 区分开来，代码如下：

```
//第3章/freeze方法.html
var data = {
  foo: 'bar'
}
var vm = new Vue({
  el: '#app',
  data: data
})

vm.$data === data // => true
vm.$el === document.getElementById("app") // => true
```

关于 Vue.js 实例自己暴露的其他 property 属性，读者可以参考官网(https://cn.vuejs.org/v2/api/)。

3.1.2 Vue.js实例的方法

开发者可以在创建 Vue.js 实例对象的时候，在传入的 JSON 可选项中自定义若干方法封装业务逻辑，以便于在需要的时候重复调用，代码如下：

```
//第3章/自定义Vue.js实例方法.html
    <div id = "app">
        {{count}}
        <br/>
        <!-- 给单击事件绑定 increment 方法 -->
        <input type = "button" v-on:click = "increment" value = "单击递增"></input>
    </div>
    <script type = "text/JavaScript">
        const data = {count:0}
        const vm = new Vue({
            el:'#app',
            data,
            methods:{
                increment(){
                    this.count++
                }
            }
        })
        //调用两次vm实例中的increment方法
        vm.increment()
        vm.increment()
    </script>
```

如上面代码所示，开发人员可以在传给 Vue.js 函数的 JSON 对象参数中用 methods 属性定义多个自定义方法，在上面的样例代码中定义的是 increment 方法。定义好后，可以将它们绑定到页面元素的事件上，对应事件触发后自动执行，也可以在 JavaScript 代码中，同调用普通方法的语法一样调用。

同 Vue. js 实例的数据属性一样，开发人员可以自定义多种方法，也可以使用 Vue. js 实例暴露的其他方法，在使用的时候，也是用 $作为前缀，用以区分开发人员的自定义方法。如 vm. $mount 方法可以实现将 vm 对象挂载到指定的 DOM 元素上。在这里对这些方法不进行详细介绍，后面章节会陆续介绍常用的方法。读者可以参考官网（https://cn. vuejs. org/v2/api/），也可以参考附录 API。

3.1.3 Vue. js 实例生命周期

每个 Vue. js 实例在被创建时都要经过一系列的初始化过程，例如，需要设置数据监听、编译模板、将实例挂载到 DOM 并在数据变化时更新 DOM 等。同时在这个过程中也会运行一些叫作生命周期钩子的函数，这给了开发人员在不同阶段添加自己代码的机会。

Vue. js 实例从创建到销毁的整个过程分为创建、绑定、更新和销毁 4 个阶段。对于每个阶段，Vue. js 里面都定义了两个对应的钩子函数，它们分别是 beforeCreate、created、beforeMount、mounted、beforeUpdate、updated、beforeDestroy 和 destroyed。

从调用 Vue. js 的构造函数开始，Vue. js 实例的生命周期的详细说明如下。

（1）创建 Vue. js 实例。

（2）初始化实例的事件和生命周期钩子函数。

（3）调用 beforeCreate 钩子函数。

（4）初始化 Vue. js 实例的注入属性和方法。

（5）调用 created 钩子函数。

（6）判断是否设置 el 选项。如果没有，就不将 vm 对象挂载到 DOM 元素上，否则进入下一步。

（7）判断是否有 el 对应的模板。如果有对应的模板，就将模板转换成 render 函数，通过 render 函数完成渲染；如果没有对应的模板，就将 el 对应的外层 html 作为模板，进行渲染。

（8）调用 beforeMount 钩子函数。

（9）在渲染的视图上，绑定 vm 中定义的数据。

（10）调用 mounted 钩子函数。

（11）当 vm 中的数据发生变化时，循环地调用 beforeUpdate 钩子函数，更新页面显示，调用 updated 钩子函数。

（12）在销毁 Vue. js 实例之前，调用 beforeDestroy 钩子函数。

（13）清除 Vue. js 实例上的所有监听器子组件等。

（14）调用 destroyed，完成 Vue. js 实例的销毁。

开发人员可以通过在代码中实现 beforeCreate、created、beforeMount、mounted、beforeUpdate、updated、beforeDestroy 和 destroyed 钩子函数，将合适的业务功能植入 Vue. js 实例的对应生命周期阶段执行。

Vue. js 实例的完整生命周期流程可以参考图 3.2。生命周期钩子方法的定义可以参考的代码如下：

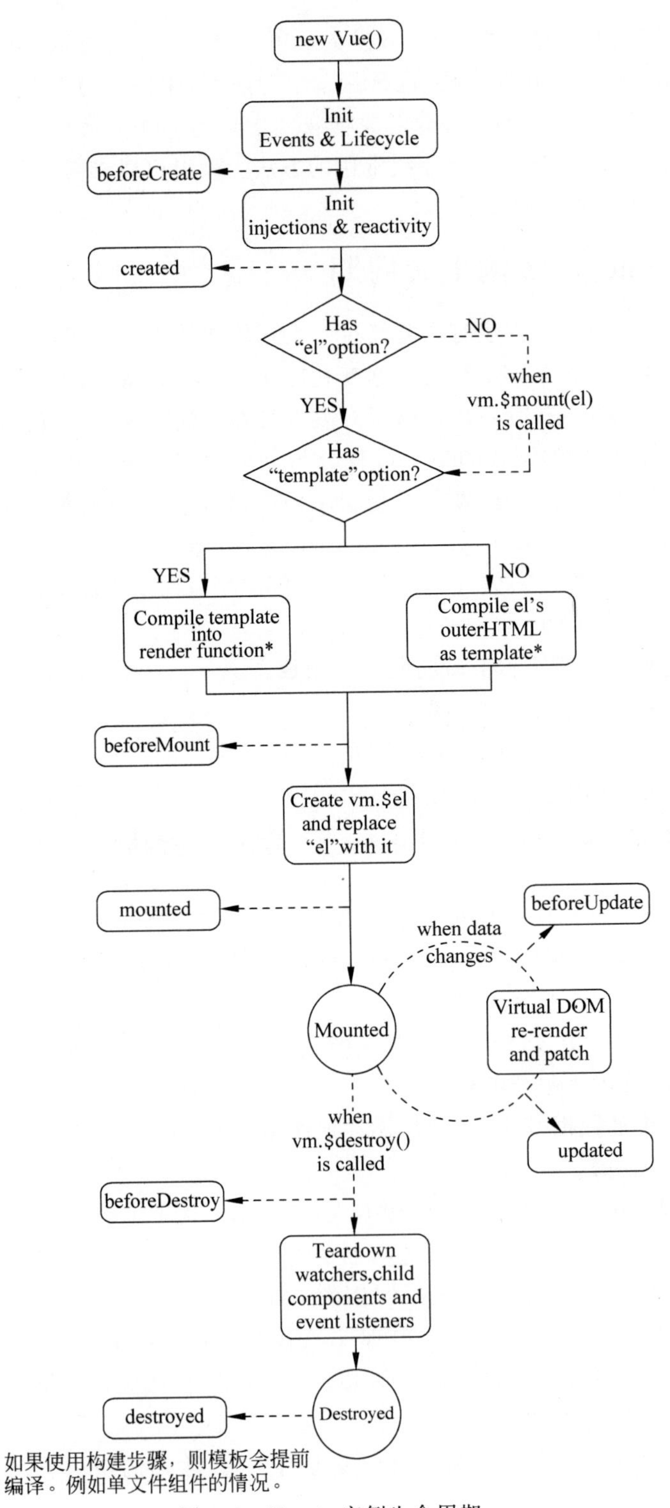

图 3.2　Vue.js 实例生命周期

```
//第 3 章/Vue.js 实例生命周期.html
<!DOCTYPE html>
<html lang="en">
    <head>
        …
        <script type="text/JavaScript" src="../static/js/Vue.js"></script>
    </head>
    <style type="text/css"></style>
    <body>
        <div id='app'>
            {{message}}
        </div>
        <script type='text/JavaScript'>
            let vm = new Vue({
                el: '#app',
                data: {
                    message: 'old message'
                },
                methods: {
                },
                beforeCreate: function(){
                    console.log('beforeCreate')
                },
                created: function(){
                    console.log('created')
                },
                beforeMount: function(){
                    console.log('beforeMount')
                },
                mounted: function(){
                    console.log('mounted')
                },
                beforeUpdate: function(){
                    console.log('beforeUpdate')
                },
                updated: function(){
                    console.log('updated')
                },
                beforeDestroy: function(){
                    console.log('beforeDestroy')
                },
                destroyed: function(){
                    console.log('destroyed')
                }
            })
            vm.$data.message = 'new message'
```

```
            setTimeout(() =>{
                const appObject = document.getElementById("app";
                document.body.removeChild(appObject)
                vm = null
            }, 2000);

        </script>
    </body>
</html>
```

3.2 插值表达式

插值表达式的作用是在页面的对应位置插入动态的数据，以便动态地显示。当被插入的数据对象在业务逻辑中发生改变时，会实时地更新到页面上。根据插入动态值的不同位置，分为文本插值、原始 html 插值、JavaScript 插值、class 属性插值和内嵌样式插值 5 种插入表达式，它们的语法和使用时的注意事项如下。

1. 文本插值表达式

文本插值表达式的作用是在标签之间插入动态的文本值。

语法如下：

```
{{表达式}}
```

案例代码如下：

```
<span>hello, {{userName}}</span>
```

说明：在 span 元素中，动态显示 userName 数据对象的值。以文本的形式显示 userName 的值。如果 userName="zhangsan"，则显示"hello，张三"；如果 userName=<b>zhangsan</b>，则显示 hello，<b>zhangsan</b>。

2. 原始 html 插值表达式

原始 html 插值表达式的作用是在标签之间插入原始的 html 内容。如果里面包含 html 标签，则浏览器会进行解析。

语法如下：

```
<标签 v-html='表达式'></标签>
```

案例代码如下：

```
<p>显示 html 内容 <span v-html='userName'></span></p>
```

说明：如果 userName 的值是<font color='red'> zhangsan </font>，则显示的结果为显示红色的"zhangsan"。

3. 属性插值表达式

属性插值表达式的作用是给 html 元素的属性绑定动态值。特别需要注意的是，Mustache 语法（双大括号语法）是不能用在 html 元素属性上的。

语法如下：

```
<标签 v-bind:属性名称 = '表达式'>...<标签>
```

案例代码如下：

```
<div v-bind:id = "divId"> div demo </div>
```

说明：对于布尔属性（只要存在就意味着其值为 true），v-bind 工作起来略有不同。示例代码如下：

```
<button v-bind:disabled = "isButtonDisabled"> Button </button>
```

如果 isButtonDisabled 的值是 null、undefined 或 false，则 disabled 属性不会被包含在渲染出来的<button>元素中，只有当 isButtonDisabled 是 true 或在 JavaScript 中能转换成 true 时，才会被渲染出来。

4. class 属性插值表达式

class 是 html 标签中的一个特殊属性，给 class 属性绑定动态值也采用 v-bind 的方式，同普通属性一样。将 active 数据属性的值绑定到 div 的 class 属性上，代码如下：

```
<div v-bind:class = "active">..</div>
```

对于 class 属性，还支持根据条件绑定数据，将多个类样式和新旧 class 值一起绑定，代码如下：

```
<div v-bind:class = "isActive:active">..</div>
```

其中 isActive 和 active 都是定义在 Vue model 中的属性名称，只有当 isActive 的值（或转换后的值）为 true 时，才将 active 属性的值绑定到 class 属性上，否则不绑定，从而实现根据条件给 class 属性绑定值，代码如下：

```
<div v-bind:class = "{clsName1, clsName2 ...}">..</div>
```

其中 clsName1 和 clsName2 都是 Vue model 的属性名称，Vue.js 的渲染结果是同时给 div 的 class 属性绑定 clsName1 和 clsName2 对应的两个类样式。当然，也可以给每个 clsName 添加条件控制，代码如下：

```
<div v-bind:class="{isClsName1:clsName1, clsName2...}">..</div>
```

最后,div 自己有个 class 属性,其值是 static,同时用 v-bind: class 给 class 属性绑定了 clsName 动态属性值,代码如下:

```
<div class="static" v-bind:class="clsName">..</div>
```

渲染结果是最后的 html 的 class 属性值既包含 static,也包含 clsName 对应的动态值,代码如下:

```
<div class="static clsValue">..</div>
```

5. 内嵌样式插值表达式

v-bind: style 的对象语法十分直观——看着非常像 CSS,但其实是一个 JavaScript 对象。CSS property 名可以用驼峰式(camelCase)或短横线分隔(kebab-case,记得用引号括起来)来命名,代码如下:

```
<div v-bind:style="{ color: activeColor, fontSize: fontSize + 'px' }"></div>

data: {
  activeColor: 'red',
  fontSize: 30
}
```

直接绑定到一个样式对象通常更好,这会让模板更清晰,代码如下:

```
<div v-bind:style="styleObject"></div>

data: {
  styleObject: {
    color: 'red',
    fontSize: '13px'
  }
}
```

v-bind: style 的数组语法可以将多个样式对象应用到同一个元素上,代码如下:

```
<div v-bind:style="[baseStyles, overridingStyles]"></div>
```

6. JavaScript 插值表达式

在页面中除了可以插入一个简单的属性值外,也可以绑定 JavaScript 表达式,代码如下:

```
{{userName + 'demo'}}
{{ok?'yes':'no'}}
{{message.split('').reverse().join('')}}
<div v-html="'hello' + userName"></div>
```

这些表达式会在所属 Vue.js 实例的数据作用域下作为 JavaScript 被解析，但有个限制，即每个绑定都只能包含单个表达式，所以下面的例子都不会生效，代码如下：

```
<!-- 这是语句,不是表达式 -->
{{ var a = 1 }} `
<!-- 流控制也不会生效,请使用三元表达式 -->
{{ if (ok) { return message } }}
```

注意，模板表达式都被放在沙盒中，只能访问全局变量的一个白名单，如 Math 和 Date。不能在模板表达式中试图访问用户定义的全局变量，全局变量的白名单如下：

```
const allowedGlobals = makeMap(
    'Infinity,undefined,NaN,isFinite,isNaN,' +
    'parseFloat,parseInt,decodeURI,decodeURIComponent,encodeURI,' +
    'encodeURIComponent,Math,Number,Date,Array,Object,Boolean,' +
    'String,RegExp,Map,Set,JSON,Intl,require' //for Webpack/Browserify
  )
```

3.3 表单输入绑定

开发人员可以用 v-model 指令在表单元素<input>、<textarea>及<select>上创建双向数据绑定。v-model 会根据控件类型自动选取正确的方法来更新元素。虽然看起来有些神奇，但 v-model 本质上不过是语法糖(Syntactic Sugar)。它负责监听用户的输入事件以更新数据，并对一些极端场景进行一些特殊处理。

v-model 会忽略所有表单元素的 value、checked、selected attribute 的初始值而总是将 Vue.js 实例的数据作为数据来源。应该通过 JavaScript 在组件的 data 选项中声明初始值。

v-model 在内部为不同的输入元素使用不同的 property 并抛出不同的事件。

(1) txt 和 textarea 元素使用 value property 和 input 事件。

(2) checkbox 和 radio 元素使用 checked property 和 change 事件。

(3) select 元素将 value 作为 prop 并将 change 作为事件。

下面从 v-model 的基本使用、值绑定和修饰符 3 方面，对使用 v-model 实现表单输入绑定进行介绍。

3.3.1 基本用法

在项目开发过程中，经常希望在表单元素中输入值后，能自动同步到对象属性中。

v-model 很好地解决了这个问题。下面通过案例，分别演示使用 v-model 改变表单元素的值后，如何自动同步到对象属性中。

1. 单行文本输入框

在单行文本输入框 input type="text"元素中，用 v-model 绑定了 Vue.js 实例中的 message 数据属性。当用户在单行文本输入框中输入值时，会自动被同步到 message 属性中，从而同步 div 中 message 的显示内容，代码如下：

```
//第 3 章/表单输入绑定之基本用法.html
...

        <div id = 'app'>
            <!-- 单行文本输入框 -->
            <input type = "text" v-model = "message" placeholder = "输入值"/><br/>

            <div>
                message is: {{message}}
            </div>
        </div>
        <script type = 'text/JavaScript'>
            const vm = new Vue({
                el: '#app',
                data: {
                    message: ''
                },
                methods: {
                }
            })
        </script>
...
```

2. 多行文本输入框

在多行文本输入框 textarea 元素中，用 v-model 绑定了 Vue.js 实例中的 message 数据属性。当用户在多行文本输入框中输入值时，会自动被同步到 message 属性中，从而同步 div 中 message 的显示内容，代码如下：

```
//第 3 章/表单输入绑定之基本用法.html
...

        <div id = 'app'>
            <!--多行文本输入框 -->
            <textarea v-model = "message" placeholder = "输入值"></textarea><br/>

            <div>
                message is: {{message}}
```

```
        </div>
    </div>
    <script type='text/JavaScript'>
        const vm = new Vue({
            el: '#app',
            data: {
                message: ''
            },
            methods: {
            }
        })
    </script>
...
```

使用 textarea 元素时需要注意，在文本区域插值(<textarea>{{text}}</textarea>)不会生效，需要用 v-model 来代替。

3. 复选框

复选框绑定的对象属性值可以是 boolean 类型的值，也可以是其他值。绑定的对象属性是 boolean 类型的值，当 checkbox 被选中时，会被同步到 checked 数据属性，这个属性是 boolean 类型的，同时会在 label 部分显示 true，否则显示 false，代码如下：

```
//第3章/表单输入绑定之基本用法.html
...
        <!-- 复选框绑定 boolean -->
        <input type="checkbox" v-model='checked' id="checkbox"/>
        <label for="checkbox">{{checked}}</label>
    </div>
    <script type='text/JavaScript'>
        const vm = new Vue({
            el: '#app',
            data: {
                checked: false
            },
            methods: {
            }
        })
    </script>
...
```

选中的那些 checkbox 会自动被同步到 checkedNames 数据属性，这个属性是字符串数组，同时会在 span 中显示所有被选中的 checkbox 的值，代码如下：

```
//第 3 章/表单输入绑定之基本用法.html
…
<div id = 'app'>
  <! -- 复选框绑定数组 -- >
  <input type = "checkbox" id = "jack" value = "Jack" v-model = "checkedNames">
  <label for = "jack"> Jack </label>
  <input type = "checkbox" id = "john" value = "John" v-model = "checkedNames">
  <label for = "john"> John </label>
  <input type = "checkbox" id = "mike" value = "Mike" v-model = "checkedNames">
  <label for = "mike"> Mike </label><br>
  <span> Checked names: {{ checkedNames }}</span>
</div>
<script type = 'text/JavaScript'>
  const vm = new Vue({
    el: '#app',
    data: {
      checkedNames: []
    },
    methods: {}
  })
</script>
…
```

4. 单选按钮

选中某个单选按钮后，会自动被同步到 picked 数据属性，在 span 元素中同步显示，代码如下：

```
//第 3 章/表单输入绑定之基本用法.html
…
<div id = 'app'>
  <! -- 单选按钮 -- >

  <input type = "radio" id = "one" value = "One" v-model = "picked">
  <label for = "one"> One </label>
  <input type = "radio" id = "two" value = "Two" v-model = "picked">
  <label for = "two"> Two </label>
  <br>
  <span> Picked: {{ picked }}</span>
</div>
<script type = 'text/JavaScript'>
  const vm = new Vue({
    el: '#app',
    data: {
      picked: ''
    },
```

```
        methods: {}
    })
</script>
...
```

5. 选择框

当选中选择框的某个选项时，会自动将选项的值同步到 selected 数据属性中，同时会被显示到 span 元素中，代码如下：

```
//第 3 章/表单输入绑定之基本用法.html
...
        <div id='app'>
            <!-- 选择框 -->
            <select v-model="selected">
                <option disabled value="">请选择</option>
                <option>A</option>
                <option>B</option>
                <option>C</option>
            </select>
            <span>Selected: {{ selected }}</span>
        </div>
        <script type='text/JavaScript'>
            const vm = new Vue({
                el: '#app',
                data: {
                    selected: ''
                },
                methods: {
                }
            })
        </script>
...
```

如果 v-model 表达式的初始值未能匹配任何选项，则<select>元素将被渲染为“未选中”状态。在 iOS 中，这会使用户无法选择第 1 个选项。因为在这样的情况下，iOS 不会触发 change 事件，因此，更推荐像上面那样提供一个值为空的禁用选项。

v-model 同时还支持多个选项的选择框，这时候可以绑定一个数组类型的数据属性。当选中多个选项时，会被自动同步到 selectedArray 数据属性中，同时对应的 span 元素会显示被选中选项的值，代码如下：

```
//第 3 章/表单输入绑定之基本用法.html
...
<div id='app'>
    <!-- 选择框 -->
```

```
    <select v-model="selectedArray" multiple style="width: 50px;">
        <option>A</option>
        <option>B</option>
        <option>C</option>
    </select>
    <br>
    <span>selectedArray: {{ selectedArray }}</span>
</div>
<script type='text/JavaScript'>
    const vm = new Vue({
        el: '#app',
        data: {
            selectedArray: []
        },
        methods: {}
    })
</script>
...
```

3.3.2 值绑定

对于单选按钮、复选框及选择框的选项，v-model 绑定的值通常是静态字符串(对于复选框也可以是布尔值)，代码如下：

```
<!-- 当选中时,`picked`为字符串 "a" -->
<input type="radio" v-model="picked" value="a">

<!-- `toggle`为 true 或 false -->
<input type="checkbox" v-model="toggle">

<!-- 当选中第 1 个选项时,`selected`为字符串 "abc" -->
<select v-model="selected">
  <option value="abc">ABC</option>
</select>
```

但是有时开发人员可能想把值绑定到 Vue.js 实例的一个动态 property 上，这时可以用 v-bind 实现，并且这个 property 的值可以不是字符串。

1. 复选框

在 checkbox 元素中，添加 true-value 和 false-value 属性，分别表示选中和没选中绑定的值。当选中的时候，v-model 绑定的 toggle 数据属性的值是 yes，否则是 no，代码如下：

```
<input type="checkbox" v-model="toggle" true-value="yes" false-value="no" />

//当选中时
```

```
vm.toggle === 'yes'
//当没有选中时
vm.toggle === 'no'
```

这里的 true-value 和 false-value 属性并不会影响输入控件的 value 属性，因为浏览器在提交表单时并不会包含未被选中的复选框。如果要确保表单中这两个值中的一个能够被提交(yes 或 no)，则应换用单选按钮。

2. 单选框

当选中 radio 时，pick 数据属性的值就是 v-bind 绑定的值 a，代码如下：

```
< input type = "radio" v - model = "pick" v - bind:value = "a">

//当选中时
vm.pick === vm.a
```

3. 选择框

当选中选项时，通过 v-model 绑定的 selected 数据属性的值会被自动同步为选项中用 v-bind 绑定的对象，代码如下：

```
< select v - model = "selected">
    <! -- 内联对象字面量 -->
  < option v - bind:value = "{ number: 123 }"> 123 </option >
</select >
//当选中时
typeof vm.selected // => 'object'
vm.selected.number // => 123
```

3.3.3 修饰符

在使用 v-model 绑定 Vue.js 实例的数据属性时，还可以添加相关的修饰符，在将数据同步到 Vue.js 实例的数据属性的时候，对数据进行必要预处理。例如转换成数字类型、去掉空格等。

1. lazy 修饰符

在默认情况下，v-model 在每次 input 事件触发后会将输入框的值与数据进行同步(除了上述输入法组合文字时)。开发人员可以添加 lazy 修饰符，从而转换为在 change 事件之后进行同步，代码如下：

```
<! -- 在"change"时而非"input"时更新 -->
< input v - model.lazy = "msg">
```

2. number 修饰符

如果想自动将用户的输入值转换为数值类型，则可以给 v-model 添加 number 修饰符，代码如下：

```
<input v-model.number = "age" type = "number">
```

这种方法通常很有用，因为即使在 type="number" 时，HTML 输入元素的值也总会返回字符串。如果这个值无法被 parseFloat()函数解析，则会返回原始的值。

3. trim 修饰符

如果要自动过滤用户输入的首尾空白字符，则可以给 v-model 添加 trim 修饰符，代码如下：

```
<input v-model.trim = "msg">
```

3.4 事件处理

在开发前端页面的时候，经常需要给 html 元素绑定事件代码，当对应的事件被触发时，能自动执行指定的业务代码。本小节将介绍 Vue.js 中对事件的处理。

3.4.1 监听事件

在 Vue.js 中，可以使用 v-on 指令监听 DOM 事件，并在触发时运行 JavaScript 代码。使用 v-on 给 button 元素绑定一段 JavaScript 代码，每单击一次按钮，就给 counter 数据属性加 1，代码如下：

```
//第 3 章/事件监听.html
...
<div id = 'app'>
    <button v-on:click = "counter += 1"> Add 1 </button>
    <p>按钮被单击了 {{ counter }} 次.</p>
</div>
<script type = 'text/JavaScript'>
    const vm = new Vue({
        el: '#app',
        data: {
            counter: 0
        }
    })
</script>
...
```

3.4.2 事件处理方法

在实际项目中，许多事件处理逻辑很复杂，所以直接把 JavaScript 代码写在 v-on 指令中是不可行的，因此 v-on 应可以接收一个需要调用的方法名称。使用 v-on 给 button 元素绑定一个定义在 Vue.js 实例中的方法，当单击按钮的时候，自动执行 Vue.js 实例中绑定的方法，代码如下：

```
//第 3 章/事件处理方法.html
...
<div id='app'>
    <!-- `greet` 是在下面定义的方法名 -->
    <button v-on:click="greet">Greet</button>
</div>
<script type='text/JavaScript'>
    const vm = new Vue({
        el: '#app',
        data: {
            name: 'Vue.js'
        },
        methods: {
            greet: function (event) {
                //`this` 在方法里指向当前 Vue.js 实例
                alert('Hello ' + this.name + '!')
                //`event` 是原生 DOM 事件
                if (event) {
                        alert(event.target.tagName)
                    }
            }
        }
    })

//也可以用 JavaScript 直接调用方法
    vm.greet() //=> 'Hello Vue.js!'
</script>
...
```

3.4.3 内联处理器中的方法

除了可以使用 v-on 给元素绑定一种方法外，也可以在内联 JavaScript 中调用方法。在 Say hi 和 Say what 两个按钮的内联 JavaScript 中调用 say 方法，代码如下：

```
//第 3 章/内联处理器中调用方法.html
...
<div id='app'>
    <button v-on:click="say('hi')">Say hi</button>
    <button v-on:click="say('what')">Say what</button>
</div>
<script type='text/JavaScript'>
```

```
    const vm = new Vue({
        el: '#app',
        methods: {
            say: function(msg){
                alert(msg)
            }
        }
    })
</script>
...
```

有时也需要在内联语句处理器中访问原始的 DOM 事件,可以用特殊变量 $event 把它传入方法。使用 $event 把事件对象传入方法中,代码如下:

```
//第 3 章/内联处理器中调用方法.html
...
        <div id='app'>
            <button v-on:click="warn('表单不能被提交')"> submit </button>
        </div>
        <script type='text/JavaScript'>
            const vm = new Vue({
                el: '#app',
                methods: {
                    warn: function (message, event) {
                        //现在可以访问原生事件对象了
                        if (event) {
                            //阻止事件提交
                            event.preventDefault()
                        }
                        alert(message)
                    }
                }
            })
        </script>
...
```

3.4.4 事件修饰符

在事件处理程序中调用 event.preventDefault()或 event.stopPropagation()方法是非常常见的需求。尽管开发人员可以在方法中轻松实现这一点,但更好的方式是方法只有纯粹的数据逻辑,而不是去处理 DOM 事件的细节。

为了解决这个问题,Vue.js 为 v-on 提供了事件修饰符。同前文提过的 v-model 修饰符一样,是由点开头的指令后缀来表示的。

(1) .stop:阻止单击事件继续传播。

（2）.prevent：提交事件不再重载页面。

（3）.capture：使用事件捕获模式添加事件。也就是说，如果内联元素中有其他事件，则先触发.capture 修饰的事件。

（4）.self：只触发发生在当前元素身上的事件，也就是说，事件不会由内部触发。

（5）.once：只触发一次。

（6）.passive：执行默认方法。浏览器只有等内核线程执行到事件监听器对应的JavaScript代码时，才能知道内部是否会调用preventDefault()方法来阻止事件的默认行为，所以浏览器本身是没有办法对这种场景进行优化的。这种场景下，用户的手势事件无法快速产生，会导致页面无法快速执行滑动逻辑，从而让用户感觉到页面卡顿。通俗地说就是每次事件产生时，浏览器都会去查询是否应由preventDefault阻止该次事件的默认动作。加上passive就是为了明确地告诉浏览器，不用查询了，这里没用preventDefault阻止默认动作。.passive修饰符一般用于滚动监听，如@scroll和@touchmove。因为在滚动监听过程中，移动每像素都会产生一次事件，每次都使用内核线程查询prevent会使滑动卡顿，通过添加passive将内核线程查询跳过，可以大大提升滑动的流畅度。

各种事件修饰符的使用和说明如下：

```
<!-- 阻止单击事件继续传播 -->
<a v-on:click.stop="doThis"></a>

<!-- 提交事件不再重载页面 -->
<form v-on:submit.prevent="onSubmit"></form>

<!-- 修饰符可以串联 -->
<a v-on:click.stop.prevent="doThat"></a>

<!-- 只有修饰符 -->
<form v-on:submit.prevent></form>

<!-- 添加事件监听器时使用事件捕获模式 -->
<!-- 即内部元素触发的事件先在此处理，然后才交由内部元素进行处理 -->
<div v-on:click.capture="doThis">...</div>

<!-- 只在 event.target 是当前元素时触发处理函数 -->
<!-- 即事件不是从内部元素触发的 -->
<div v-on:click.self="doThat">...</div>

<!-- 单击事件将只会触发一次 -->
<a v-on:click.once="doThis"></a>

<!-- 滚动事件的默认行为（滚动行为）将会立即触发 -->
<!-- 而不会等待 `onScroll` 完成 -->
<!-- 这其中包含 `event.preventDefault()` 的情况 -->
<div v-on:scroll.passive="onScroll">...</div>
```

使用事件修饰符时的注意事项如下。

(1) .passive 和.prevent 冲突，不能同时绑定在同一个监听器上。

(2) 使用修饰符时，顺序很重要，相应的代码会以同样的顺序产生。

用 v-on：click.prevent.self 会阻止所有的单击，而 v-on：click.self.prevent 只会阻止对元素自身的单击。

(3) .once 修饰符还能被用到自定义的组件事件上。

3.4.5 按键修饰符

在监听键盘事件时，经常需要检查详细的按键。Vue.js 允许为 v-on 在监听键盘事件时添加按键修饰符，对按键进行详细检查后再执行绑定的方法。当所在的键是 Enter 键的时候，调用 vm.submit()方法，代码如下：

```
<!-- 只有在 `key` 是 `Enter` 时调用 `vm.submit()` -->
< input v - on:keyup.enter = "submit">
```

Vue.js 提供了绝大多数常用的按键的别名，它们是：

(1) .enter。

(2) .tab。

(3) .delete(捕获“删除”和“退格”键)。

(4) .esc。

(5) .space。

(6) .up。

(7) .down。

(8) .left。

(9) .right。

除了这些 Vue.js 提供的按键别名外，Vue.js 还提供了一个机制，允许开发人员通过全局 config.keyCodes 对象自定义按键修饰符别名。通过 config.keyCodes 对象，定义 v、f1、"media-play-pause"和 up4 个按键别名，代码如下：

```
Vue.config.keyCodes = {
  v: 86,
  f1: 112,
  //camelCase 不可用
  mediaPlayPause: 179,
  //取而代之的是 kebab - case 且用双引号括起来
  "media - play - pause": 179,
  up: [38, 87]
}
```

自定义按键别名的代码如下：

```
<input type="text" v-on:keyup.media-play-pause|v|f1|up="method">
```

3.4.6 系统修饰符

除了前面的事件、按键修饰符外，Vue.js 还提供了系统修饰符，实现按特殊键配合鼠标或键盘事件才能触发事件。比较常用的键是 Ctrl、Alt、Shift 和 Meta 键，它们对应的修饰符是.ctrl、.alt、.shift 和.meta。注意：在 Mac 系统的键盘上，Meta 键对应 command 键(⌘)。在 Windows 系统的键盘上 Meta 键对应 Windows 徽标键(⊞)。在 Sun 操作系统的键盘上，Meta 键对应实心宝石键(◆)。它们同其他键盘和鼠标的联合使用代码如下：

```
<!-- Alt + C -->
<input v-on:keyup.alt.67="clear">

<!-- Ctrl + Click -->
<div v-on:click.Ctrl="doSomething">Do something</div>
```

需要注意的是，修饰键与常规按键不同，在和 keyup 事件一起使用时，当事件触发时修饰键必须处于按下状态。换句话说，只有在按住 Ctrl 键的情况下释放其他按键，才能触发 keyup.ctrl，而单单释放 Ctrl 键不会触发事件。如果想要这样的行为，应将 Ctrl 键换作 keyCode：keyup.17。

Vue 2.5 以后，新增了.exact 修饰符，能精准地控制系统修饰符组合事件，代码如下：

```
<!-- 即使 Alt 或 Shift 键被一同按下时也会触发 -->
<button v-on:click.Ctrl="onClick">A</button>

<!-- 有且只有 Ctrl 键被按下的时候才触发 -->
<button v-on:click.Ctrl.exact="onCtrlClick">A</button>

<!-- 没有任何系统修饰符被按下的时候才触发 -->
<button v-on:click.exact="onClick">A</button>
```

最后，Vue.js 还提供了 3 个仅响应特定的鼠标按钮的修饰符，它们分别是.left、.right 和.middle，代码如下：

```
<span v-on:mousedown.left='mouseLeft'>鼠标左击</span>
<span v-on:mousedown.right='mouseRight'>鼠标右击</span>
<span v-on:mousedown.middle='mouseMiddle'>鼠标中键</span>
```

3.5 指令

指令 (Directives) 是带有 v- 前缀的特殊 attribute。指令的职责是，当表达式的值改变时，将其产生的连带影响响应式地作用于 DOM。熟悉使用指令，能方便开发人员快速地实现页面逻辑。

3.5.1 v-text 和 v-html 指令

v-text 和 v-html 的作用是在元素之间插入动态值，不同的是，v-html 会识别动态内容中的网页信息，而 v-text 是将网页信息当纯文本处理。

语法如下：

```
<标签 v-text/v-html="表达式"></标签>
```

案例代码如下：

```
<div id="app">
    <p v-text="htmlName"></p>
    <p v-html="htmlName"></p>
</div>
<script>
    const vm = new Vue({
        el:"#app",
        data:{
            htmlName:'<font color="red">张三</font>'
        }
    })

</script>
```

3.5.2 v-bind 指令

v-bind 指令的作用是动态地绑定一个或多个 attribute，或将一个组件 prop 绑定到表达式。

在绑定 class 或 style attribute 时，支持其他类型的值，如数组或对象。

在绑定 prop 时，prop 必须在子组件中声明。可以用修饰符指定不同的绑定类型。

当没有参数时，可以绑定到一个包含键值对的对象。注意此时 class 和 style 绑定不支持数组和对象。

语法如下：

```
<标签 v-bind:属性='表达式'>...</标签>
```

案例代码如下：

```
<!-- 绑定一个 attribute -->
<img v-bind:src="imageSrc">

<!-- 动态 attribute 名 (2.6.0+) -->
```

```
<button v-bind:[key] = "value"></button>

<!-- 缩写 -->
<img :src = "imageSrc">

<!-- 动态 attribute 名缩写 (2.6.0+) -->
<button :[key] = "value"></button>

<!-- 内联字符串拼接 -->
<img :src = "'/path/to/images/' + fileName">

<!-- class 绑定 -->
<div :class = "{ red: isRed }"></div>
<div :class = "[classA, classB]"></div>
<div :class = "[classA, { classB: isB, classC: isC }]">

<!-- style 绑定 -->
<div :style = "{ fontSize: size + 'px' }"></div>
<div :style = "[styleObjectA, styleObjectB]"></div>

<!-- 绑定一个全是 attribute 的对象 -->
<div v-bind = "{ id: someProp, 'other-attr': otherProp }"></div>

<!-- 通过 prop 修饰符绑定 DOM attribute -->
<div v-bind:text-content.prop = "text"></div>

<!-- prop 绑定."prop"必须在 my-component 中声明. -->
<my-component :prop = "someThing"></my-component>

<!-- 通过 $props 将父组件的 props 一起传给子组件 -->
<child-component v-bind = " $props"></child-component>

<!-- XLink -->
<svg><a :xlink:special = "foo"></a></svg>


<div id = "app">
    <div v-bind:id = "divId"> div demo </div>
    <div v-bind:disabled = "isdisabled"> disabled </div>
</div>
<script>
    const vm = new Vue({
            el:"#app",
            data:{
                    divId:'demoDiv',
                    isdisabled:true
```

```
        }
    })

    console.log(document.getElementById('demoDiv'))
</script>
```

说明：如果 isButtonDisabled 的值是 null、undefined 或 false，则 disabled attribute 甚至不会被包含在渲染出来的 <button> 元素中。

v-bind 可以缩写成冒号，代码如下：

```
<标签 :属性 = '表达式'>...</标签>
```

3.5.3 v-once 指令

v-once 指令的作用是控制插值只会在 UI 上渲染一次，当再改变属性的值时此值不会同步到 UI 上。

语法如下：

```
<标签 v-once ...>...</标签>
```

案例代码如下：

```
<div id="app">
    <p v-once>{{userName}}</p>
    <p>{{userName}}</p>
    <p v-once v-bind:id="'id' + age"> age{{age}}</p>
    <p v-bind:id="'id' + age"> age{{age}}</p>
</div>
<script>
    const vm = new Vue({
        el:"#app",
        data:{
            userName:'zhangsan',
            age:12
        }
    })
    //修改值后,不再同步到 UI
    vm.$data.age = 14;
    vm.$data.userName = 'lisi';
</script>
```

3.5.4 v-model 指令

v-text 和 v-html 只能实现单向绑定：对象属性能同步到 UI，但是在 UI 上改变值后，不会同步到对象属性。v-model 指令可以实现双向绑定，经常用在表单元素上，也可以用在 Vue.js 的自定义 Component 上。

语法如下：

```
<表单元素|VueComponent v-mode="表达式" ...>...</表单元素>
```

案例代码如下：

```
<div id="app">
    <input v-model="message" type="text" name="msg" placeholder="请输入消息" /><br/>
    <textarea v-model="message"></textarea><br/>
    <input v-model="edu" type="text" placeholder="输入学历:1,2,3">
    <select v-model="edu">
        <option value="1">大专</option>
        <option value="2">本科</option>
        <option value="3">博士</option>
    </select><br/>
    <input v-model="sex" placeholder="输入性别:0,1" />
    <input v-model="sex" type="radio" value="0" name="sex"/>男
    <input v-model="sex" type="radio" value="1" name="sex"/>女<br/>
    <input v-model="likes" placeholder="输入值" />
    <input v-model="likes" type="checkbox" value="0" name="likes0"/>阅读<br/>
</div>
<script>
    const vm = new Vue({
        el:"#app",
        data:{
            message:'demo',
            edu:2,
            sex:1,
            likes:false
        }
    })
</script>
```

3.5.5 v-if、v-else-if 和 v-else 指令

v-if、v-else-if 和 v-else 指令的作用是条件性地渲染一部分内容。

语法如下：

```
<div v-if="type === 'A'">
  A
```

```
</div>
<div v-else-if="type === 'B'">
  B
</div>
<div v-else-if="type === 'C'">
  C
</div>
<div v-else>
  Not A/B/C
</div>
```

案例代码如下：

```
<div id="app">
    <div v-if="type > 0.8">
     A
    </div>
    <div v-else-if="type > 0.6">
     B
    </div>
    <div v-else-if="type > 0.4">
     C
    </div>
    <div v-else>
     Not A/B/C
    </div>
    <button v-on:click="changeType"> changeType </button>
</div>
<script>
    const vm = new Vue({
        el:"#app",
        data:{
            type:Math.random()
        },
        methods:{
                    changeType:function(){
                        this.type = Math.random;
            }
        }
    })
</script>
```

说明：在分支语句和 v-for 语句中，可能会存在相同元素需要渲染。Vue.js 会尽可能高效地渲染元素，通常会复用已有元素而不是从头开始渲染。这么做除了使 Vue.js 运行时变得非常快之外，还有其他一些好处。例如，应用允许用户在不同的登录方式之间切换，代码如下：

```
<template v-if="loginType === 'username'">
  <label>Username</label>
  <input placeholder="Enter your username">
</template>
<template v-else>
  <label>E-mail</label>
  <input placeholder="Enter your email address">
</template>
```

在上面的代码中切换 loginType 将不会清除用户已经输入的内容。因为两个模板使用了相同的元素,<input>不会被替换掉——仅仅替换了它的 placeholder。这样不总是符合实际需求,所以 Vue.js 为开发人员提供了一种方式来表达"这两个元素是完全独立的,不要复用它们"。只需添加一个具有唯一值的 key attribute,代码如下:

```
<template v-if="loginType === 'username'">
  <label>Username</label>
  <input placeholder="Enter your username" key="username-input">
</template>
<template v-else>
  <label>E-mail</label>
  <input placeholder="Enter your email address" key="email-input">
</template>
```

每次切换时,输入框都将被重新渲染。

3.5.6 v-show 指令

v-show 指令的作用是根据条件展示元素。带有 v-show 的元素始终会被渲染并保留在 DOM 中。v-show 只是简单地切换元素的 CSS property display。

语法如下:

```
<标签 v-show="表达式">...</标签>
```

案例代码如下:

```
<div id="app">
    <div v-show="ok">ok</div>
    <button v-on:click="toChange">Change</button>
</div>
<script>
    const vm = new Vue({
        el:"#app",
        data:{
            ok:true
```

```
            },
            methods:{
                toChange:function(){
                    this.ok = !this.ok;
                }
            }
        })
</script>
```

注意：v-show 不支持<template>元素，也不支持 v-else。

v-if 同 v-show 的对比：

v-if 是“真正”的条件渲染，因为它会确保在切换过程中条件块内的事件监听器和子组件适当地被销毁和重建。

v-if 也是惰性的：如果在初始渲染时条件为假，则什么也不做——直到条件第一次变为真时才会开始渲染条件块。v-show 就简单得多——不管初始条件是什么，元素总是会被渲染，并且只是简单地基于 CSS 进行切换。

一般来讲，v-if 有更高的切换开销，而 v-show 有更高的初始渲染开销，因此，如果需要非常频繁地切换，则使用 v-show 较好；如果在运行时条件很少改变，则使用 v-if 较好。

3.5.7 v-for 指令

v-for 指令的作用是遍历数组元素，并且渲染到 ui。

语法如下：

```
<标签 v-for="item in|of items" :key="item">...</标签>
or
<标签 v-for="(item, index) in|of items" :key="index">...</标签>
```

1）遍历数组

案例代码如下：

```
<div id="app">
    <div v-for="item in items" :key="item.msg">
            <span v-text="item.msg"></span>
    </div>
    <div v-for="(item, index) in items" :key="index">
            <p v-text="item.msg + '->' + index"></p>
    </div>

    <div v-for="item of items" :key="item.msg">
            <span v-text="item.msg"></span>
    </div>
    <div v-for="(item, index) of items" :key="index">
```

```
        <p v-text="item.msg + '->' + index"></p>
    </div>
</div>
<script>
    const vm = new Vue({
            el:"#app",
            data:{
                items:[
                    {msg:'m1'},
                    {msg:'m2'},
                    {msg:'m3'},
                    {msg:'m4'},
                ]
            }
    })
</script>
```

2）遍历对象

案例代码如下：

```
<div id="app">
    <div v-for="value in user" :key='value'>
            {{value}}
    </div>

    <div v-for="(value, name) in user" :key='name'>
            {{name}}->{{value}}
    </div>

    <div v-for="(value, name, index) in user" :key='name+index'>
            {{index}}:{{name}}->{{value}}
    </div>
</div>
<script>
    const vm = new Vue({
            el:"#app",
            data:{
                user:{
                    name:'zhangsan',
                    age:12,
                    sex:'男',
                }
            }
    })
</script>
```

3.5.8 v-on 指令

v-on 指令的作用是绑定事件监听器。事件类型由参数指定。表达式可以是一种方法的名字或一个内联语句，如果没有修饰符，则可以省略。

当用在普通元素上时，只能监听原生 DOM 事件。当用在自定义元素组件上时，也可以监听子组件触发的自定义事件。

在监听原生 DOM 事件时，方法以事件为唯一的参数。如果使用内联语句，则语句可以访问一个 $event property：v-on：click="handle('ok'，$event)"

语法如下：

```
<标签 v-on:事件名称[.修饰符] = "表达式"> ... </标签>
```

v-on 指令可以缩写为 @。

v-on 支持的修饰符如下。

. stop：调用 event. stopPropagation()。

. prevent：调用 event. preventDefault()。

. capture：添加事件侦听器时使用 capture 模式。

. self：只当事件是从侦听器绑定的元素本身触发时才触发回调。

. {keyCode | keyAlias}：只当事件是从特定键触发时才触发回调。

. native：监听组件根元素的原生事件。

. once：只触发一次回调。

. left：(2.2.0)只当单击鼠标左键时触发。

. right：(2.2.0)只当右击时触发。

. middle：(2.2.0)只当单击鼠标中键时触发。

. passive：(2.3.0)以{passive：true}模式添加侦听器。

案例代码如下：

```
<!-- 方法处理器 -->
<button v-on:click = "doThis"></button>

<!-- 动态事件 (2.6.0+) -->
<button v-on:[event] = "doThis"></button>

<!-- 内联语句 -->
<button v-on:click = "doThat('hello', $event)"></button>

<!-- 缩写 -->
<button @click = "doThis"></button>

<!-- 动态事件缩写 (2.6.0+) -->
```

```
<button @[event] = "doThis"></button>

<!-- 停止冒泡 -->
<button @click.stop = "doThis"></button>

<!-- 阻止默认行为 -->
<button @click.prevent = "doThis"></button>

<!-- 阻止默认行为,没有表达式 -->
<form @submit.prevent></form>

<!-- 串联修饰符 -->
<button @click.stop.prevent = "doThis"></button>

<!-- 键修饰符,键别名 -->
<input @keyup.enter = "onEnter">

<!-- 键修饰符,键代码 -->
<input @keyup.13 = "onEnter">

<!-- 单击回调只会触发一次 -->
<button v-on:click.once = "doThis"></button>

<!-- 对象语法 (2.4.0+) -->
<button v-on = "{ mousedown: doThis, mouseup: doThat }"></button>
```

3.6 Vue.js 响应原理

Vue.js 的一个重要特色是实现了 View 和 Model 的自动同步响应，接下来介绍一下 Vue.js 的响应原理和特殊数据的检测响应。

3.6.1 响应式原理

当项目代码把一个普通的 JavaScript 对象传入 Vue.js 实例作为 data 选项时，Vue.js 将遍历此对象所有的 property，并使用 Object.defineProperty 把这些 property 全部转换为 getter/setter。Object.defineProperty 是 ES5 中一个无法 shim（低版本不兼容）的特性，这也就是 Vue.js 不支持 IE 8 及更低版本浏览器的原因。

这些 getter/setter 对用户来讲是不可见的，但是在内部它们让 Vue.js 能够追踪依赖，在 property 被访问和修改时通知变更。这里需要注意的是不同浏览器在控制台打印数据对象时对 getter/setter 的格式化并不同。

每个组件实例都对应一个 watcher 实例，它会在组件渲染的过程中把“接触”过的数据 property 记录为依赖。之后当依赖项的 setter 被触发时，会通知 watcher，从而使与它关联的组件重新被渲染,如图 3.3 所示。

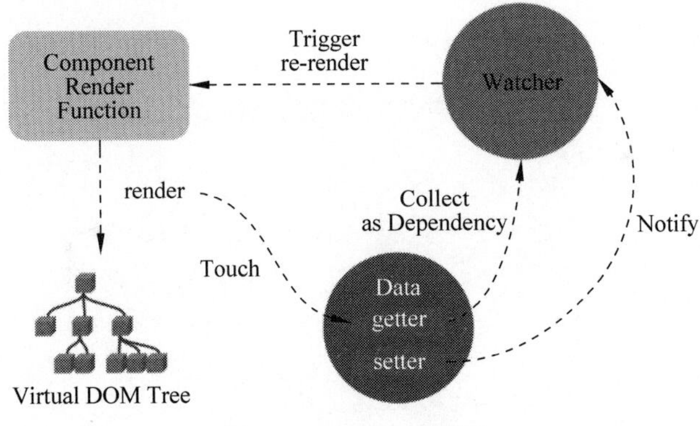

图 3.3 响应式原理图

由于 JavaScript 的限制,Vue.js 不能检测数组和对象的变化。尽管如此,状态管理工具 Vuet 提供了一些办法来回避这些限制并保证它们的响应性。

3.6.2 对象的检测响应

Vue.js 无法检测对象 property 的添加或移除。由于 Vue.js 会在初始化实例时对 property 执行 getter/setter 转换,所以 property 必须在 data 对象上存在才能让 Vue.js 将它转换为响应式的。

同时 Vue.js 不允许动态添加根级响应式 property,所以开发人员必须在初始化实例前声明所有根级响应式 property,哪怕只是一个空值。如果未在 data 选项中声明 message,则 Vue.js 将警告渲染函数正在试图访问不存在的 property,代码如下:

```
var vm = new Vue({
  data: {
    //将 message 声明为一个空值字符串
    message: ''
  },
  template: '<div>{{ message }}</div>'
})
//之后设置 `message`
vm.message = 'Hello!'
```

这样的限制在背后是有其技术原因的,它消除了在依赖项跟踪系统中的一类边界情况,也使 Vue.js 实例能更好地配合类型检查系统工作,但与此同时在代码可维护性方面也有一点重要的考虑:data 对象就像组件状态的结构(schema)。提前声明所有的响应式 property,可以让组件代码在未来修改或给其他开发人员阅读时更易于理解。

对于已经创建的实例,Vue.js 不允许动态添加根级别的响应式 property,但是,可以使用 Vue.set(object,propertyName,value)方法向嵌套对象添加响应式 property,代码如下:

```
Vue.set(vm.someObject, '新属性名称', 属性值)
```

还可以使用 vm. $set 实例方法，这也是全局 Vue.set()方法的别名，代码如下：

```
this.$set(this.someObject,'b',2)
```

如果需要添加多个属性，则可以使用 Object.assign()，将新对象的属性（可以是多个属性）与原对象的属性一起混合进一个新对象，代码如下：

```
//第 3 章/检测对象.html
…
<div id = "app">
    <p>{{a}}</p>
    <p>{{user.userName}}</p>
    <p>{{user.age}}</p>
    <p>{{user.sex}}</p>
    <p>{{user.code}}</p>
    <p>{{user.name}}</p>
</div>
<script>
    const _address = {code:'123'}
    const vm = new Vue({
            el:"#app",
            data:{
                a:1,
                user:{
                    userName:'zhangsan',
                }
            }
    })
    vm.user.age = 12;
    //能输出 age 的值,但是不能响应到 View
    console.log(vm.user.age + "," + vm.user.sex)
    Vue.set(vm.user, 'sex', '男')
    //输出用户对象,并且能响应到 View
    console.log(vm.user)
    //全部合并成 vm.user 对象,响应到 View
    vm.user = Object.assign({},vm.user, {code:'001',name:'张三'})
</script>
…
```

3.6.3 数组的检测响应

Vue.js 不能检测以下数组的变动：

（1）当利用索引直接设置一个数组项时，Vue.js 检测不到数组的变动。

例如：vm. items[indexOfItem]=newValue。

(2) 当修改数组的长度时，Vue. js 检测不到数组的变动。

例如：vm. items. length=newLength。

Vue. js 提供了两种方式都可以实现和 vm. items[indexOfItem]=newValue 相同的效果，同时也将在响应式系统内触发状态更新。一种方式是调用 Vue. js 的 set()方法，替代数组指定下标的值；另一种是调用数组的 splice()方法，替换数组指定下标的值。可以修改数组指定索引对应元素的值，Vue. js 检测不到变动的问题，代码如下：

```
//Vue.set
Vue.set(vm.items, indexOfItem, newValue)
//Array.prototype.splice
//splice(index,howmany,item1…itemn) 删除从 index 开始的 howmany 个元素
//用 item1..n 替代这些删除的元素,并且返回删除的元素
vm.items.splice(indexOfItem, 1, newValue)
```

也可以使用 vm. $set()实例方法，该方法是全局方法 Vue. set()的一个别名，代码如下：

```
vm.$set(vm.items, indexOfItem, newValue)
```

为了修改数组长度，Vue. js 检测不到变动的问题可以使用 splice，代码如下：

```
vm.items.splice(newLength)
```

综合示例代码如下：

```
//第 3 章/检测数组.html
<div id="app">
    <p>{{ages}}</p>
</div>
<script>
    const vm = new Vue({
            el:"#app",
            data:{
                ages:[1,2,3]
            }
    })
    vm.ages[1] = 20;
    //输出修改,但不响应到 View
    console.log(vm.ages)
    //响应到 View
    Vue.set(vm.ages, 1, 21)
    vm.$set(vm.ages, 1, 22)
    vm.$set(vm.ages, vm.ages.length,4)
    console.log(vm.ages)
```

```
    //修改数组长度,响应到 View
    vm.ages.splice(vm.ages.length - 1)
</script>
```

除了前面介绍的 set()方法和 splice()方法外,Vue.js 还提供了一些其他方法,实现变更数组和替换数组,同时还能检测到数组的变化。

1. 变更数组方法

变更方法会改变以前的数组(原数组)。Vue.js 将被侦听的数组的变更方法进行了包裹,所以它们也将会触发视图更新。被包裹的变更数组的方法如下。

1) push()

在数组中添加一个新元素,并返回数组的新长度值,代码如下:

```
arrayObj.push([item1 [item2 [. . . [itemN ]]]])
```

2) pop()

移除数组中的最后一个元素并返回该元素,代码如下:

```
arrayObj.pop()
```

3) shift()

移除数组中的第 1 个元素并返回该元素,代码如下:

```
arrayObj.shift()
```

4) unshift()

将指定的元素插入数组的开始位置并返回新数组的长度,代码如下:

```
arrayObj.unshift([item1[, item2 [, . . . [, itemN]]]])
```

5) splice()

从一个数组中移除一个或多个元素,如果必要,则在所移除元素的位置上插入新元素,并且返回所移除的元素,代码如下:

```
arrayObj.splice(start, deleteCount,[item1[, item2[, . . . [,itemN]]]])
```

6) sort()

返回一个对元素完成排序的数组,代码如下:

```
arrayobj.sort(sortfunction)
```

如果为 sortfunction 参数提供了一个函数,则表示在执行 sort 时,基于 sortfunction 对元素进行大小比较。该函数必须返回下列值之一:

(1) 负值，如果所传递的第 1 个参数比第 2 个参数小。

(2) 零，如果两个参数相等。

(3) 正值，如果第 1 个参数比第 2 个参数大。

7) reverse()

返回一个数组元素被反转的数组，代码如下：

```
arrayObj.reverse()
```

2. 替换数组方法

替换返回，不会改变以前的数组，但是会返回一个新的数组。

1) filter()

根据 filter 条件返回新数组，代码如下：

```
array.filter(function(currentValue, indedx, arr), thisValue)
```

currentValue：必须，当前元素的值；index：可选，当前元素的索引值；arr：可选，当前元素属于的数组对象；thisValue：可选，对象作为该执行回调时使用，传递给函数，用作"this"的值。如果省略了 thisValue，则"this"的值为"undefined"。以下代码演示了 filter 方法的使用：

```
var arr = [3,9,4,3,6,0,9];
//返回大于 5 的值
function max5(arr){
    return arr.filter(x) = (> x > 5);
}
//移除数组中与 item 相同的值并返回
function remove(arr, item) {
    return arr.filter(val => val != item);
}
//移除重复值
var r = arr.filter(function (element, index, self) {
        return self.indexOf(element) == index;
});
//查找所有同 item 值相同的元素的位置
function findAllOccurrences(arr, target) {
    var res = [];
    arr.filter(function(ele, index){
        return (ele === target)&&res.push(index);
    })
    return res;
}
```

2）concat()

返回一个新数组，这个新数组是由两个或更多个数组组合而成的，代码如下：

```
array1.concat([item1[, item2[, . . . [, itemN]]]])
```

3）slice()

截取一段子数组并返回，代码如下：

```
arrayObject.slice(start,end)            //包含开始，不包含结尾
```

3.6.4 异步更新问题

Vue.js 在更新 DOM 时是异步执行的。只要侦听到数据变化，Vue.js 将开启一个队列，并缓冲在同一事件循环中发生的所有数据变更。如果同一个 watcher 被多次触发，则只会被推入队列中一次。这种在缓冲时去除重复数据对于避免不必要的计算和 DOM 操作是非常重要的，然后，在下一个事件的循环 tick 中，Vue.js 将刷新队列并执行实际（已去重的）工作。Vue.js 在内部对异步队列尝试使用原生的 Promise.then、MutationObserver 和 setImmediate，如果执行环境不支持，则会采用 setTimeout(fn,0)代替。

例如，如果设置 vm.someData='new value'，则该组件不会立即被重新渲染。当刷新队列时，组件会在下一个事件循环 tick 中更新。多数情况下开发人员不需要关心这个过程，但是如果开发人员想基于更新后的 DOM 状态来做点什么，这就可能会有些棘手。虽然 Vue.js 通常鼓励开发人员使用"数据驱动"的方式思考，避免直接接触 DOM，但是有时开发人员必须这么做。为了在数据变化之后等待 Vue.js 完成更新 DOM 操作，可以在数据变化之后立即使用 Vue.nextTick(callback)。这样回调函数将在 DOM 更新完成后被调用，代码如下：

```
<div id = "app">
    <div>{{message}}</div>
</div>
<script>
    const vm = new Vue({
            el:"#app",
            data:{
                        message:"123"
        }
    })
    vm.message = 'new message' //更改数据
    let bool = vm.$el.textContent === 'new message' //false
    console.log(bool)
    Vue.nextTick(function () {
```

```
    let bool = vm. $el.textContent === 'new message' //true
    console.log(bool)
  })
</script>
```

在组件内使用 vm. $nextTick() 实例方法特别方便，因为它不需要全局 Vue.js，并且回调函数中的 this 将自动被绑定到当前的 Vue.js 实例上，代码如下：

```
Vue.component('example', {
  template: '<span>{{ message }}</span>',
  data: function ) {
    return {
      message: '未更新'
    }
  },
  methods: {
    updateMessage: function () {
      this.message = '已更新'
      console.log(this. $el.textContent) // => '未更新'
      this. $nextTick(function () {
        console.log(this. $el.textContent) // => '已更新'
      })
    }
  }
})
```

因为 $nextTick() 会返回一个 Promise 对象，所以开发人员可以使用新的 ES2017 async/await 语法完成相同的事情，代码如下：

```
methods: {
  updateMessage: async function () {
    this.message = '已更新'
    console.log(this. $el.textContent) // => '未更新'
    await this. $nextTick()
    console.log(this. $el.textContent) // => '已更新'
  }
}
```

第4章

compute属性和watch侦听器

第3章介绍了Vue.js实现View同Vue.js实例对象中数据属性的同步，当Vue.js实例对象的对象属性值发生改变的时候，能自动更新并渲染到View，但是在实际项目中，Vue.js实例对象中的数据，会依赖其他数据的改变而改变，View层需要实时感知到底层数据的变化，从而及时渲染到View层并显示。Vue.js提供了compute和watch属性，可以满足这样的项目需求。本章介绍compute和watch侦听器的特点、使用方法和注意事项。

4.1 compute属性

模板内的表达式在使用时非常便利，但是设计它们的初衷是进行简单运算。在模板中放入太多的逻辑会让模板负担过重且难以维护。以下代码比较难理解和维护：

```
<div id = "example">
  {{ message.split('').reverse().join('') }}
</div>
```

在这个地方，模板不再是简单的声明式逻辑。开发人员必须看一段时间才能意识到，这里是想显示变量message的翻转字符串。当开发人员在模板中的多处包含此翻转字符串时，就会更加难以处理。

对于类似这样的复杂逻辑，开发人员应使用计算属性实现，代码如下：

```
//第4章/计算属性.html
…
<div id = 'app'>
    <p>基本属性 message: "{{ message }}"</p>
    <p>计算属性 reversedmessage: "{{ reversedMessage }}"</p>
</div>
<script type = 'text/JavaScript'>
    const vm = new Vue({
        el: '#app',
        data: {
```

```
            message: 'Hello'
        },
        computed: {
            //计算属性的 getter
            reversedMessage: function () {
            //`this` 指向 vm 实例
            return this.message.split('').reverse().join('')
            }
        }
    })
</script>
...
```

在上面的代码中声明了一个计算属性 reversedMessage，并且定义了一个函数，将这个计算属性赋给这个函数，相当于 vm.reversedMessage 计算属性的 getter()函数。可以直接通过这个函数获取计算属性 reversedMessage 的值。因为 reversedMessage 计算属性的值是基于 message 属性计算出来的，所以也可以通过改变 message 属性的值，然后重新计算，得到 reversedMessage 的新值，代码如下：

```
console.log(vm.reversedMessage) // => 'olleH'
vm.message = 'Goodbye'
console.log(vm.reversedMessage) // => 'eybdooG'
```

开发人员可以打开浏览器的控制台，自行修改例子中的 vm。vm.reversedMessage 的值始终取决于 vm.message 的值，如果单独修改 vm.reversedMessage 的值，控制台会提示计算属性不能设置值，因为没有 setter()方法。需要说明一下，计算属性默认只有 getter()方法，而没有 setter()方法，所以这样直接改变计算属性的值会抛出异常，但是计算属性是可以添加 setter()方法的，后面再专题介绍。

开发人员可以像绑定普通 property 一样在模板中绑定计算属性。Vue.js 知道 vm.reversedMessage 依赖于 vm.message，因此当 vm.message 发生改变时，所有依赖 vm.reversedMessage 的绑定也会更新，而且最妙的是开发人员已经以声明的方式创建了这种依赖关系：计算属性的 getter()方法是没有副作用(Side Effect)的，这使它更易于测试和理解。

4.1.1 compute 属性的 setter()方法

compute(计算)属性默认只有 getter()方法，开发人员是无法直接修改计算属性的值的。如果开发人员需要直接修改计算属性的值，则可以给计算属性添加 setter()方法。同时定义 fullName 计算属性的 getter()和 setter()方法，代码如下：

```
//第 4 章/计算属性的 setter 方法.html
...
```

```
< div id = 'app'>
    firstName:< input type = "text" v - model = "firstName">< br/>
    lastName:< input type = "text" v - model = "lastName">< br/>
    fullName:< input type = "text" v - model = "fullName">< br/>
</div >
< script type = 'text/JavaScript'>
    const vm = new Vue({
        el: '#app',
        data: {
            firstName: '',
            lastName: ''
        },
        computed:{
            fullName: {
                get: function(){
                    return this.firstName + ' ' + this.lastName;
                },
                set: function(newValue){
                    let names = newValue.split(' ')
                    this.firstName = names[0]
                    this.lastName = names[1]
                }
            }
        }
    })
</script >
...
```

这样就可以在控制台直接给 fullName 计算属性赋予新的值了,如 vm.fullName='zhao liu'。

4.1.2 compute 属性同方法的对比

在 Vue.js 中,可以将计算属性绑定到 View 中进行渲染显示,同样可以将一个 Vue.js 实例对象的方法,绑定到 View 中显示方法返回的结果,在页面上都会显示,代码如下:

```
//第 4 章/计算属性同方法的对比.html
...
< div id = 'app'>
    < div >
        < p >计算属性:< br/>
            {{computedNow}}< br/>
        </p >
        < p >方法:< br/>
            {{methodNow()}}< br/>
        </p >
```

```
        </div>
    </div>
    <script type='text/JavaScript'>
        const vm = new Vue({
            el: '#app',
            methods: {
                methodNow:function(){
                    console.log('调用了 methodNow 方法')
                    return Date.now()
                }
            },
            computed: {
                computedNow:function(){
                    console.log('执行了 computedNow 计算属性')
                    return Date.now()
                }
            }
        })
    </script>
    ...
```

但是如果在 View 中同时绑定多次计算属性和方法，则虽然它们的显示效果一样，但是查看控制台时会发现计算属性函数中的打印语句只执行了一次，而函数中的打印语句执行了多次（每绑定一次就执行一次），代码如下：

```
//第 4 章/计算属性同方法的对比.html
...
<div id='app'>
    <div>
        <p>计算属性:<br/>
            {{computedNow}}<br/>
            {{computedNow}}<br/>
            {{computedNow}}<br/>

        </p>
        <p>方法:<br/>
            {{methodNow()}}<br/>
            {{methodNow()}}<br/>
            {{methodNow()}}<br/>

        </p>
    </div>
</div>
<script type='text/JavaScript'>
    const vm = new Vue({
```

```
        el: '#app',
        methods: {
            methodNow:function(){
                console.log('调用了 methodNow 方法')
                return Date.now()
            }
        },
        computed: {
            computedNow:function(){
                console.log('执行了 computedNow 计算属性')
                return Date.now()
            }
        }
    })
</script>
...
```

其原因是计算属性是基于它们的响应式依赖进行缓存的。也就是说，计算属性会将它们计算的值保存在缓存，只有当它们依赖的响应式数据发生改变后，计算属性才会重新计算，否则就直接从缓存中获取值进行渲染，而方法不一样，每次调用都执行一次。

计算属性适合于计算时性能开销比较大的数据，这样就避免每次渲染的时候都要重新计算，浪费资源。如果开发人员不希望使用缓存，则可改成使用方法。

4.2　watch 侦听器

当需要在数据发生变化且需要执行异步代码或开销比较大的逻辑时使用 compute 属性不合适，开发者可以自定义一个 watch 侦听器，来侦听数据的变化，代码如下：

```
//第 4 章/watch 侦听器.html
<!DOCTYPE html>
<html>
    <head>
        <meta charset="utf-8">
        <title>test Vue.js</title>
         <script src="https://cdn.jsdelivr.net/npm/lodash@4.13.1/lodash.min.js">
</script>
        <script type="text/JavaScript" src="../static/js/Vue.js"></script>
    </head>
    <body>
        <div id="app">
            <p>
                输入问题:
                <input v-model="question">
```

```
            </p>
            <p>{{ answer }}</p>
        </div>
        <script>

            const vm = new Vue({
                el:"#app",
                data:{
                    question:'',
                    answer:'请输入问题,我才好给你答案'
                },
                watch:{
                    //侦听 question 的变化,自动调用
                    question:function(){
                        this.answer = '等待你输入问题...'
                        this.debouncedGetAnswer;
                    }
                },
                created:function(){
                    //500ms 后调用 getAnswer 方法,控制调用频率
                    this.debouncedGetAnswer = _.debounce(this.getAnswer, 500)
                },
                methods:{
                    getAnswer:function(){
                        this.answer = "思索中……"
                        window.setTimeout("getResult)", 500)
                    }
                }
            })

            function getResult(){
                console.log('111')
                vm.$data.answer = '答案是:' + Math.random();
            }
        </script>
    </body>
</html>
```

在上面的代码中,定义了一个 question 侦听器,用于侦听 question 数据属性。当 question 数据属性发生变化时,会自动执行 question 侦听器后面的函数内容。在 question 侦听器的函数逻辑中执行的是一个异步延时代码。

4.3 计算属性同 watch 侦听器的对比

watch 侦听器是 Vue.js 提供的一种用来观察和响应 Vue.js 实例上的数据变动的一种方式,一般只用于变动的开销比较大或异步方法调用的情况。否则使用计算属性要比使用

watch 侦听器合适。虽然可以用 watch 侦听器侦听 firstName1 和 lastName1 的变化，但是用计算属性计算 fullName2 要简单得多，代码如下：

```
//第 4 章/watch 同 compute 的对比.html
...
<div id = 'app'>
    FirstName1 < input type = "text" v - model = "firstName1"> LastName1 < input type = "text"
v - model = "lastName1">< br/>
    FullName1 {{fullName1}}< br/>< br/>
    FirstName2 < input type = "text" v - model = "firstName2"> LastName2 < input type = "text"
v - model = "lastName2">< br/>
    FullName2 {{fullName2}}< br/>< br/>
</div >
< script type = 'text/JavaScript'>
    const vm = new Vue({
        el: '#app',
        data: {
            firstName1: '',
            lastName1: '',
            fullName1: '',
            firstName2: '',
            lastName2: ''
        },
        watch:{
            firstName1: function(){
                this.fullName1 = this.firstName1 + '' + this.lastName1;
            },
            lastName1: function(){
                this.fullName1 = this.firstName1 + '' + this.lastName1;
            }
        },
        computed:{
            fullName2:function(){
                return this.firstName2 + " " + this.lastName2
            }
        },
        methods: {
        }
    })
</script >
...
```

第二篇　Vue.js组件化编程

第5章

组件化编程

组件(Component)是 Vue.js 最强大的功能之一。通过组件,可以扩展 HTML 元素,以及封装可重用的代码。有了组件后,开发人员就可以使用独立可复用的小组件,构建大型的前端应用。绝大多数应用的界面可以抽象成一个组件树。例如一个页面可能会有页头、侧边栏、内容区等部分,这些都可以封装成一个个独立的组件,然后按关系组合起来,形成一个完整的界面,如图5.1所示。

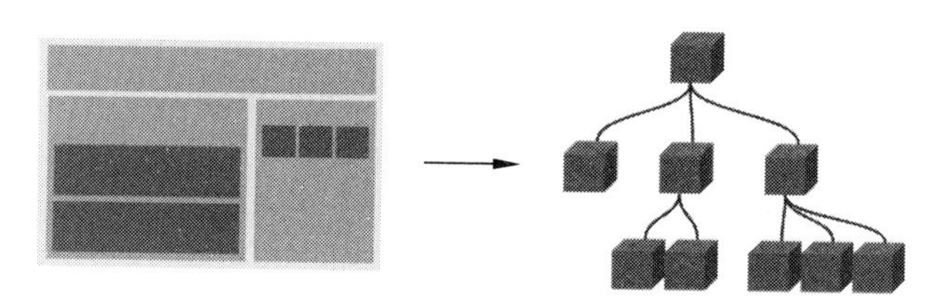

图5.1　应用界面和组件树

本章将介绍怎样定义和使用组件,以及使用组件过程中的各种方式和技巧。

5.1　第1个组件

定义一个计算器组件,实现单击后自动累计按钮被单击的次数。在一个页面上就可以重复地使用这个计算器组件,各种累计自己被单击的次数,代码如下:

```
<div id = 'app'>
    <! -- 重复使用 ButtonCounter 组件 -- >
    < button - counter ></button - counter >
    < button - counter ></button - counter >
    < button - counter ></button - counter >
    < button - counter ></button - counter >
</div >
< script type = 'text/JavaScript'>
    //定义一个计算器组件 -- ButtonCounter
    Vue.component("ButtonCounter", {
```

```
        data(){
            return {
                count: 0
            }
        },
        template:`<button v-on:click="count++">单击了我{{count}}次</button`
    })
    const vm = new Vue({
        el: '#app',
        data: {
        },
        methods: {
        }
    })
</script>
```

5.2 使用自定义组件

自定义组件的步骤分为三步：定义并创建组件、注册组件和使用组件。接下来以计算器按钮组件为案例，介绍怎样自定义组件。

5.2.1 自定义组件

一个组件一般由视图、数据和业务逻辑三部分组成，所以在所定义的组件对象中，一般包含 template、data 和 methods。创建一个组件对象，通过 template、data 和 methods 属性定义组件包含的视图、数据和业务逻辑，代码如下：

```
//第5章/计算器按钮组件.html
...
<script type='text/JavaScript'>
    const buttonCounter = {
        template: `<button v-on:click="counter">我被单击了{{count}}次</button>`,
        data(){
            return {
                count: 0
            }
        },
        methods:{
            counter(){
                this.count++
            }
        }
    }
```

```
    const vm = new Vue({
        el: '#app',
        data: {
        },
        methods: {
        }
    })
</script>
...
```

作为子组件(在一个页面中,有且只能有一个根 Vue.js 实例),data 属性必须是一个函数,返回包含子组件实例所有数据属性的对象。也就是说,一定要写成如上代码的形式,而不能写成如下的形式:

```
...
  data: {
     count: 0
  }
...
```

如果 templat 的视图内容比较复杂,直接写在组件对象的 template 数字后面会太复杂,不方便理解和维护,这时候可以使用< template >元素单独定义,再在组件对象的 template 属性中使用,代码如下:

```
...
<template id = "buttonCounterTemplate">
    <button v - on:click = "counter">我被单击了{{count}}次</button>
</template>
<script type = 'text/JavaScript'>
    const buttonCounter = {
        template: "#buttonCounterTemplate", //引用单独定义的模板
        ...
    }
    ...
</script>
...
```

使用< template >单独定义模板时需要注意,里面的内容有且只能有一个根元素。如果视图中有很多元素,则需要用一个大元素(例如 div)包含全部小元素。

5.2.2 全局注册组件

定义好组件对象后,在使用时需要先注册。注册组件有两种方式:全局注册和局部注册。这里先介绍全局注册的特点和方式。

全局注册的特点是，注册一个组件，在应用的其他组件中都可以使用。其优势是只要注册一次，其他要使用的地方不用再注册，可直接使用。

全局注册方式调用的是 Vue.js 的 component 方法，带两个参数，第 1 个参数是注册的组件名称，第 2 个参数是要注册的组件对象，代码如下：

```
Vue.component(组件名称, 组件对象)
```

使用 Vue.component()方法，全局注册前面定义的 buttonCounter 组件对象，组件的名称是 ButtonCounter，代码如下：

```
...
<script>
  ...
  Vue.component("ButtonCounter", buttonCounter)
  ...
</script
...
```

如果组件直接在 DOM 中使用，则组件名称的命名，除了遵循见名思意的原则外，强烈推荐遵循 W3C 规范中的自定义组件名（字母全小写且必须包含一个连字符）。这样可更好地避免在使用组件的时候，同当前及未来的 HTML 元素相冲突。

定义组件名的方式有两种。

1. 使用 kebab-case

```
Vue.component('my-component-name', { /* ... */ })
```

当使用 kebab-case（短横线分隔命名）定义一个组件名时，开发人员必须在引用这个自定义元素时使用 kebab-case，例如 <my-component-name>。

2. 使用 PascalCase

```
Vue.component('MyComponentName', { /* ... */ })
```

当使用 PascalCase（首字母大写命名）定义一个组件名时，开发人员在引用这个自定义元素时两种命名法都可以使用。也就是说<my-component-name>和<MyComponentName>都是可接受的。注意，尽管如此，直接在 DOM(非字符串的模板)中使用时只有 kebab-case 是有效的。

5.2.3 局部注册组件

全局注册是一次注册，在任何组件中都可以用，在方便使用的同时，也有劣势存在。例如，如果开发者使用一个像 webpack 这样的构建系统，则全局注册所有的组件意味着即便不再使用一个组件了，它仍然会被包含在最终的构建结果中。这造成了用户下载的

JavaScript 中包含了没必要的内容。

在这些情况下，开发人员可以通过一个普通的 JavaScript 对象来定义组件，代码如下：

```
var ComponentA = { /* ... */ }
var ComponentB = { /* ... */ }
var ComponentC = { /* ... */ }
```

然后在 components 选项中定义开发人员想要使用的组件，代码如下：

```
new Vue({
  el: '#app',
  components: {
    'component-a': ComponentA,
    'component-b': ComponentB
  }
})
```

对于 components 对象中的每个 property 来讲，其 property 名就是自定义元素的名字，其 property 值就是这个组件的选项对象。

注意局部注册的组件在其子组件中是不可用的。例如，如果希望 ComponentA 在 ComponentB 中可用，则需要写成如下代码的形式：

```
var ComponentA = { /* ... */ }

var ComponentB = {
  components: {
    'component-a': ComponentA
  },
  ...
}
```

或者通过 Babel 和 webpack 使用 ES2015 模块，这样代码看起来更科学，代码如下：

```
import ComponentA from './ComponentA.vue'

export default {
  components: {
    ComponentA
  },
  ...
}
```

注意在 ES2015+中，在对象中放一个类似 ComponentA 的变量名其实是 ComponentA: ComponentA 的缩写，即这个变量名同时表示变量名和将对象赋给变量名。

在 Vue.js 根实例组件中，使用局部注册的方式注册 ButtonCounter 组件，代码如下：

```
//第 5 章/局部注册.html
…
<div id='app'>
    <button-counter></button-counter>
</div>
<template id="buttonCounterTemplate">
    <button v-on:click="counter">我被单击了{{count}}次</button>
</template>
<script type='text/JavaScript'>
    const ButtonCounter = {
        template: "#buttonCounterTemplate",
        data(){
            return {
                count: 0
            }
        },
        methods:{
            counter(){
                this.count++
            }
        }
    }
    const vm = new Vue({
        el: '#app',
        components:{
            ButtonCounter
        },
        data: {
        },
        methods: {
        }
    })
</script>
…
```

在模块式开发前端应用的项目中，还可以使用 import/require 方式完成组件的局部注册。

5.2.4 使用组件

组件注册完后，就可以用标签的方式，在其他组件中使用被注册过的组件，注意全局注册的组件，可以用在任何组件的视图中，而局部注册的组件，则只能用在注册的当前组件中。定义 ComponentA、ComponentB 和 ComponentC 3 个组件，使用全局注册的方式注册 ComponentA 组件，在 vm 中局部注册 ComponentB 和 ComponentC 两个组件。这样 ComponentA 组件，既可以在 vm 中使用，也可以在 ComponentC 中使用，而 ComponentB

组件，只能在 vm 中使用，当在 ComponentC 组件中使用的时候，就会抛出异常，代码如下：

```
//第 5 章/使用子组件.html
…
<div id='app'>
    <component-a></component-a>
    <component-b></component-b><br/>
    <component-c></component-c>
</div>
<template id="componentC">
    <div>
        <!-- 使用 ComponentA 按钮正常 -->
        <component-a></component-a><br/>
        <!-- 使用 ComponentB 按钮抛出异常 -->
        <component-b></component-b>
    </div>
</template>
<script type='text/JavaScript'>
    const ComponentA = {
        template:`<button>ComponentA 按钮</button>`
    }
    //全局注册 ComponentA,所以可以在 vm 和 ComponentC 中使用
    Vue.component('ComponentA', ComponentA)

    const ComponentB = {
        template:`<button>ComponentB 按钮</button>`
    }
    const ComponentC = {
        template:'#componentC'
    }

    const vm = new Vue({
        el: '#app',
        components:{
            ComponentB,
            ComponentC
        },
        data: {
        },
        methods: {
        }
    })
</script>
…
```

5.3 父组件将值传到子组件

在 Vue.js 中，可以在定义子组件中定义多个 prop 属性，用来接收父组件传过来的数据。也就是说，父组件可以通过子组件的 prop 属性，给子组件传递值。定义一个 ViewCount 组件，显示通过 propCount 传入的值。在 vm 根实例组件中，注册并且使用 ViewCount 组件，显示单击按钮累计的单击次数，代码如下：

```
//第 5 章/父组件通过 prop 属性给子组件传值.html
…
<div id='app'>
    <button v-on:click="clickMe">单击我</button>
    <!-- 使用 ViewCount 组件显示单击次数 -->
    <view-count :prop-count='count'></view-count>
</div>
<template id="viewCountTemplate">
    <div>{{propCount}}</div>
</template>
<script type='text/JavaScript'>
    const ViewCount = {
        props:['propCount'],
        template:"#viewCountTemplate"
    }
    const vm = new Vue({
        el: '#app',
        components:{
            ViewCount
        },
        data: {
            count:0
        },
        methods: {
            clickMe(){
                this.count++
            }
        }
    })
</script>
…
```

传值的语法如下：

```
<子组件名称 :prop 属性名称="表达式"></子组件名称>
```

或

```
<子组件名称 v-bind:prop属性名称="表达式"></子组件名称>
```

5.3.1 prop的大小写

HTML中的attribute名对大小写不敏感，所以浏览器会把所有大写字符解释为小写字符。这意味着当开发人员使用DOM中的模板时，用camelCase（驼峰命名法）命名的prop名需要使用其等价的kebab-case（短横线分隔命名法）命名，代码如下：

```
Vue.component('sub-component', {
  //在 JavaScript 中使用的是 camelCase
  props: ['postTitle'],
  template: '<h3>{{ postTitle }}</h3>'
})

<!-- 在 HTML 中使用的是 kebab-case -->
<sub-component post-title="hello!"></sub-component>
```

5.3.2 prop的数据类型

prop除了支持数字和string类型外，还支持其他类型，代码如下：

```
props: {
  title: String,
  likes: Number,
  isPublished: Boolean,
  commentIds: Array,
  author: Object,
  callback: Function,
  contactsPromise: Promise //or any other constructor
}
```

这不仅为组件提供了使用参考文档，还会在它们遇到错误的类型时从浏览器的JavaScript控制台提示用户。使用方式和说明的样例代码如下。

1. 传入一个数字

```
<!-- 即便 '42' 是静态的,仍然需要 'v-bind' 来告诉 Vue.js -->
<!-- 这是一个 JavaScript 表达式而不是一个字符串. -->
<sub-component v-bind:likes="42"></sub-component>

<!-- 用一个变量进行动态赋值. -->
<sub-component v-bind:likes="post.likes"></sub-component>
```

2. 传入一个布尔值

```
<!-- 包含该 prop 没有值的情况在内,都意味着 'true'. -->
<sub-component is-published></sub-component>
```

```
<!-- 即便 'false' 是静态的,仍然需要 'v-bind' 来告诉 Vue.js -->
<!-- 这是一个 JavaScript 表达式而不是一个字符串. -->
<sub-component v-bind:is-published="false"></sub-component>

<!-- 用一个变量进行动态赋值. -->
<sub-component v-bind:is-published="post.isPublished"></sub-component>
```

3. 传入一个数组

```
<!-- 即便数组是静态的,我们仍然需要 'v-bind' 来告诉 Vue.js -->
<!-- 这是一个 JavaScript 表达式而不是一个字符串. -->
<sub-component v-bind:comment-ids="[234, 266, 273]"></sub-component>

<!-- 用一个变量进行动态赋值. -->
<sub-component v-bind:comment-ids="post.commentIds"></sub-component>
```

4. 传入一个对象

```
<!-- 即便对象是静态的,仍然需要 'v-bind' 来告诉 Vue.js -->
<!-- 这是一个 JavaScript 表达式而不是一个字符串. -->
<sub-component
  v-bind:author="{
    name: 'Veronica',
    company: 'Veridian Dynamics'
  }"
></sub-component>

<!-- 用一个变量进行动态赋值. -->
<sub-component v-bind:author="post.author"></sub-component>
```

5. 传入一个对象的所有 property

如果开发人员要将一个对象的所有 property 都作为 prop 传入,则可以使用不带参数的 v-bind(取代 v-bind: prop-name)。将 user 对象的所有属性传递到组件中,代码如下:

```
<template id="subUserId">
 <div>
 <p>name:{{userAttr.name}}</p>
 <p>age:{{userAttr.age}}</p>
 </div>
</template>
<div id="app">
 <!-- 传入用户对象 -->
 <user-component :user-attr="user"></user-component>
</div>
```

```
<script>
 Vue.component("user-component",{
   props:['userAttr'],
   template:"#subUserId"
 })
 const vm = new Vue({
   el:"#app",
   data:{
   user:{
       name:'张三',
       age:12
     }
   }
 })
</script>
```

5.3.3 prop 单向数据流

所有的 prop 都使其父子 prop 之间形成了一个单向下行绑定：父级 prop 的更新会向下流动到子组件中，但是反过来则不行。这样会防止从子组件意外变更父级组件的状态，从而导致应用的数据流向难以理解。

另外，每次父级组件发生变更时，子组件中所有的 prop 都将被刷新为最新的值。这意味着开发人员不应该在一个子组件的内部改变 prop。如果这样做了，Vue.js 则会在浏览器的控制台中发出警告。

这里有两种常见的试图变更一个 prop 的情形：

(1) 使用 prop attribute 将一个初始值传递给子组件的本地属性，子组件直接操作本地属性。

(2) 在子组件中定义计算属性，基于 prop attribute 传入的值进行计算处理。

代码如下：

```
//第 5 章/prop 数据流向.html
…
<!-- 定义模板 -->
<template id="template1">
    <div>
        <!-- 初始值方式 -->
        <p>
            <span>传入的 counter:{{counter}}</span><br>
            subCounter:<input v-model="subCounter" /><br>
            <span>子组件更新 counter:{{subCounter}}</span>
        </p>
```

```
        <! -- 计算属性方式 -->
        <p>
            <span>传入的 size:{{size}}</span><br>
            subSize:<input v-model="computeSize" /><br>
            <span>子组件更新 size:{{computeSize}}</span>
        </p>
    </div>
</template>

<div id="app">
    counter:<input v-model="counter"/><br>
    size:<input v-model="size" />
    <sub-test :counter="counter" :size="size"></sub-test>
</div>
<script>
    const subVue = {
        props:['counter', 'size'],          //定义 prop 属性接收父组件的值
        data:function(){
            return {
                subCounter:this.counter     //给本地属性赋值
            };
        },
        computed:{
            computeSize:{               //定义计算属性,处理 prop 属性传入的值后给子组件使用
                get:function(){
                    return 'result->' + this.size;
                },
                set:function(newValue){
                    //不能响应到父组件,会抛出异常
                    this.size = newValue.slice("result->".length)
                }
            }
        },
        template:"#template1"
    }
    const vm = new Vue({
        el:"#app",
        data:{
            counter:1,
            size:10
        },
        components:{subTest:subVue}
    })
</script>
...
```

5.3.4 prop 属性验证

开发人员还可以为组件的 prop 指定验证要求。如果有一个需求没有被满足，则 Vue.js 会在浏览器控制台中警告。这在开发一个会被别人用到的组件时尤其有帮助。

为了定制 prop 的验证方式，开发人员可以为 props 中的值提供一个带有验证需求的对象，而不是一个字符串数组，代码如下：

```
//第 5 章/验证 prop 属性.html
...
<template id = "template1">
    <div>
        <span>{{attr1}}</span><br>
        <span>{{attr2}}</span><br>
        <span>{{attr3}}</span><br>
        <span>{{attr4}}</span><br>
        <span>{{attr5.message}}</span><br>
        <span>{{attr6}}</span><br>
    </div>
</template>
<div id = "app">
<son-component
:attr1 = "a1"
:attr2 = "a2"
:attr3 = "a3"
:attr4 = "a4"
:attr5 = "a5"
:attr6 = "a6"></son-component>
</div>
<script>
    const SonComponent = {
        props:{
            attr1:Number,
            attr2:[String,Number],
            attr3:{
                type:String,
                required:true
            },
            attr4:{
                type:Number,
                default:10
            },
            attr5:{
                type:Object,
                default:function(){
```

```
                    return {message:'hello'}
                }
            },
            attr6:{
                type:String,
                validator:function(value){
                    //这个值必须匹配下列字符串中的一个
                    return ['success', 'warning', 'danger'].indexOf(value) !== -1
                }
            }
        },
        template:"#template1"
    }
    const vm = new Vue({
        el:"#app",
            components:{
            SonComponent
        },
        data:{
            a1:1,
            a2:'hello',
            a3:'world',
            a4:11,
            a5: {message:'hai'},
            a6:'success'
        }
    })
</script>
…
```

当 prop 验证失败时，Vue.js 将会发出一个控制台警告。

实例的属性是在对象创建之前进行验证的，所以实例的属性（如 data 和 computed）在 default 和 validator 函数中不可用。

prop 支持的类型包括：String、Number、Boolean、Array、Object、Date、Function、Symbol，同时支持自定义的构造函数，能使用 instanceof 进行确认，代码如下：

```
//第 5 章/instanceof 的使用.html
…
<template id="template1">
    <div>
        {{personAttr.firstName}} {{personAttr.lastName}}
    </div>
</template>
```

```
<div id = "app">
    <son-component :person-attr = "person"></son-component>
</div>
<script>
    function Person(first, last){
        this.firstName = first;
        this.lastName = last;
    }
    const SonComponent = {
        template:"#template1",
        props:{
            personAttr:{
                type: Person,
                validator:function(value){
                    return value instanceof Person;
                }
            }
        }
    }
    const vm = new Vue({
        el:"#app",
        components:{
            SonComponent
        },
        data:{
            person:new Person("san", "zhang")
        }
    })
</script>
...
```

5.3.5 非 prop 的 attribute

组件可以接受任意的 attribute，而这些 attribute 会被添加到这个组件的根元素上。

显式定义的 prop 适用于向一个子组件传入信息，这也是 Vue.js 中推荐的做法，即向子组件传值的方式，然而组件库的作者并不总能预见组件会被用于怎样的场景。这也是为什么组件可以接受任意的 attribute，而这些 attribute 会被添加到这个组件的根元素上。

如下例子中的 son-component 组件的 notprop 属性，没有在 son-component 的 props 属性中定义，但是会被直接渲染到子组件的根元素（div 元素）中，代码如下：

```
//第 5 章/非 prop 属性.html
...
<template id = "template1">
    <div class = "subClass" name = "subName">子组件</div>
```

```
</template>
<div id = "app">
    <son - component
        :notprop = "notPropValue"
        :class = "clsValue"
        :name = "nameValue"></son - component>
</div>
<script>
    const SonComponent = {
        template:"#template1"
    }
    const vm = new Vue({
        el:"#app",
        data:{
            notPropValue:'hello',
            clsValue:'parentClass',
            nameValue:'parentName'
        },
        components:{
            SonComponent
        }
    })
    vm.$data.notPropValue = "hai"
</script>
...
```

渲染代码如下：

```
<div notprop = "hai"
class = "parentClass subClass"
name = "parentName">子组件</div>
```

div 中的 notprop＝"hai"是从< son-component：notprop＝"notPropValue"></son-component>传递过去的。

1. 替换/合并已有的 attribute

如果在子组件中也定义了非 prop 的 attribute，同时在使用组件的时候也定义了该 attribute，这时候最后的值，存在替换/合并问题。class 和 style 的值会被合并，其他属性值会被替换。

如上面的代码中，在 son-component 组件中定义了 class＝"subClass" name＝"subName"，同时在组件使用的时候定义了：class＝"clsValue"和：name＝"nameValue"，最后渲染的结果是 class＝"parentClass subClass"和 name＝"parentName"。class 属性的值被合并了，而 name 属性的值只有一个：外面的值被替换了组件里面的 name 值。

2. 禁用 Attribute 继承

如果开发人员不希望组件的根元素继承 attribute，则可以在组件的选项中设置 inheritAttrs：false

因为继承的 attribute 只能作用到根元素上，如果需要将 attribute 继承到子组件的非根元素上，则可以使用 v-bind=" $attrs"将 attribute 绑定到子元素的非根元素上，代码如下：

```
<div id = "app">
  <son-component test = 'tValue' required
placeholder = '请输入姓名'></son-component>
</div>
<script>
  Vue.component("SonComponent",{
      inheritAttrs:false,
      template:`<div><input type = 'text' v-bind = " $attrs"/></div>`
  })
  const vm = new Vue({
      el:"#app"
  })
</script>
```

注意：class 和 style 属性不在作用范围。

5.4 子组件将值传到父组件

组件的 prop 属性只能实现父组件向子组件传值，在实际的前端项目中，需要实现子组件将值传给父组件。Vue.js 提供了 3 种机制，实现子组件将值传给父组件。

5.4.1 使用 $emit 方法调用父组件方法传值

在 Vue.js 的父组件中，可以通过 v-on 指令，给子组件的指定事件绑定一个函数，在子组件中，用 $emit 方法触发自己的事件，从而执行被绑定的函数。$emit 方法的第 1 个参数是一个字符串，对应 v-on 指定的事件名称，父组件中使用 v-on 给 son-component 组件的 parent-method 事件绑定了定义在父组件中 parentMethod 函数，在 son-component 的 toTest 函数中，使用 this. $emit('parent-method')方式触发 parent-method 事件，执行 parentMethod 方法，实现父组件中的 count 自增，代码如下：

```
//第 5 章/子组件基于 $emit 调用父组件的方法.html
...
<div id = 'app'>
    <son-component v-on:parent-method = "parentMethod"></son-component>
    <br/>
    <div>{{count}}</div>
```

```
</div>
<template id="sonComponent">
    <button v-on:click='toTest'>单击子组件</button>
</template>
<script type='text/JavaScript'>
    const SonComponent = {
        template:'#sonComponent',
        methods:{
            toTest(){
                this.$emit('parent-method')
            }
        }
    }
    const vm = new Vue({
        el: '#app',
        components:{
            SonComponent
        },
        data: {
            count:0
        },
        methods: {
            parentMethod(){
                this.count++
            }
        }
    })
</script>
...
```

$emit 方法必须有一个参数指定要触发的事件，同时支持更多的可选参数，通过这些参数，子组件可以将自己的数据传递给事件绑定的方法，而绑定的方法是定义在父组件中的，所以就可以间接地使用 $emit 方法，将子组件中的数据传递给父组件。在 son-component 子组件的 toTest 方法中，通过第 2 个参数给父组件中绑定的 parentMethod 方法传递 count 的递增幅度 step，代码如下：

```
//第5章/使用$emit方法传值.html
...
<div id='app'>
    <son-component v-on:parent-method="parentMethod"></son-component>
    <br/>
    <div>{{count}}</div>
</div>
<template id="sonComponent">
```

```
    <button v-on:click='toTest'>单击子组件</button>
</template>
<script type='text/JavaScript'>
    const SonComponent = {
        template:'#sonComponent',
        methods:{
            toTest(){
                this.$emit('parent-method', 2)
            }
        }
    }
    const vm = new Vue({
        el: '#app',
        components:{
            SonComponent
        },
        data: {
            count:0
        },
        methods: {
            parentMethod(step){
                this.count += step
            }
        }
    })
</script>
...
```

5.4.2 调用父组件的方法传值

prop属性的数据类型支持Function，利用这个特点，开发人员可以在父组件中定义一个Function类型的prop属性，给子组件传递一个函数对象，在子组件中调用这个函数，通过函数的参数，可以将子组件中的数据传递给父组件。父组件基于子组件的funcData prop属性，给子组件son-component传递increment函数对象，在子组件son-component的toTest方法中，调用传入的increment函数，并且传入参数step的值，代码如下：

```
//第5章/调用父组件的方法传值.html
...
<div id='app'>
    <son-component v-bind:func-data="increment"></son-component><br/>
    {{count}}
</div>
<template id="sonComponentTemplate">
    <button v-on:click="toTest">单击递增</button>
```

```
</template>
<script type='text/JavaScript'>
    const SonComponent = {
        template: '#sonComponentTemplate',
        props:{
            funcData:{
                type:Function
            }
        },
        methods:{
            toTest(){
                this.funcData(2)
            }
        }
    }
    const vm = new Vue({
        el: '#app',
        components:{
            SonComponent
        },
        data: {
            count: 0
        },
        methods: {
            increment(step){
                this.count += step
            }
        }
    })
</script>
...
```

5.4.3 使用 v-model 实现父子组件的数据同步

v-model 指令可以实现 input 输入框同组件数据属性双向同步，改变输入框的值，此值能自动被同步到 Vue.js 实例对象中。同样，改变 Vue.js 实例对象的数据属性，此数据属性也能自动被同步到 input 输入框，代码如下：

```
//第5章/使用v-model同步数据.html
...
<div id='app'>
    name:{{name}}<br/>
    <input v-model='name'/><br/>
</div>
```

```
<script type='text/JavaScript'>
    const vm = new Vue({
        el: '#app',
        data: {
            name: ''
        },
    })
</script>
...
```

实际上，v-model 是 v-bind：value 和 v-on：input 两个指令的组合。v-bind：value 指令将 Vue.js 实例对象的数据属性绑定到 input 元素的 value 属性。v-on：input 指令给 input 元素的 input 事件绑定一个函数，该函数将 input 输入框的 value 属性值赋给 Vue.js 实例对象的数据属性。<input>元素中的 b-bind：value='name'，将 name 数据属性的值绑定到 input 的 value 属性上，这样 input 输入框就可以实时显示 name 数据属性的值了。<input>元素中的 v-on：input="demoInputChange($event)"将 demoInputChange 函数绑定到 input 元素的 input 事件上，并且传入了当前的事件对象，当 input 事件触发时自动执行 demoInputChange 函数，将 input 的 value 属性的值赋给 name 数据属性，从而实现了 input 元素中的 value 同 Vue.js 实例对象中的 name 数据属性的双向绑定，代码如下：

```
//第 5 章/使用 v-model 同步数据.html
...
<div id='app'>
    name:{{name}}<br/>
    <input v-bind:value='name' v-on:input="demoInputChange($event)"><br/>
</div>
<script type='text/JavaScript'>
    const vm = new Vue({
        el: '#app',
        data: {
            name: ''
        },
        methods: {
            demoInputChange(event){
                this.name = event.target.value
            }
        }
    })
</script>
...
```

既然 v-bind 和 v-on 的组合可以实现 input 元素的 value 属性同 Vue.js 实例对象的数据属性的双向绑定，同样可以用在子组件上，实现子组件的 value 和数据属性的双向绑定，代码如下：

```
//第5章/使用v-model同步数据.html
...
<div id='app'>
    <!-- 子组件使用v-bind和v-on:input的组合 -->
    age:{{age}}<br/>
    <son-component v-bind:age="age" v-on:input="sonChange"></son-component>
</div>
<template id="sonComponentTemplate">
    <input type='text' v-bind:value="age" v-on:input="toChange($event)">
</template>
<script type='text/JavaScript'>
    const SonComponent = {
        template: '#sonComponentTemplate',
        props:['age'],
        methods:{
            toChange(event){
                this.$emit('input',event.target.value)
            }
        }
    }
    const vm = new Vue({
        el: '#app',
        components:{
            SonComponent
        },
        data: {
            age: 0
        },
        methods: {
            sonChange(age){
                this.age = age
            }
        }
    })
</script>
...
```

使用v-model合并子组件的v-bind和v-on指令，代码如下：

```
//第5章/使用v-model同步数据.html
...
<div id='app'>
    <!-- 子组件使用v-bind和v-on:input的组合 -->
    age:{{age}}<br/>
    <son-component v-model="age"></son-component>
</div>
```

```
<template id="sonComponentTemplate">
    <input type='text' v-bind:value="age" v-on:input="toChange($event)">
</template>
<script type='text/JavaScript'>
    const SonComponent = {
        template: '#sonComponentTemplate',
        props:['age'],
        methods:{
            toChange(event){
                this.$emit('input',event.target.value)
            }
        }
    }
    const vm = new Vue({
        el: '#app',
        components:{
            SonComponent
        },
        data: {
            age: 0
        }
    })
</script>
...
```

5.5 Vue.js 组件对象的常用属性

在 Vue.js 中，给 Vue.js 组件对象定义了很多属性，比较常用的有 $data、$props、$parent、$root、$children 和 $refs 等属性。Vue.js 组件对象的属性都是以 $ 为前缀的，用来区分定义在组件里面的数据属性。这些属性的作用分别如下。

(1) $data：获取 Vue.js 组件的数据属性对象，包含自定义的所有数据属性。

(2) $props：获取 Vue.js 组件的 props 属性对象，包含所有的 prop 属性。

(3) $parent：获取 Vue.js 组件对象的父组件。

(4) $root：获取 Vue.js 组件对象的根对象，否则就是自己。

(5) $children：获取 Vue.js 组件对象的所有子组件数组。

(6) $refs：获取组件对象里面的所有 ref 组件数组，可以根据 ref 名称获取指定的子组件。

这些组件属性的使用，参见如下代码：

```
//第 5 章/Vue.js 实例对象的常用属性.html
...
<div id='app'>
```

```
<son1></son1>

    <hr>
    <son2 ref="second"></son2>
    <br/>
    <button v-on:click="toTest">parent test</button>
</div>
<template id="sonTemplate">
    <div>
        name:{{name}} --- value:{{value}}<br/>
        <button v-on:click="toTest">测试</button>
    </div>
</template>
<script type='text/JavaScript'>
    const Son1 = {
        name: 'Son1',
        template:'#sonTemplate',
        data(){
            return {
                name:'son1',
                value:'value1'
            }
        },
        methods:{
            toTest(){
                console.log(this.$parent.$data.value)
            }
        }
    }

    const Son2 = {
        name: 'Son2',
        template:'#sonTemplate',
        data(){
            return {
                name:'son2',
                value:'value2'
            }
        },
        methods:{
            toTest(){
                console.log(this.$parent.$data.value)
            }
        }
    }
```

```
    const vm = new Vue({
        el: '#app',
        components:{
            Son1,
            Son2
        },
        data: {
            value: 'parentValue'
        },
        methods: {
            toTest(){
                this.$children.forEach(element => {
                    console.log(element.$data.name + ',,,,' + element.$data.value)
                });
                console.log(this.$refs.second.$data.value)
            }
        }
    })
</script>
...
```

5.6 事件总线

在 Vue.js 中实现组件之间数据的传递，大概有以下几种方式：

(1) 通过 props 将父组件的数据传递给子组件。

(2) 通过 $emit 方法将子组件中的数据，通过绑定的函数传递给父组件。

(3) 在父组件中定义方法，通过@(v-bind)自定义方法='自定义函数名称'传递给子组件，在子组件中可以直接调用。

(4) 通过 Vue.js 组件实例的 $parent、$root、$children、$refs 等属性，获取相关的组件对象。

还可以使用事件总线的方式和 Vuex 的 state 的属性实现任意组件之间的数据共享。接下来介绍 Vue.js 中的事件总线，并用它实现组件之间的数据共享传递。

事件总线又称为 EventBus。在 Vue.js 中可以使用 EventBus 来作为沟通的桥梁，就像是所有组件共用相同的事件中心，可以向该中心注册发送事件或接收事件，所以组件都可以上下平行地通知其他组件，但由于太方便，所以若使用不慎，就会造成难以维护的灾难(所以由 Vuex 作为管理状态中心，将通知的概念提升为状态共享)。

使用事件总线，可以分三步实现。

1. 注册事件总线

创建一个全局的 Vue.js 对象，用来充当事件总线中心，以及用来传递事件。该事件总线只能用在当前页面的组件中，也可以将这个 Vue.js 对象保存到 Vue.js 的原型属性中，从而注册一个全局的事件总线中心，代码如下：

```
//创建事件总线
const eventBus = new Vue()
//或者注册全局事件中心
//Vue.prototype.$bus = eventBus
```

2. 接收事件总线

定义一个组件，使用 $on(事件名称，回调函数)方法，给指定的总线事件绑定一个函数，监听事件总线中的事件，代码如下：

```
//第 5 章/事件总线.html
…
<!-- 第 1 个组件模板，显示 age -->
<template id="view1">
    <div>view1 age:{{age}}</div>
</template>

<script type="text/JavaScript">
    //创建 MyView1 对象
    const MyView1 = {
        data() {
            return {
                age: 1
            }
        },

        mounted() {
            //接收 bus 总线中的 bindEvent 事件
            //this.$bus.$on('bindEvent', user => {
            eventBus.$on('bindEvent', user => {
                this.age = user.age + this.age
            })
        },
        template: '#view1'
}
…
</script>
…
```

3. 发送事件总线

在组件代码中，用 $emit(事件名称，参数)方法，向事件总线中心发送一个事件，并传递参数内容，代码如下：

```
//第 5 章/事件总线.html
…
<script type="text/JavaScript">
    …
    const vm = new Vue({
        …
        methods: {
            publishEvent() {
                this.value++
                //往 bus 总线发布 bindEvent 事件(全局)
                //this.$bus.$emit('bindEvent', {name:'zhangsan', age:this.value})
                //局部
                eventBus.$emit('bindEvent', {name:'zhangsan', age:this.value})
            }
        }
    })
</script>
…
```

整合后的代码如下：

```
//第 5 章/事件总线.html
…
<div id="app">
    <my-view1></my-view1>
    <my-view2></my-view2>

    <button v-on:click='publishEvent'>测试</button>
</div>

<!--第 1 个组件模板,显示 age-->
<template id="view1">
    <div>view1 age:{{age}}</div>
</template>

<!--第 2 个组件目标,显示 age-->
<template id="view2">
    <div>view2 age:{{age}}</div>
</template>

<script type="text/JavaScript">
    //创建 MyView1 对象
    const MyView1 = {
        data() {
            return {
                age: 1
```

```
            }
        },
        mounted() {
            //接收 bus 总线中的 bindEvent 事件
            //this. $bus. $on('bindEvent', user => {
            eventBus. $on('bindEvent', user => {
                this.age = user.age + this.age
            })
        },
        template: '#view1'
    }

    //创建 MyView2 对象
    const MyView2 = {
        data() {
            return {
                age: 10
            }
        },
        mounted() {
            //接收 bus 总线中的 bindEvent 事件
            //this. $bus. $on('bindEvent', user => {
            // this.age = this.age + user.age
            //})
            //this. $bus. $once('bindEvent', user => {
            // this.age = this.age + user.age
            //})
            eventBus. $on('bindEvent', user => {
                this.age = this.age + user.age
            })
        },
        template: '#view2'
    }

    //事件总线
    //Vue.prototype. $bus = new Vue()
    const eventBus = new Vue()

    const vm = new Vue({
        el:'#app',
        data: {
            value: 1
        },
        components: {
            MyView1,
            MyView2
```

```
        },
        methods: {
            publishEvent() {
                this.value++
                //往 bus 总线发布 bindEvent 事件
                //this.$bus.$emit('bindEvent', {name:'zhangsan', age:this.value})
                eventBus.$emit('bindEvent', {name:'zhangsan', age:this.value})
            }
        }
    })
</script>
...
```

5.7 插槽

项目中有很多组件显示的内容，在不同的地方使用的时候，有些内容需要变化，这时候可以用插槽内容分发功能实现。例如开发人员可以定义一个组件，专门用来显示提示信息。只是显示的信息包含在插槽 slot 元素中，满足在不同的地方，提示的内容不一样。MessageAlert 组件中包含了 slot 插槽，运行的时候，就会将组件元素包含的内容动态地渲染到视图中，代码如下：

```
//第 5 章/简单插槽.html
...
<div id="app">
    <message-alert>提醒</message-alert>
    <message-alert>警告</message-alert>
    <message-alert>错误</message-alert>
</div>
<script>
    const MessageAlert = {
            template:`<div>
                <strong>msg:</strong>
                <slot></slot>
                </div>`
    }
    Vue.component("MessageAlert",MessageAlert;
    const vm = new Vue({
            el:"#app"
    })
</script>
...
```

5.7.1 插槽的缺省内容和编译作用域

在实战中，slot 中往往希望有缺省值，实现起来很简单，直接在子元素的 slot 之间包含缺省值即可，调用的时候，如果有新值，则自动会将缺省值覆盖，如< slot >缺省值</slot >。

当然，分发到 slot 中的内容，可以包含动态内容，此时动态的变量只能使用父组件里面的 property，代码如下：

```
//第 5 章/插槽缺省值和编译作用域.html
...
<div id = "app">
    <! -- 使用缺省值 -- >
    < slot - dft - content ></slot - dft - content >
    <! -- 使用静态内容 -- >

    < slot - dft - content >提交</slot - dft - content >
    <! -- 编译作用域 -- >

    < slot - dft - content >{{parentValue}}</slot - dft - content >
    <! --  放开会报异常,sonValue 未定义

    < slot - dft - content >{{sonValue}}</slot - dft - content >
    -- >

</div >
< script >
    Vue.component("SlotDftContent",{
        template:` < button type = 'submit'>
                            < slot > submit </slot >
                  </button >`,
        data:function(){
              return {
                            sonValue:'sonContent'
              }
        }
    })
    const vm = new Vue({
        el:"#app",
        data:{
            parentValue:'parent'
        }
    })
</script >
...
```

5.7.2　具名插槽

有时候在一个子元素中需要定义多个插槽，使用子元素的时候，将不同内容分发到不同插槽中去。这时候，开发人员可以使用 slot 的 name 属性，给每个 slot 命名(没有命名的为 default)，使用子组件时，用 template 元素包含要分发到每个 slot 的内容，使用“v-slot：目标插槽的名称”来指定对应的分发插槽，代码如下：

```
//第 5 章/具名插槽.html
...
<div id = "app">
    <base - layout >
        <template v - slot:header >头</template >
        <template v - slot:footer >尾</template >
        <p>缺省</p >
        <p>内容</p >
    </base - layout >
</div >
<script >
    Vue.component("BaseLayout",{
        template:`<div class = "container">
            <header >
                <slot name = "header"></slot >
            </header >
            <main >
                <slot ></slot >
            </main >
            <footer >
                <slot name = "footer"></slot >
            </footer >
            </div >`
    })
    const vm = new Vue({
        el:"#app"
    })
</script >
...
```

5.7.3　作用域插槽

有时候，需要获取子元素的数据，进行整理后再分发给插槽，这时候可以使用作用域插槽。开发人员可以在 slot 元素中，使用 v-bind 指令将对象绑定后，传递到组件外面，如：<slot v-bind：userAttr="user" >...</slot >就是将子元素中的 user 数据，通过 v-bind 指令绑定到 userAttr 属性上。userAttr 属性，我们称它为 slot property。在父组件中，可以在

template 元素中使用指定的插槽属性变量操作 userAttr 绑定的 user 数据，代码如下：

```
<div id = "app">
    <! -- v - slot:default 指定分发给 default slot -- >
    < range - slot v - slot:default = "slotProps">
        {{slotProps.userAttr.userName + "整理好了"}}
    </range - slot >
</div >
< script >
    //作用域插槽
    Vue.component("RangeSlot",{
            template:`< div >
                        < slot v - bind:userAttr = "user">
                            {{user.name}}
                        </slot >
                      </div >`,
                        data:function){
                            return {
                                    user:{
                                        name:'李四',
                                        userName:'lisi'
                                    }
                            }
                        }
    })
    const vm = new Vue({
        el:"#app"
    })
</script >
```

5.7.4 动态插槽名

动态指令参数也可以用在 v-slot 上，用来定义动态的插槽名。特别要注意的是，这里不支持驼峰命名的变量名，语法如下：

```
< range - slot >
    < template v - slot:[动态的插槽名称]>
        ...
    </template >
</range - slot >
```

5.7.5 具名插槽的缩写

跟 v-on 和 v-bind 一样，v-slot 也有缩写，即把参数之前的所有内容（v-slot：）替换为字符＃。v-slot：header 可以被重写为＃header，代码如下：

```
<base-layout>
  <template #header>
    <h1>Here might be a page title</h1>
  </template>

  <p>A paragraph for the main content.</p>
  <p>And another one.</p>

  <template #footer>
    <p>Here's some contact info</p>
  </template>
</base-layout>
```

然而，和其他指令一样，该缩写只在有参数的时候才可用，这意味着以下语法是无效的：

```
<!-- 这样会触发一个警告 -->
<current-user #="{ user }">
  {{ user.firstName }}
</current-user>
```

如果开发人员希望使用缩写，则必须始终以明确的插槽名取而代之，代码如下：

```
<current-user #default="{ user }">
  {{ user.firstName }}
</current-user>
```

5.8 动态组件和异步组件

Vue.js 提供了两种特殊组件：动态组件和异步组件。

5.8.1 动态组件

有时候，一个组件内部需要根据不同的选择，显示不同的组件。如在标签卡布局里面，单击不同的标签，内容块需要显示该标签对应的内容。Vue.js 中提供了 component 元素组件，可以动态地显示指定的组件。

在 component 元素中有个 is 属性，通过给 is 属性绑定一个组件名称，component 就可以显示该组件名称对应的组件，语法如下：

```
<component v-bind:is="currTabComponent"><component>
```

如果 currTabComponent 的值是 test-comp1，就表示显示名称为 test-comp1 的组件。样例代码如下：

```
//第 5 章/动态组件.html
...
<div id = "app">
    <! -- 显示标签 Tabs -->
    <button
        v - for = "tab in tabs"
        v - bind:key = "tab"
        v - bind:class = "['tab - button', { active: currentTab === tab }]"
        v - on:click = "currentTab = tab"
        >
        {{ tab }}
        </button>
    <! -- 显示 tab 对应的组件 -->
        <! -- <keep - alive> -->
        <component v - bind:is = "currentTabComponent" class = "tab"></component>
        <! -- </keep - alive> -->
</div>
<script>
    //<! -- 定义 Home 组件 -->
    Vue.component("dync - comp0",{
        template:`<div> Component Home </div>`
    };
    //<! -- 定义组件 1 -->
    Vue.component("dync - comp1",{
        template:`<div> Component 111 </div>`
    };
    //<! -- 定义组件 2 -->
    Vue.component("dync - comp2",{
        template:`<div> Component 222 </div>`
    })
    const vm = new Vue({
        el:"#app",
        data:{
            currentTab:'Comp1',
            tabs:["Comp0","Comp1","Comp2"]
        },
        computed:{
            currentTabComponent:function(){
                //返回 tab 对应组件的名称,以便在 component 中动态显示
                return 'dync - ' + this.currentTab.toLocaleLowerCase;
            }
        }
    })
</script>
...
```

5.8.2　异步组件

在大型应用中，开发人员可能需要将应用分割成小的代码块，并且只在需要的时候才从服务器加载所需的模块。为了简化，Vue.js 允许开发人员以一个工厂函数的方式定义组件，这个工厂函数会异步解析组件的定义。Vue.js 只有在这个组件需要被渲染的时候才会被触发，并且会把结果缓存起来供未来重渲染，代码如下：

```
Vue.component('async-example', function (resolve, reject) {
  setTimeout(function () {
    //在此定义组件
    resolve({
      template: `
        <div>
          我是异步加载的哦
        </div>
      `
    })
  }, 1000);
});
```

这个工厂函数会收到一个 resolve 回调，这个回调函数会在从服务器得到组件定义的时候被调用。开发人员也可以调用 reject(reason)来表示加载失败。这里的 setTimeout 是为了演示用的，如何获取组件取决于开发人员自己。

5.8.3　keep-alive

在前面的动态组件案例中，用户单击 tab 切换组件后，回到以前 tab 对应的组件上时，该组件会被重新渲染。有时候希望回到上次单击的状态，这时候，开发人员可以在 component 的外面包含 keep-alive 元素，这样就可以保持这些组件的状态，以避免反复重渲染导致的性能问题，代码如下：

```
<!-- 失活的组件将会被缓存! -->
<keep-alive>
  <component v-bind:is="currentTabComponent"></component>
</keep-alive>
```

5.9　处理组件边界问题

接下来介绍 Vue.js 中处理与边界情况有关的功能，即需要对 Vue.js 的规则做一些小调整的特殊情况。不过应注意这些功能都是有好有坏的，开发人员需要根据实际情况斟酌选择。

5.9.1 访问元素的 & 组件

在绝大多数情况下，开发人员最好不要触达另一个组件实例内部或手动操作 DOM 元素。不过在有些情况下，需要直接操作组件实例的内部数据和手动操作 DOM 元素，Vue.js 为 Vue.js 实例对象提供了很多 & 组件，开发人员用这些方式可以直接操作组件内部的其他组件。

1. 访问根实例

在每个 Vue.js 实例中都定义了 $root 属性，开发人员可以使用 $root 获取整个组件树的根组件，从而操作根组件实例的内容。创建一个根组件对象，代码如下：

```
//Vue.js 根实例
new Vue({
  data: {
    foo:1
  },
  computed: {
    bar:function () { /* ... */ }
  },
  methods: {
    baz:function () { /* ... */ }
  }
})
```

所有的子组件都可以将这个实例作为一个全局 store 访问或使用，代码如下：

```
//获取根组件的数据
this.$root.foo

//写入根组件的数据
this.$root.foo = 2

//访问根组件的计算属性
this.$root.bar

//调用根组件的方法
this.$root.baz()
```

注意：对于 demo 或非常小型的有少量组件的应用来讲这个模式是很方便的，不过将这个模式扩展到中大型应用就不然了。因此在绝大多数情况下，我们强烈推荐使用 Vuex 来管理应用的状态。

2. 访问父级组件实例

和 $root 类似，$parent property 可以用来从一个子组件访问父组件的实例。它提供了一种机制，可以在后期随时触达父级组件，以替代将数据以 prop 的方式传入子组件的方式。

在绝大多数情况下，触达父级组件会使应用更难调试和理解，尤其是隔了一段时间后，很难找出哪个变更是从哪里发起的。

<google-map>组件可以定义一个 map property，所有的子组件都需要访问它。在这种情况下<google-map-markers>可以通过类似 this.$parent.getMap 的方式访问那个地图，以便为其添加一组标记，代码如下：

```
<google-map>
  <google-map-markers v-bind:places="iceCreamShops"></google-map-markers>
</google-map>
```

但是，通过这种模式构建出来的组件的内部仍然容易出现问题。例如，在<google-map>和<google-map-markers>之间添加一个新的 <google-map-region> 组件时，代码如下：

```
<google-map>
  <google-map-region v-bind:shape="cityBoundaries">
    <google-map-markers v-bind:places="iceCreamShops"></google-map-markers>
  </google-map-region>
</google-map>
```

在 <google-map-markers> 内部就需要编写一些类似的代码：

```
var map = this.$parent.map || this.$parent.$parent.map
```

这样它很快就会失控，变得难以理解和追踪。

3. 访问子组件实例或子元素

尽管存在 prop 和事件，有的时候开发人员仍可能需要在 JavaScript 里直接访问一个子组件。为了达到这个目的，可以通过 ref 的 attribute 为子组件赋予一个 ID 引用，代码如下：

```
<base-input ref="usernameInput"></base-input>
```

这样在 JavaScript 代码中，就可以用以下方式访问 <base-input> 实例，代码如下：

```
this.$refs.usernameInput
```

当然，也可以使用一个类似的 ref 提供对内部这个指定元素的访问，示例代码如下：

```
<input ref="input">
```

甚至可以通过其父级组件定义方法，代码如下：

```
methods: {
  //用来从父级组件聚焦输入框
  focus:function() {
```

```
    this.$refs.input.focus()
  }
}
```

这样就允许父级组件通过下面的代码聚焦 <base-input> 里的输入框，代码如下：

```
this.$refs.usernameInput.focus()
```

当 ref 和 v-for 一起使用的时候，开发人员得到的 ref 将会是一个包含了对应数据源的子组件的数组。

注意： $refs 只会在组件渲染完成之后生效，并且不是响应式的。它仅作为一个用于直接操作子组件的"逃生舱"——开发人员应该避免在模板或计算属性中访问 $refs。

4. 依赖注入

前面介绍的 $parent 属性，可以为开发人员提供一种直接引用父组件的方式，但是在更深层级的嵌套组件上使用的时候，很难跟踪，特别是在中间插入新的组件的时候。Vue.js 提供了依赖注入的机制，可以很好地解决这个问题。

所谓依赖注入，就是在父组件中，可以使用 provide 选项指定想要提供给后代子组件使用的数据和方法，然后在任何后代子组件中都可以使用 inject 选项来接收想要添加到当前实例上的属性。

在<google-map>组件中使用 provide 选项，暴露 getMap 方法，代码如下：

```
provide:function () {
  return {
    getMap: this.getMap                //暴露 getMap 方法
  }
}
```

在<google-map>的任何后代组件中，使用 inject 选项添加到自己的属性中，代码如下：

```
inject: ['getMap']
```

本质上，依赖注入是一部分大范围有效的 prop 属性，以前说的 props 属性只能将数据传递给直接子组件，而依赖注入则可以将数据传递给任何后代子组件。满足以下两个条件的数据传递，可以使用依赖注入方式实现。

(1) 祖先组件不需要知道哪些后代组件使用它提供的 property。

(2) 后代组件不需要知道被注入的 property 来自哪里。

然而，依赖注入还是有负面影响的。它将应用程序中的组件与它们当前的组织方式耦合起来，使重构变得更加困难。同时所提供的 property 是非响应式的。这是出于设计的考虑，因为使用它们来创建一个中心化规模化的数据和使用 $root 做这件事都是不够好的。如果开发人员想要共享的这个 property 是应用特有的，而不是通用化的，或者如果想在祖先组件中更新所提供的数据，开发人员则可以使用像Vuex这样真正的状态管理方案实现。

5.9.2 程序化的事件侦听

在Vue.js中，可以使用$emit触发一个事件，v-on指令可以在侦听到事件后，执行绑定在事件上的函数。另外，Vue.js还提供了以下几种方式侦听$emit发送的事件。

(1) 通过$on(eventName,eventHandler)侦听一个事件。

(2) 通过$once(eventName, eventHandler)一次性侦听一个事件。

(3) 通过$off(eventName,eventHandler)停止侦听一个事件。

项目中一般不会用到这些，但是当项目需要在一个组件实例上手动侦听事件时，这些就可以派上用场。在EventComponent组件的mounted回调函数中，使用$on的方式给自己添加了一个侦听myclick事件的函数，当单击EventComponent组件的button按钮时，在执行的事件代码中使用$emit方式触发myclick事件，从而实现count属性的递增，代码如下：

```
//第5章/程序化侦听.html
...
<div id='app'>
    <event-component></event-component>
</div>
<template id="eventComponent">
    <div>
    count: {{count}}<br/>
    <button v-on:click="toClick">单击</button>
    </div>
</template>
<script type='text/JavaScript'>
    const EventComponent = {
        template:'#eventComponent',
        data(){
            return {
                count: 0
            }
        },
        mounted:function(){
            this.$on('myclick', function(step){
                this.count += step
            })
        },
        methods:{
            toClick(){
                this.$emit('myclick', 2)
            }
```

```
            }
        }
        const vm = new Vue({
            el: '#app',
            components:{
                EventComponent
            },
        })
</script>
...
```

5.9.3 循环引用组件

Vue.js 中的组件支持自己引用自己的循环引用和自己同其他组件之间的相互循环引用。

1. 递归循环引用

Vue.js 中的组件可以自己引用自己。因为组件在使用之前,需要进行注册。递归引用自己的组件,一样支持局部注册和全局注册。全局注册和使用一个递归组件的样例,代码如下:

```
//第 5 章/递归组件.html
...
<div id='app'>
    <recursion-component :count="count"></recursion-component>
</div>
<template id="recursionComponent">
    <span>&gt;
        <span v-if="count>1">
            <!--递归引用自己-->
            <recursion-component :count="count-1"></recursion-component>
        </span>
    </span>
</template>
<script type='text/JavaScript'>
    const RecursionComponent = {
        name:'RecursionComponent',
        template:'#recursionComponent',
        props:{
            count:{
                type:Number
            }
        },
        mounted:function(){
```

```
                console.log(this.count)
            }
    }
    //全局注册
        Vue.component("RecursionComponent", RecursionComponent)
        const vm = new Vue({
            el: '#app',
            data: {
                count: 5
            },
            methods: {
            }
        })
</script>
...
```

因为是全局注册，所以在任何组件里面，包括自己都可以使用。同样，递归组件支持局部注册，只是需要注意的是，在组件中注册自己是通过 name 选项完成的。RecursionComponent 组件在内部注册了自己，代码如下：

```
const RecursionComponent = {
        name:'RecursionComponent',
        ...
}
```

在递归引用组件的时候，一定要用 v-if 等手段，避免无限循环，否则会抛出无限循环错误。

2. 组件之间循环引用

以下代码，定义了两个组件，即 TreeNode 和 TreeNodeChild。TreeNode 用来显示当前节点的名称，如果包含子节点，就循环 TreeNodeChild 显示子节点。TreeNodeChild 组件用来显示一个新的 TreeNode，这样就形成了组件之间的循环调用。

需要注意的是，组件之间的循环引用，要注意条件判断，避免出现死循环。另外，因为组件之间要相互引用，引用前要相互注册，而目前注册的前提是被注册的组件要先被初始化才能被注册，这样就会造成 TreeNode 和 TreeNodeChild 组件注册的悖论，所以暂时循环引用的组件，只适用于全局注册，代码如下：

```
//第 5 章/循环组件 tree.html
...
<style type="text/css">
    * {
        margin: 0;
```

```
        padding: 0;
    }
    li{
        list - style: none;
    }
    ul li{
        margin - left: 20px;
    }
</style>
<div id = 'app'>
    <tree - node :node = "treeData"></tree - node>
</div>
<! -- 显示树节点内容 -->
<template id = "treeNode">
    <! -- 如果有子节点 -->
    <ul v - if = "node.children">
        <! -- 显示自己的名称 -->
        <span>{{node.name}}</span>
        <! -- 便利列出子节点 -->
        <li v - for = "(child, index) in node.children" :key = "index">
            <tree - node - child :node = "child"></tree - node - child>
        </li>
    </ul>
    <! -- 如果没有子节点 -->
    <span v - else>{{node.name}}</span>
</template>
<! -- 显示子节点 -->
<template id = "treeNodeChild">
    <tree - node :node = "node"></tree - node>
</template>

<script type = 'text/JavaScript'>
    const TreeNode = {
        name: 'TreeNode',
        template: '#treeNode',
        props:{
            node: {
                type: Object
            }
        }
    }
    const TreeNodeChild = {
        name: 'TreeNodeChild',
        template: '#treeNodeChild',
        props: {
            node: {
```

```
                type: Object
            }
        }
    }

    Vue.component('TreeNode', TreeNode)
    Vue.component('TreeNodeChild', TreeNodeChild)

    const vm = new Vue({
        el: '#app',
        data: {
            treeData:{
                name: 'TP2012 班',
                children: [
                    {
                        name: '组 1',
                        children: [
                            {
                                name: '1 组员 1'
                            },
                            {
                                name: '1 组员 2'
                            },
                            {
                                name: '1 组员 3'
                            },
                            {
                                name: '1 组员 4'
                            }
                        ]
                    },
                    {
                        name: '组 2',
                        children: [
                            {
                                name: '2 组员 1'
                            },
                            {
                                name: '2 组员 2'
                            },
                            {
                                name: '2 组员 3'
                            },
                            {
                                name: '2 组员 4'
                            }
```

```
                        ]
                    }
                    ,{
                        name: '组 3',
                        children: [
                            {
                                name: '3 组员 1'
                            },
                            {
                                name: '3 组员 2'
                            },
                            {
                                name: '3 组员 3'
                            },
                            {
                                name: '3 组员 4'
                            }
                        ]
                    }
                ]
            }
        },
        methods: {
        }
    })
</script>
...
```

5.9.4 其他模板

在 Vue.js 中,除了可以使用 template 定义组件模板外,还支持内联模板和 X-Template 模板。

1. 内联模板

在使用组件的时候,在组件中添加 inline-template 特殊属性,就是告知 Vue.js,使用组件包含的内容作为模板。使用 inline-template 标记,并且使用< inner-template-component >元素中包含的内容作为模板,代码如下:

```
//第 5 章/其他模板.html
...
<div id='app'>
    <inner-template-component inline-template>
        <div>
            <span v-html="desc"></span>:{{count}}
```

```
        </div>
    </inner-template-component>
</div>
<script type='text/JavaScript'>

    const InnerTemplateComponent = {
        name: 'InnerTemplateComponent',
        data(){
            return {
                count: 1,
                desc: '内联模板组件案例'
            }
        }
    }
    const vm = new Vue({
        el: '#app',
        components:{
            InnerTemplateComponent
        },
        data: {
        },
        methods: {
        }
    })
</script>
...
```

在 Vue.js 所示的 DOM 中定义内联模板，使模板的撰写工作更加灵活，但是 inline-template 会让模板的作用域变得更加难以理解，所以作为最佳实践，应在组件内优先选择 template 选项或.vue 文件里的一个 template 元素来定义模板。

2. X-Template 模板

X-Template 模板定义的方式是在一个 script 元素中，为其带上 text/x-template 的类型，然后通过一个 id 将模板引用过去，代码如下：

```
//第 5 章/其他模板.html
...
<div id='app'>
    <xtemplate-component></xtemplate-component>
</div>
<script type='text/JavaScript'>
    const XtemplateComponent = {
        name: 'XTemplateComponent',
        template: '#XTemplateComponent',
        data(){
```

```
            return {
                count: 100,
                desc: 'XTemplate 模板组件案例'
            }
        }
    }
    const vm = new Vue({
        el: '#app',
        components:{
            XtemplateComponent
        },
        data: {
        },
        methods: {
        }
    })
</script>
...
```

X-Template 模板需要定义在 Vue. js 所属的 DOM 元素外，可以用于模板特别大的 demo 或极小型的应用，但是其他情况下应避免使用，因为这会将模板和该组件的其他定义分离开。

5.9.5 控制组件的更新

Vue. js 作为一个响应式系统，它自动实现更新，不过还有特殊情况，如开发人员希望能强制更新，又或者如开发人员希望阻止不必要的更新。Vue. js 也提供了对应的支持方式。

1. 强制更新

开发人员可以通过调用 Vue. js 实例的 $forceUpdate 方法，迫使 Vue. js 实例重新渲染。注意它仅仅影响实例本身和插入插槽内容的子组件，而不是所有子组件。

需要强制更新的情况比较少，绝大部分是开发人员的编码失误，例如没有留意数组或对象的变更检测事项，或者依赖了未被 Vue. js 响应式系统跟踪的状态。

2. 降低更新

渲染普通的 HTML 元素在 Vue. js 中是非常快速的，但有的时候可能有一个组件，这个组件包含了大量静态内容。在这种情况下，可以在根元素上添加 v-once attribute 以确保这些内容只计算一次，然后缓存起来，代码如下：

```
Vue.component('terms-of-service', {
  template: `
    <div v-once>
      <h1>Terms of Service</h1>
      ... a lot of static content ...
```

```
    </div>
  `
})
```

注意：不要过度使用这个模式。当需要渲染大量静态内容时，极少数的情况下它会给你带来便利，除非当前功能非常明显渲染变慢了，不然它完全是没有必要的——再加上它在后期会带来很多困惑。例如，设想另一个开发者并不熟悉 v-once 或漏看了它在模板中，他们可能会花很多个小时去找出模板为什么无法正确更新。

第6章 组件的过渡和动画

Vue.js 在插入、更新或者移除 DOM 时，提供了多种不同方式的应用过渡效果，包括以下工具：

（1）在 CSS 过渡和动画中自动应用 class。

（2）可以配合使用第三方 CSS 动画库，如 Animate.css。

（3）在过渡钩子函数中使用 JavaScript 直接操作 DOM。

（4）可以配合使用第三方 JavaScript 动画库，如 Velocity.js。

本章介绍进入/离开和列表的过渡。

6.1 进入/离开和列表过渡

Vue.js 提供了 transition 的封装组件，在下列情形中，可以给任何元素和组件添加进入/离开过渡。

（1）条件渲染（使用 v-if）。

（2）条件展示（使用 v-show）。

（3）动态组件。

（4）组件根节点。

使用 transition 控制<p>hello</p>动态隐藏和显示，代码如下：

```
//第6章/第1个动画.html

<!DOCTYPE html>
<html lang="en">
<head>
    <meta charset="UTF-8">
    <meta http-equiv="X-UA-Compatible" content="IE=edge">
    <meta name="viewport" content="width=device-width, initial-scale=1.0">
    <!-- 引入 Vue.js -->
    <script src="../static/js/Vue.js" type="text/JavaScript"></script>
    <title>Demo vue</title>
```

```
</head>
<body>
    <!-- 定义 Vue 的视图 -->
    <div id="app">
        <button v-on:click="show = !show">
            Toggle
        </button>
        <transition name="fade">
          <p v-if="show">hello</p>
        </transition>
    </div>
    <script type="text/JavaScript">
        //创建 Vue.js 对象
        const vm = new Vue({
            el: "#app",
            data:{
                show: true
            }
        })
    </script>
    <style type="text/css">
      .fade-enter-active, .fade-leave-active {
        transition: opacity .5s;
      }
      .fade-enter, .fade-leave-to
/* .fade-leave-active below version 2.1.8 */ {
        opacity: 0;
      }
    </style>
</body>
</html>
```

6.1.1　单元素/组件过渡

在进入/离开的过渡中，会有 6 个 class 切换，它们分别如下。

（1）v-enter：定义进入过渡的开始状态。在元素被插入之前生效，在元素被插入之后的下一帧移除。

（2）v-enter-active：定义进入过渡生效时的状态。在整个进入过渡的阶段中应用，在元素被插入之前生效，在过渡/动画完成之后移除。这个类可以被用来定义进入过渡的过程时间、延迟和曲线函数。

（3）v-enter-to：2.1.8 版及以上版本定义进入过渡的结束状态。在元素被插入之后下一帧生效(与此同时 v-enter 被移除)，在过渡/动画完成之后移除。

（4）v-leave：定义离开过渡的开始状态。在离开过渡被触发时立刻生效，下一帧被移除。

(5) v-leave-active：定义离开过渡生效时的状态。在整个离开过渡的阶段中应用，在离开过渡被触发时立刻生效，在过渡/动画完成之后移除。这个类可以被用来定义离开过渡的过程时间、延迟和曲线函数。

(6) v-leave-to：2.1.8版及以上版本定义离开过渡的结束状态。在离开过渡被触发之后下一帧生效(与此同时 v-leave 被删除)，在过渡/动画完成之后移除。

这6个 class 的关系如图6.1所示。

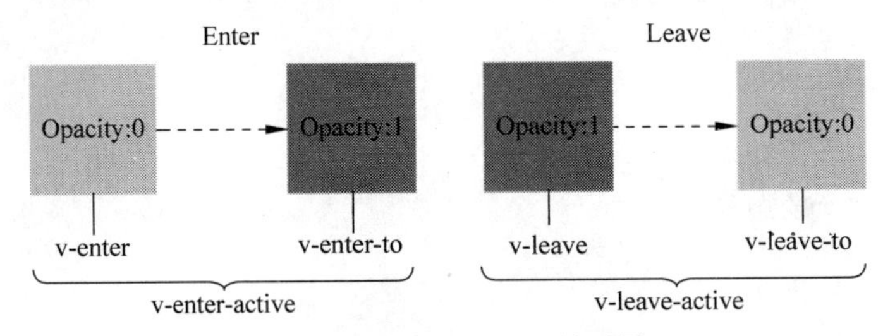

图6.1 transition class 关系图

这些过渡中切换的类名，默认为以 v-为前缀，例如 v-enter、v-enter-active 等，如果给 transition 组件的 name 赋值了，则类名就以 name 的值加-为前缀。例如在第6章第1个动画.html 中 transition 的 name 属性值是 fade，那么 class 的名称就是 fade-enter、fade-enter-active 等。

常用的过渡是使用 CSS 过渡的，v-enter-active 和 v-leave-active 可以控制进入/离开过渡的不同的缓和曲线，代码如下：

```
//第6章/CSS过渡.html

<!DOCTYPE html>
<html lang = "en">
<head>
    <meta charset = "UTF - 8">
    <meta http - equiv = "X - UA - Compatible" content = "IE = edge">
    <meta name = "viewport" content = "width = device - width, initial - scale = 1.0">
    <!-- 引入 Vue.js -->
    <script src = "../static/js/Vue.js" type = "text/JavaScript"></script>
    <title>Demo vue</title>
</head>
<body>
    <!-- 定义 Vue 的视图 -->
    <div id = "app">
        <button @click = "show = !show">
            Toggle render
        </button>
        <transition name = "slide - fade">
            <p v - if = "show"> hello </p>
```

```
        </transition>
    </div>
    <script type="text/JavaScript">
        //创建 Vue.js 对象
        const vm = new Vue({
            el: "#app",
            data:{
                show: true
            }
        })
    </script>
    <style type="text/css">
        /* 可以设置不同的进入和离开动画 */
        /* 设置持续时间和动画函数 */
        .slide-fade-enter-active {
          transition: all .3s ease;
        }
        .slide-fade-leave-active {
          transition: all .8s cubic-bezier(1.0, 0.5, 0.8, 1.0);
        }
        .slide-fade-enter, .slide-fade-leave-to
        /* .slide-fade-leave-active for below version 2.1.8 */ {
          transform: translateX(10px);
          opacity: 0;
        }
    </style>
</body>
</html>
```

CSS 动画用法同 CSS 过渡一样，其区别是在动画中 v-enter 类名在节点插入 DOM 后不会立即被删除，而是在 animationend 事件触发时被删除，代码如下：

```
///第 6 章/CSS 动画.html

<!DOCTYPE html>
<html lang="en">
<head>
    <meta charset="UTF-8">
    <meta http-equiv="X-UA-Compatible" content="IE=edge">
    <meta name="viewport" content="width=device-width, initial-scale=1.0">
    <!--引入 Vue.js-->
    <script src="../static/js/Vue.js" type="text/JavaScript"></script>
    <title>Demo vue</title>
</head>
<body>
```

```
        <!-- 定义 Vue 的视图 -->
        <div id = "app">
            <button @click = "show = !show"> Toggle show </button>
            <transition name = "bounce">
              <p v-if = "show">测试动画...</p>
            </transition>
        </div>
        <script type = "text/JavaScript">
            //创建 Vue.js 对象
            const vm = new Vue({
                el: "#app",
                data:{
                    show: true
                }
            })
        </script>
        <style>
            .bounce-enter-active {
            animation: bounce-in .5s;
            }
            .bounce-leave-active {
            animation: bounce-in .5s reverse;
            }
            @keyframes bounce-in {
            0% {
                transform: scale(0);
            }
            50% {
                transform: scale(1.5);
            }
            100% {
                transform: scale(1);
            }
            }
        </style>
    </body>
    </html>
```

根据需求，开发人员可以使用 transition 的相关属性，自定义过渡类名，这些属性的名称分别是 enter-class、enter-active-class、enter-to-class（2.1.8+）、leave-class、leave-active-class 和 leave-to-class（2.1.8+）。

它们的优先级高于普通的类名，这对于 Vue.js 的过渡系统和其他第三方 CSS 动画库（如 Animate.css）结合使用十分有用，代码如下：

```
//第 6 章/自定义类名.html

<!DOCTYPE html>
<html lang="en">
<head>
    <meta charset="UTF-8">
    <meta http-equiv="X-UA-Compatible" content="IE=edge">
    <meta name="viewport" content="width=device-width, initial-scale=1.0">
    <!-- 引入 Vue.js -->
    <script src="../static/js/Vue.js" type="text/JavaScript"></script>
    <link href="../static/css/animate.css" rel="stylesheet" type="text/css">
    <title>Demo vue</title>
</head>
<body>
    <!-- 定义 Vue.js 的视图 -->
    <div id="app">
        <button @click="show = !show">
            Toggle render
        </button>
        <transition
            name="custom-classes-transition"
            enter-active-class="animated tada"
            leave-active-class="animated bounceOutRight"
        >
            <p v-if="show">hello</p>
        </transition>
    </div>
    <script type="text/JavaScript">
        //创建 Vue.js 对象
        const vm = new Vue({
            el: "#app",
            data:{
                show: true
            }
        })
    </script>
</body>
</html>
```

Vue.js 通过设置 transitioned 或 animationend 事件监听器，来判断元素的过渡或动画是否完成，具体设置哪个监听器监听，Vue.js 可以根据元素的 CSS 规则自动判断。如果在样式规则中同时存在 transition 和 animation，则 Vue.js 无法知道应该添加哪个监听器。这种情况下，需要开发人员使用 transition 的 type 属性，指定当前需要添加哪个监听器。type 属性的值选项是 transition 和 animation。前者表示添加过渡监听器，后者表示添加动画监听器。

在很多情况下，Vue.js 可以自动得出过渡效果的完成时机。默认情况下，Vue.js 会等待其在过渡效果的根元素的第 1 个 transitionend 或 animationend 事件，然而也可以不这样设定——例如，开发者可以开发一个精心编排的一系列过渡效果，其中一些嵌套的内部元素相比于过渡效果的根元素有延迟或更长的过渡效果。

在这种情况下可以用<transition>组件上的 duration prop 定制一个显性的过渡持续时间（以毫秒计），代码如下：

```
<transition :duration="1000">...</transition>
```

也可以定制进入和移出的持续时间，代码如下：

```
<transition :duration="{ enter: 500, leave: 800 }">...</transition>
```

在元素的过渡和动画过程中，可以通过 transition 的属性，绑定 JavaScript 回调函数，从而在必要的时候，可以给过渡和动画过程添加业务处理，代码如下：

```
<transition
  v-on:before-enter="beforeEnter"
  v-on:enter="enter"
  v-on:after-enter="afterEnter"
  v-on:enter-cancelled="enterCancelled"

  v-on:before-leave="beforeLeave"
  v-on:leave="leave"
  v-on:after-leave="afterLeave"
  v-on:leave-cancelled="leaveCancelled"
>
  <!-- ... -->
</transition>
...
methods: {
  //--------
  //进入中
  //--------

  beforeEnter: function (el) {
    ...
  },
  //当与 CSS 结合使用时
  //回调函数 done 是可选的
  enter: function (el, done) {
    ...
    done()
  },
```

```
    afterEnter: function (el) {
      ...
    },
    enterCancelled: function (el) {
      ...
    },

    //--------
    //离开时
    //--------

    beforeLeave: function (el) {
      ...
    },
    //当与 CSS 结合使用时
    //回调函数 done 是可选的
    leave: function (el, done) {
      ...
      done()
    },
    afterLeave: function (el) {
      ...
    },
    //leaveCancelled 只用于 v-show 中
    leaveCancelled: function (el) {
      ...
    }
  }
```

这些钩子函数可以结合 CSS transitions/animations 使用，也可以单独使用。

注意：当只用 JavaScript 过渡的时候，在 enter 和 leave 中必须使用 done 进行回调。否则，它们将被同步调用，过渡会立即完成。

推荐对于仅使用 JavaScript 过渡的元素添加 v-bind：css＝"false"，Vue.js 会跳过 CSS 的检测。这也可以避免过渡过程中 CSS 的影响。

6.1.2 初始渲染的过渡

在 transition 组件中，可以通过 appear attribute 设置节点在初始渲染的过渡，代码如下：

```
<transition appear>
  <!-- ... -->
</transition>
```

这里默认和进入/离开过渡一样，同样也可以自定义 CSS 类名，代码如下：

```
<transition
  appear
  appear-class="custom-appear-class"
  appear-to-class="custom-appear-to-class" (2.1.8+)
  appear-active-class="custom-appear-active-class"
>
  <!-- ... -->
</transition>
```

同样可以自定义 JavaScript 钩子，代码如下：

```
<transition
  appear
  v-on:before-appear="customBeforeAppearHook"
  v-on:appear="customAppearHook"
  v-on:after-appear="customAfterAppearHook"
  v-on:appear-cancelled="customAppearCancelledHook"
>
  <!-- ... -->
</transition>
```

在上面的例子中，无论是 appear attribute 还是 v-on：appear 钩子都会生成初始渲染过渡。

6.1.3 多元素过渡

在实战项目中，可以使用 v-if/v-else 控制多个元素标签的过渡。最常见的多标签过渡是一个列表和描述这个列表为空消息的元素，代码如下：

```
<transition>
  <table v-if="items.length > 0">
    <!-- ... -->
  </table>
  <p v-else>Sorry, no items found.</p>
</transition>
```

但是有一点需要注意，当有相同标签名的元素进行切换时，需要通过 key attribute 设置唯一的值来标记，以让 Vue.js 区分它们，否则 Vue.js 为了效率只会替换相同标签内部的内容。即使在技术上没有必要，给在< transition >组件中的多个元素设置 key 是一个更好的实践，代码如下：

```
<transition>
  <button v-if="isEditing" key="save">
    Save
  </button>
```

```
    <button v-else key="edit">
      Edit
    </button>
  </transition>
```

在一些场景中，也可以通过给同一个元素的 key attribute 设置不同的状态来代替 v-if 和 v-else，上面的例子可以重写，代码如下：

```
<transition>
  <button v-bind:key="isEditing">
    {{ isEditing ? 'Save' : 'Edit' }}
  </button>
</transition>
```

使用多个 v-if 的多个元素的过渡可以重写为绑定了动态 property 的单个元素的过渡，代码如下：

```
<transition>
  <button v-if="docState === 'saved'" key="saved">
    Edit
  </button>
  <button v-if="docState === 'edited'" key="edited">
    Save
  </button>
  <button v-if="docState === 'editing'" key="editing">
    Cancel
  </button>
</transition>
```

可以重写，代码如下：

```
<transition>
  <button v-bind:key="docState">
    {{ buttonMessage }}
  </button>
</transition>
...
computed: {
  buttonMessage: function() {
    switch (this.docState) {
      case 'saved': return 'Edit'
      case 'edited': return 'Save'
      case 'editing': return 'Cancel'
    }
  }
}
```

在实战中，经常需要实现在一个位置上两个元素可以相互替换（一个隐藏，另一个显示），Vue.js 对这两个元素的进入和离开是同时控制的，代码如下：

```
//第 6 章/过渡模式.html

<!DOCTYPE html>
<html lang="en">
<head>
    <meta charset="UTF-8">
    <meta http-equiv="X-UA-Compatible" content="IE=edge">
    <meta name="viewport" content="width=device-width, initial-scale=1.0">
    <!-- 引入 Vue.js -->
    <script src="../static/js/Vue.js" type="text/JavaScript"></script>
    <title>Demo vue</title>
</head>
<body>
    <!-- 定义 Vue.js 的视图 -->
    <div id="app">
      <transition>
          <button @click="show=!show" v-if="show" :key="1" style="position:
absolute;">on</button>
          <button @click="show=!show" v-else :key="2" style="position: absolute;">
off</button>
      </transition>
    </div>
    <script type="text/JavaScript">
        //创建 Vue.js 对象
        const vm = new Vue({
            el: "#app",
            data:{
                show: true
            }
        })
    </script>
    <style type="text/css">
        .v-enter-active {
          transition: all .3s ease;
        }
        .v-leave-active {
          transition: all .3s cubic-bezier(1.0, 0.5, 0.8, 1.0);
        }
        .v-enter /* .v-leave-active for below version 2.1.8 */ {
          transform: translateX(40px);
          opacity: 0;
        }
        .v-leave-to{
```

```
            transform: translateX( - 40px);
            opacity: 0;
        }
    </style>
</body>
</html>
```

单击 On/Off 按钮的时候，On/Off 按钮的离开和进入是同时进行的。为了提高用户体验，需要控制离开和进入的顺序，例如先离开，再进入。这时候就可以使用 transition 元素的 mode 属性，指定过渡模式，以此来控制是先离开再进入，还是先进入再离开。mode 属性有两个值可以选择，分别是 in-out 和 out-in，分别表示先进入再离开模式和先离开再进入模式，代码如下：

```
...
    <!-- 定义 Vue.js 的视图 -->
    <div id = "app">
      <transition>
          <button @click = "show = ! show" v - if = "show" :key = "1" style = "position:
absolute;"> on </button>
          <button @click = "show = ! show" v - else :key = "2" style = "position: absolute;">
off </button>
      </transition>
</div>
...
```

6.1.4 多组件过渡

Vue.js 同样支持多组件的过渡。实现多组件的过渡比较简单，直接使用 transition 元素，结合动态组件实现，代码如下：

```
//第 6 章/多组件过渡.html

<!DOCTYPE html>
<html lang = "en">
<head>
    <meta charset = "UTF - 8">
    <meta http - equiv = "X - UA - Compatible" content = "IE = edge">
    <meta name = "viewport" content = "width = device - width, initial - scale = 1.0">
    <!-- 引入 Vue.js -->
    <script src = "../static/js/Vue.js" type = "text/JavaScript"></script>
    <title> Demo vue </title>
    <style>
        .tab - button {
```

```
            padding: 6px 10px;
            border-top-left-radius: 3px;
            border-top-right-radius: 3px;
            border: 1px solid #ccc;
            cursor: pointer;
            background: #f0f0f0;
            margin-bottom: -1px;
            margin-right: -1px;
          }
          .tab-button:hover {
            background: #e0e0e0;
          }
          .tab-button.active {
            background: #da9797;
          }
          .tab {
            border: 1px solid #ccc;
            padding: 10px;
          }

          .v-enter-active {
            transition: all .3s ease;
          }
          .v-leave-active {
            transition: all .3s cubic-bezier(1.0, 0.5, 0.8, 1.0);
          }
          .v-enter /* .v-leave-active for below version 2.1.8 */ {
            transform: translateX(40px);
            opacity: 0;
          }
          .v-leave-to{
            transform: translateX(-40px);
            opacity: 0;
          }
        </style>
</head>
<body>
    <!--定义 Vue.js 的视图-->
    <div id="app">
        <button v-for="tab in tabs"
                :key="tab"
                :class="['tab-button', {active:currTab === tab}]"
                @click="currTab = tab">
            {{tab}}
        </button>
        <keep-alive>
```

```
                < transition mode = 'in - out'>
                    < component :is = "componentName" style = "position: absolute;"></component >
                </transition >
                </keep - alive >
            </div >
            < script type = "text/JavaScript">
                //创建 Vue.js 对象
                const vm = new Vue({
                    el: "#app",
                    data: {
                        tabs: ['Com1','Com2', 'Com3'],
                        currTab: 'Com1'
                    },
                    computed:{
                        componentName(){
                            return 'dync - ' + this.currTab.toLowerCase()
                        }
                    },
                    components:{
                        'dync - com1':{
                            template: '< div > Component 1 </div >'
                        },
                        'dync - com2':{
                            template: '< div > Component 2 </div >'
                        },
                        'dync - com3':{
                            template: '< div > Component 3 </div >'
                        },
                    }
                })
            </script >
        </body >
        </html >
```

6.1.5 列表过渡

通过 transition 元素，可以实现单个节点和多个节点在同一时间渲染过渡，同样，也可以使用 transition-group 组件，实现一个列表中多项的过渡。transition-group 相对 transition 有以下几个特点：

(1) 与 transition 不同，它会以一个真实元素呈现，默认为 span，也可以通过 tag 属性指定。

(2) 在 transition-group 中不存在过渡模式，因为在列表过渡中不存在切换的离开/进入。

(3) 内部元素,必须有一个唯一的 key 属性值。

(4) CSS 过渡的类,将会应用到内部的元素中,而不是组/容器本身。

使用 CSS 类名,实现列表的进入/离开过渡,代码如下:

```
//第 6 章/CSS 类名列表进入/离开过渡.html

<!DOCTYPE html>
<html lang="en">
<head>
    <meta charset="UTF-8">
    <meta http-equiv="X-UA-Compatible" content="IE=edge">
    <meta name="viewport" content="width=device-width, initial-scale=1.0">
    <!--引入 Vue.js-->
<script src="../static/js/Vue.js" type="text/JavaScript"></script>
<script src="../static/js/lodash.min.js" type="text/JavaScript"></script>
    <title>Demo vue</title>
</head>
<body>
    <!--定义 Vue.js 的视图-->
    <div id="app">
        <button v-on:click="add">Add</button>
        <button v-on:click="remove">Remove</button>
        <button @click="shuffle">shuffle</button>
        <transition-group name="list" tag="div">
          <span v-for="item in items" v-bind:key="item" class="list-item">
            {{ item }}
          </span>
        </transition-group>
    </div>
    <script type="text/JavaScript">
        //创建 Vue.js 对象
        const vm = new Vue({
            el: "#app",
            data: {
                items: [1,2,3,4,5,6,7,8,9],
                nextNum: 10
            },
            methods: {
                randomIndex: function () {
                  return Math.floor(Math.random() * this.items.length)
                },
                add: function () {
                  this.items.splice(this.randomIndex(), 0, this.nextNum++)
                },
                remove: function () {
```

```
                    this.items.splice(this.randomIndex(), 1)
                },
                shuffle: function () {
                    this.items = _.shuffle(this.items)
                }
            }
        })
    </script>
    <style>
        .list - item {
        display: inline - block;
         margin - right: 10px;
        }
        .list - enter - active, .list - leave - active {
         transition: all 1s;
        }
        .list - enter, .list - leave - to
        /* .list - leave - active for below version 2.1.8 */ {
        opacity: 0;
          transform: translateY(30px);
        }
    </style>
</body>
</html>
```

通过代码可以了解，列表的过渡实现同单元素的过渡实现基本类似，除了将过渡元素改成了 transition-group 外。

单击上面样例的 Shuffle 案例，会将数字随机打乱，但是很快就替换完成，没有动画的感觉。

在 transition-group 中有个 v-move 属性，能改变定位。开发人员通过给 v-move 属性设置过渡的切换时机和过渡曲线，就可以实现动态的定位过渡。v-move 名称的命名规则同 v-enter、v-enter-active 等属性的规则一样（默认为 v-前缀，否则就以 transition-group 的 name 属性的值作为前缀）。在上面样例的 style 中，添加如下代码的 CSS 类，单击 Shuffle 按钮，就可以体会到定位的动态过渡效果，代码如下：

```
...
    .list - move {
      transition: transform 3s;
    }
...
```

Vue.js 使用了一个叫作 FLIP 的简单的动画队列，使用 transforms 将元素从之前的位置平滑过渡到新的位置。使用 FLIP 技术，可完成位置完美的平滑过渡，代码如下：

```
//第 6 章/FLIP 动画技术.html

<!DOCTYPE html>
<html lang="en">
<head>
    <meta charset="UTF-8">
    <meta http-equiv="X-UA-Compatible" content="IE=edge">
    <meta name="viewport" content="width=device-width, initial-scale=1.0">
    <!--引入 Vue.js-->
    <script src="../static/js/Vue.js" type="text/JavaScript"></script>
    <script src="../static/js/lodash.min.js" type="text/JavaScript"></script>
    <title>Demo vue</title>
</head>
<body>
    <!--定义 Vue.js 的视图-->
    <div id="app">
        <button v-on:click="shuffle">Shuffle</button>
        <button v-on:click="add">Add</button>
        <button v-on:click="remove">Remove</button>
        <transition-group name="list-complete" tag="p">
          <span
            v-for="item in items"
            v-bind:key="item"
            class="list-complete-item"
          >
            {{ item }}
          </span>
        </transition-group>
    </div>
    <script type="text/JavaScript">
        //创建 Vue.js 对象
        const vm = new Vue({
            el: "#app",
            data: {
                items: [1,2,3,4,5,6,7,8,9],
                nextNum: 10
            },
            methods: {
                randomIndex: function () {
                return Math.floor(Math.random() * this.items.length)
                },
                add: function () {
                this.items.splice(this.randomIndex(), 0, this.nextNum++)
                },
                remove: function () {
                this.items.splice(this.randomIndex), 1)
```

```
                },
                shuffle: function () {
                this.items = _.shuffle(this.items)
                }
            }
        })
    </script>
    <style>
        .list-complete-item {
          transition: all 1s;
           display: inline-block;
         margin-right: 10px;
        }
        .list-complete-enter, .list-complete-leave-to
        /* .list-complete-leave-active for below version 2.1.8 */ {
          opacity: 0;
          transform: translateY(30px);
        }
        .list-complete-leave-active {
          position: absolute;
        }
    </style>
</body>
</html>
```

需要注意的是使用 FLIP 过渡的元素不能被设置为 display：inline。作为替代方案，可以设置为 display：inline-block 或者放置于 flex 中。

FLIP 不仅可以实现单列过渡，还可以实现多维网格过渡，代码如下：

```
//第 6 章/FLIP 多维网格过渡.html

<!DOCTYPE html>
<html>
  <head>
    <title>List Move Transitions Sudoku Example</title>
    <script src="../static/js/Vue.js" type="text/JavaScript"></script>
    <script src="../static/js/lodash.min.js" type="text/JavaScript"></script>
    <style>
        .container {
            display: flex;
            flex-wrap: wrap;
            width: 238px;
            margin-top: 10px;
        }
        .cell {
```

```
            display: flex;
            justify-content: space-around;
            align-items: center;
            width: 25px;
            height: 25px;
            border: 1px solid #aaa;
            margin-right: -1px;
            margin-bottom: -1px;
        }
        .cell:nth-child(3n) {
            margin-right: 0;
        }
        .cell:nth-child(27n) {
            margin-bottom: 0;
        }
        .cell-move {
            transition: transform 1s;
        }
    </style>
</head>
<body>
  <div id="sudoku-demo" class="demo">
    <h1>Lazy Sudoku</h1>
    <p>Keep hitting the shuffle button until you win.</p>

    <button @click="shuffle">
      Shuffle
    </button>
    <transition-group name="cell" tag="div" class="container">
      <div v-for="cell in cells" :key="cell.id" class="cell">
        {{ cell.number }}
      </div>
    </transition-group>
  </div>

  <script>
    new Vue({
      el: "#sudoku-demo",
      data: {
        cells: Array.apply(null, { length: 81 }).map(function(_, index) {
          return {
            id: index,
            number: (index % 9) + 1
          };
        })
      },
```

```
            methods: {
              shuffle: function() {
                this.cells = _.shuffle(this.cells);
              }
            }
          };
        </script>
      </body>
    </html>
```

通过 data attribute 与 JavaScript 通信，可以实现列表的交错过渡，代码如下：

```
<script src="https://cdnjs.cloudflare.com/ajax/libs/velocity/1.2.3/velocity.min.js">
</script>

<div id="staggered-list-demo">
  <input v-model="query">
  <transition-group
    name="staggered-fade"
    tag="ul"
    v-bind:css="false"
    v-on:before-enter="beforeEnter"
    v-on:enter="enter"
    v-on:leave="leave"
  >
    <li
      v-for="(item, index) in computedList"
      v-bind:key="item.msg"
      v-bind:data-index="index"
    >{{ item.msg }}</li>
  </transition-group>
</div>
new Vue({
  el: '#staggered-list-demo',
  data: {
    query: '',
    list: [
      { msg: 'Bruce Lee' },
      { msg: 'Jackie Chan' },
      { msg: 'Chuck Norris' },
      { msg: 'Jet Li' },
      { msg: 'Kung Fury' }
    ]
  },
  computed: {
```

```
    computedList: function () {
      var vm = this
      return this.list.filter(function (item) {
        return item.msg.toLowerCase().indexOf(vm.query.toLowerCase) !== -1
      })
    }
  },
  methods: {
    beforeEnter: function (el) {
      el.style.opacity = 0
      el.style.height = 0
    },
    enter: function (el, done) {
      var delay = el.dataset.index * 150
      setTimeout(function () {
        Velocity(
          el,
          { opacity: 1, height: '1.6em' },
          { complete: done }
        )
      }, delay)
    },
    leave: function (el, done) {
      var delay = el.dataset.index * 150
      setTimeout(function () {
        Velocity(
          el,
          { opacity: 0, height: 0 },
          { complete: done }
        )
      }, delay)
    }
  }
})
```

6.1.6 可复用的过渡

过渡可以通过 Vue.js 的组件系统实现复用。要创建一个可复用的过渡组件，需要将 transition 或者 transition-group 作为根组件，然后将任何子组件放置在其中就可以了。使用 template 实现，代码如下：

```
Vue.component('my-special-transition', {
  template: '\
    <transition\
```

```
      name = "very - special - transition"\
      mode = "out - in"\
      v - on:before - enter = "beforeEnter"\
      v - on:after - enter = "afterEnter"\
    >\
      < slot ></slot >\
    </transition >\
  ',
  methods: {
    beforeEnter: function (el) {
      ...
    },
    afterEnter: function (el) {
      ...
    }
  }
})
```

函数式组件更适合完成这个任务,代码如下:

```
Vue.component('my - special - transition', {
  functional: true,
  render: function (createElement, context) {
    var data = {
      props: {
        name: 'very - special - transition',
        mode: 'out - in'
      },
      on: {
        beforeEnter: function (el) {
          ...
        },
        afterEnter: function (el) {
          ...
        }
      }
    }
    return createElement('transition', data, context.children)
  }
})
```

6.1.7 动态过渡

在 Vue.js 中即使是过渡也是由数据驱动的,动态过渡最基本的例子是通过 name attribute 来绑定动态值,代码如下:

```
<transition v-bind:name="transitionName">
  <!-- ... -->
</transition>
```

当开发人员想用 Vue.js 的过渡系统来定义 CSS 过渡/动画在不同过渡间切换会非常有用。

所有过渡 attribute 都可以动态绑定，但不仅只有 attribute 可以利用，还可以通过事件钩子获取上下文中的所有数据，这是因为事件钩子都是方法。这意味着，根据组件状态的不同，JavaScript 过渡会有不同的表现，代码如下：

```
<script src="https://cdnjs.cloudflare.com/ajax/libs/velocity/1.2.3/velocity.min.js">
</script>

<div id="dynamic-fade-demo" class="demo">
  Fade In: <input type="range" v-model="fadeInDuration" min="0" v-bind:max=
"maxFadeDuration">
  Fade Out: <input type="range" v-model="fadeOutDuration" min="0" v-bind:max=
"maxFadeDuration">
  <transition
    v-bind:css="false"
    v-on:before-enter="beforeEnter"
    v-on:enter="enter"
    v-on:leave="leave"
  >
    <p v-if="show">hello</p>
  </transition>
  <button
    v-if="stop"
    v-on:click="stop = false; show = false"
  >Start animating</button>
  <button
    v-else
    v-on:click="stop = true"
  >Stop it!</button>
</div>
new Vue({
  el: '#dynamic-fade-demo',
  data: {
    show: true,
    fadeInDuration: 1000,
    fadeOutDuration: 1000,
    maxFadeDuration: 1500,
    stop: true
  },
```

```
    mounted: function ) {
      this.show = false
    },
    methods: {
      beforeEnter: function (el) {
        el.style.opacity = 0
      },
      enter: function (el, done) {
        var vm = this
        Velocity(el,
          { opacity: 1 },
          {
            duration: this.fadeInDuration,
            complete: function () {
              done()
              if !vm.stop) vm.show = false
            }
          }
        )
      },
      leave: function (el, done) {
        var vm = this
        Velocity(el,
          { opacity: 0 },
          {
            duration: this.fadeOutDuration,
            complete: function () {
              done()
              vm.show = true
            }
          }
        )
      }
    }
  })
```

创建动态过渡的最终方案是使组件通过接受 props 来动态修改之前的过渡。

6.2 状态过渡

Vue.js 的过渡系统提供了非常多简单的方法用于设置进入/离开和列表的动效。对那些本身是数值形式的数据，或可以转化成数值形式的数据，Vue.js 也提供了一套机制，结合第三方库实现切换元素的过渡状态。

6.2.1 状态动画与侦听器

通过侦听器侦听数值属性的更新。用侦听器侦听 number 值的变化，从而实现 tweenedNumber 属性的状态过渡，代码如下：

```
//第 6 章/状态动画和侦听.html

<!DOCTYPE html>
<html lang="en">
<head>
    <meta charset="UTF-8">
    <meta http-equiv="X-UA-Compatible" content="IE=edge">
    <meta name="viewport" content="width=device-width, initial-scale=1.0">
    <!-- 引入 Vue.js -->
    <script src="../static/js/Vue.js" type="text/JavaScript"></script>
    <script src="../static/js/gsap.min.js"></script>
    <title>Demo vue</title>
</head>
<body>
    <!-- 定义 Vue 的视图 -->
    <div id="app">
        <input v-model.number="number" type="number" step="10">
        <p>{{ animatedNumber }}</p>
    </div>
    <script type="text/JavaScript">
        //创建 Vue.js 对象
        const vm = new Vue({
            el: "#app",
            data: {
                number: 0,
                tweenedNumber: 0
            },
            computed: {
                animatedNumber: function() {
                    //小数位为 0
                return this.tweenedNumber.toFixed(0);
                }
            },
            watch: {
                number: function(newValue) {
                  //将 this.$data 中的 tweenedNumber 在 0.5s 内变成 newValue
                  gsap.to(this.$data, { duration: 0.5, tweenedNumber: newValue });
                }
            }
        })
```

```
    </script>
</body>
</html>
```

上面的代码，当改变 number 值时，能看到动态效果，但是对于不能像数字一样直接存储的数据，例如颜色，就需要用第三方库协助实现。用 tween.js 和 color.js 实现 CSS 中 color 的状态过渡，代码如下：

```
//第 6 章/Color 属性过渡.html

<!DOCTYPE html>
<html lang = "en">
<head>
    <meta charset = "UTF - 8">
    <meta http - equiv = "X - UA - Compatible" content = "IE = edge">
    <meta name = "viewport" content = "width = device - width, initial - scale = 1.0">
    <!-- 引入 Vue.js -->
    <script src = "../static/js/Vue.js" type = "text/JavaScript"></script>
    <script src = "../static/js/tween.js"></script>
    <script src = "../static/js/color.js"></script>
    <title> Demo vue </title>
</head>
<body>
    <!-- 定义 Vue.js 的视图 -->
    <div id = "app">
        <input
            v - model = "colorQuery"
            v - on:keyup.enter = "updateColor"
            placeholder = "Enter a color"
        >
        <button v - on:click = "updateColor"> Update </button>
        <p> Preview:</p>
        <span
            v - bind:style = "{ backgroundColor: tweenedCSSColor }"
            class = "example - 7 - color - preview"
        ></span>
        <p>{{ tweenedCSSColor }}</p>
    </div>
    <script type = "text/JavaScript">
        var Color  = net.brehaut.Color
    //创建 Vue.js 对象
        const vm  = new Vue({
            el: "#app",
            data: {
                                colorQuery: '',
```

```
                    color: {
                    red: 0,
                    green: 0,
                    blue: 0,
                    alpha: 1
                    },
                    tweenedColor: {}
        },
        created: function () {
            this.tweenedColor = Object.assign({}, this.color)
        },
        watch: {
            color: function () {
            function animate () {
                if (TWEEN.update()) {
                requestAnimationFrame(animate)
                }
            }

            new TWEEN.Tween(this.tweenedColor)
                .to(this.color, 750)
                .start()

            animate()
            }
        },
        computed: {
            tweenedCSSColor: function () {
            return new Color({
                    red: this.tweenedColor.red,
                    green: this.tweenedColor.green,
                    blue: this.tweenedColor.blue,
                    alpha: this.tweenedColor.alpha
            }).toCSS()
            }
        },
        methods: {
                updateColor: function () {
                this.color = new Color(this.colorQuery).toRGB)
                this.colorQuery = ''
                }
        }
    })
</script>
<style>
    .example-7-color-preview {
```

```
            display: inline-block;
            width: 50px;
            height: 50px;
        }
    </style>
</body>
</html>
```

6.2.2　把过渡放在组件中

管理太多的状态过渡会很快地增加 Vue.js 实例或者组件的复杂性，为了解决这样的问题，可以将动画封装到专用的子组件中，代码如下：

```
//第 6 章/组件封装动画.html

<!DOCTYPE html>
<html lang="en">
<head>
    <meta charset="UTF-8">
    <meta http-equiv="X-UA-Compatible" content="IE=edge">
    <meta name="viewport" content="width=device-width, initial-scale=1.0">
    <!-- 引入 Vue.js -->
    <script src="../static/js/Vue.js" type="text/JavaScript"></script>
    <script src="../static/js/tween.js"></script>
    <title>Demo vue</title>
</head>
<body>
    <!-- 定义 Vue.js 的视图 -->
    <div id="app">
        <input v-model.number="firstNumber" type="number" step="20"> +
        <input v-model.number="secondNumber" type="number" step="20"> =
        {{ result }}
        <p>
          <animated-integer v-bind:value="firstNumber"></animated-integer> +
          <animated-integer v-bind:value="secondNumber"></animated-integer> =
          <animated-integer v-bind:value="result"></animated-integer>
        </p>
    </div>
    <script type="text/JavaScript">
        //这种复杂的补间动画逻辑可以被复用
        //任何整数都可以执行动画
        //组件化使我们的界面十分清晰
        //可以支持更多更复杂的动态过渡策略
        Vue.component('animated-integer', {
        template: '<span>{{ tweeningValue }}</span>',
```

```
props: {
    value: {
    type: Number,
    required: true
    }
},
data: function () {
    return {
    tweeningValue: 0
    }
},
watch: {
    value: function (newValue, oldValue) {
    this.tween(oldValue, newValue)
    }
},
mounted: function () {
    this.tween(0, this.value)
},
methods: {
    tween: function (startValue, endValue) {
    var vm = this
    function animate () {
        if (TWEEN.update) {
        requestAnimationFrame(animate)
        }
    }

    new TWEEN.Tween({ tweeningValue: startValue })
        .to({ tweeningValue: endValue }, 500)
        .onUpdate(function() {
        vm.tweeningValue = this.tweeningValue.toFixed(0)
        })
        .start()

    animate()
    }
}
})
//创建 Vue.js 对象
const vm = new Vue({
    el: "#app",
    data: {
    firstNumber: 20,
    secondNumber: 40
},
```

```
            computed: {
                result: function () {
                return this.firstNumber + this.secondNumber
                }
            }
            })
        </script>
    </body>
    </html>
```

第 7 章

复用和组合

在实际项目中，有很多组件的功能或特征是重复的。Vue.js 提供了复用和组合机制，能将这些公共的特征和功能抽取出来，再混入(集成)不同的对象中去。本章将介绍 Vue.js 组件的直接复用和组合。

7.1 混入

混入(mixin)提供了一种非常灵活的方式，来分发 Vue.js 组件中的可复用功能。一个混入对象可以包含任意组件选项。当组件使用混入对象时，所有混入对象的选项将被“混合”进入该组件本身的选项。

在 myMixin 中定义 created 构造函数，同时在 methods 中定义一个 hello 函数。通过 Vue.extend()函数，将 myMixin 对象混入到一个新的 Component 对象中，在 Component 的实例中，就包含了 myMixin 中定义的内容，代码如下：

```
//第 7 章/混入基础.html

<!DOCTYPE html>
<html lang="en">
<head>
    <meta charset="UTF-8">
    <meta http-equiv="X-UA-Compatible" content="IE=edge">
    <meta name="viewport" content="width=device-width, initial-scale=1.0">
    <!-- 引入 Vue.js -->
    <script src="../static/js/Vue.js" type="text/JavaScript"></script>
    <title>Demo vue</title>
</head>
<body>
    <!-- 定义 Vue.js 的视图 -->
    <div id="app">

    </div>
```

```
    <script type="text/JavaScript">
        let myMixin = {
            created:function(){
                this.hello()
            },
            methods:{
                hello(){
                    console.log('hello, my mixins')
                }
            }
        }
        let Component = Vue.extend({
            mixins: [myMixin],

        })
        let instance = new Component()
    </script>
</body>
</html>
```

浏览器的console会输出"hello,my mixins"内容,说明Component对象中,混入了定义在myMixin中的created钩子函数和定义在methods中的hello函数。

7.1.1 选项合并

当组件和混入对象含有同名选项时,这些选项将以恰当的方式进行“合并”。

对于数据属性,会进行递归合并,如果有属性名称重复(冲突),则以组件的数据属性优先。在mixin中定义message、foo两个数据属性,并且在组件里面定义message和bar两个数据属性,混入的结果是,一共有message、foot和bar 3个数据属性,其中message数据的值是组件里面message的值,代码如下:

```
var mixin = {
  data: function() {
    return {
      message: 'hello',
      foo: 'abc'
    }
  }
}


new Vue({
  mixins: [mixin],
  data: function() {
    return {
```

```
      message: 'goodbye',
      bar: 'def'
    }
  },
  created: function() {
    console.log(this.$data)
    // => { message: "goodbye", foo: "abc", bar: "def" }
  }
})
```

如果在混入对象和被混入对象中都定义了同名的钩子函数,则混入后将会合并成一个数组,也就是同一个钩子会绑定两个函数,钩子事件触发的时候,会调用多个函数,而且混入对象的钩子函数将在组件自身钩子之前调用,代码如下:

```
var mixin = {
  created: function() {
    console.log('混入对象的钩子被调用')
  }
}

new Vue({
  mixins: [mixin],
  created: function() {
    console.log('组件钩子被调用')
  }
})

// => "混入对象的钩子被调用"
// => "组件钩子被调用"
```

作为对象的选项,如 methods、components 和 directives 等,会被合并到一个对象中。如果存在键名冲突,则保留组件自己的键值对,代码如下:

```
var mixin = {
  methods: {
    foo: function() {
      console.log('foo')
    },
    conflicting: function() {
      console.log('from mixin')
    }
  }
}
```

```
var vm = new Vue({
  mixins: [mixin],
  methods: {
    bar: function() {
      console.log('bar')
    },
    conflicting: function() {
      console.log('from self')
    }
  }
})

vm.foo() // => "foo"
vm.bar() // => "bar"
vm.conflicting() // => "from self"
```

注意：Vue.extends()的合并策略也是一样的。

7.1.2 全局混入

混入也可以进行全局注册，但使用时应格外小心。一旦使用全局混入，它将影响每个之后创建的 Vue.js 实例。使用恰当时，可以用来为自定义选项注入处理逻辑，代码如下：

```
//为自定义的选项 'myOption' 注入一个处理器
Vue.mixin({
  created: function() {
    var myOption = this.$options.myOption
    if (myOption) {
      console.log(myOption)
    }
  }
})

new Vue({
  myOption: 'hello!'
})
// => "hello!"
```

7.2 自定义指令

在项目实战中，有可能需要对普通的 DOM 元素进行底层操作，这时候可以用自定义指令的方式实现。定义一个 v-demo 指令，封装控制台输出元素的样例业务。在 input 元素中使用 v-demo 指令，页面在运行的时候，控制台上会输出 input 元素，代码如下：

```
    <!-- 定义 Vue.js 的视图 -->
    <div id="app">
        <input type="text" v-demo/>
    </div>
    <script type="text/JavaScript">
        Vue.directive('demo', {
            inserted: function(el){
                console.log(el)
            }
        })
        //创建 Vue.js 对象
        const vm = new Vue({
            el: "#app",
        })
</script>

//控制台输出 <input type="text">
```

同样,可以在组件内部注册一个局部指令,在组件里面使用。在 MyComponent 组件中定义 mydemo 指令,在组件的 div 和 input 元素中使用 mydemo 指令,代码如下:

```
    <!-- 定义 Vue.js 的视图 -->
    <div id="app">
        <my-component />
    </div>
    <script type="text/JavaScript">
        let MyComponent = {
            template: `<div v-mydemo><input v-mydemo type='text' value='default value'/
></div>`,
            directives:{
                mydemo: {
                    inserted: function(el){
                        console.log('mydemo->' + el)
                    }
                }
            }
        }
        //创建 Vue.js 对象
        const vm = new Vue({
            el: "#app",
            components:{
                MyComponent
            }
        })
    </script>
//控制台输出了 div 对象和 input 对象
```

7.2.1 钩子函数

在指令对象中包含 5 个钩子函数，它们分别如下。

(1) bind：只调用一次，指令在第一次绑定到元素时调用。在这里可以进行一次性的初始化设置。

(2) inserted：被绑定元素在插入父节点时调用(仅保证父节点存在，但不一定已被插入文档中)。

(3) update：所在组件的 VNode 更新时调用，但是可能发生在其子 VNode 更新之前。指令的值可能发生了改变，也可能没有，但是可以通过比较更新前后的值来忽略不必要的模板更新(详细的钩子函数的参数见 7.2.2 节)。

(4) componentUpdated：指令所在组件的 VNode 及其子 VNode 全部更新后调用。

(5) unbind：只调用一次，指令与元素解绑时调用。

7.2.2 钩子函数参数

钩子函数有 el、binding、vnode 和 oldVnode 等 4 个参数，它们的特点和意义分别如下。

(1) el：指令所绑定的元素，可以用来直接操作 DOM。

(2) binding：一个对象，包含以下 property。

- name：指令名，不包括 v-前缀。
- value：指令的绑定值，例如在 v-my-directive="1+1"中，绑定值为 2。
- oldValue：指令绑定的前一个值，仅在 update 和 componentUpdated 钩子中可用。无论值是否改变都可用。
- expression：字符串形式的指令表达式。例如在 v-my-directive="1+1"中，表达式为"1+1"。
- arg：传给指令的参数，可选。例如在 v-my-directive：foo 中，参数为"foo"。
- modifiers：一个包含修饰符的对象。例如在 v-my-directive. foo. bar 中，修饰符对象为{foo：true，bar：true}。

(3) vnode：Vue. js 编译生成的虚拟节点。移步 VNode API 来了解更多详情。

(4) oldVnode：上一个虚拟节点，仅在 update 和 componentUpdated 钩子中可用。

注意：除了 el 之外，其他参数都是只读的，切勿修改。

获取钩子参数的代码如下：

```
<!-- 定义 Vue.js 的视图 -->
<div id = "app">
    <div v-demo:foo.a.b = "message"> hello,how are you </div>
</div>
<script type = "text/JavaScript">
    Vue.directive('demo', {
```

```
            bind: function (el, binding, vnode) {
                var s = JSON.stringify
                el.innerHTML =
                'name: ' + s(binding.name) + '<br>' +
                'value: ' + s(binding.value) + '<br>' +
                'expression: ' + s(binding.expression) + '<br>' +
                'argument: ' + s(binding.arg) + '<br>' +
                'modifiers: ' + s(binding.modifiers) + '<br>' +
                'vnode keys: ' + Object.keys(vnode).join(', ')
            }
        })
        //创建 Vue.js 对象
        const vm = new Vue({
            el: "#app",
            data: {
                message: 'hello!'
            }
        })
    </script>
```

指令的参数可以是动态的。例如，在 v-mydirective:[argument]="value"中，argument 参数可以根据组件实例的数据进行更新，这使自定义指令可以在应用中被灵活使用。

通过动态参数，控制元素是固定在 left 还是固定在 top，代码如下：

```
<div id="dynamicexample">
  <h3>Scroll down inside this section ↓</h3>
  <p v-pin:[direction]="200">I am pinned onto the page at 200px to the left.</p>
</div>
Vue.directive('pin', {
  bind: function (el, binding, vnode) {
    el.style.position = 'fixed'
    //根据动态参数判断是固定在 left 还是固定在 top
    var s = (binding.arg == 'left'? 'left' : 'top')
    el.style[s] = binding.value + 'px'
  }
})

new Vue({
  el: '#dynamicexample',
  data: function () {
    return {
      direction: 'left'
    }
  }
})
```

7.2.3 函数简写

如果 bind 和 update 触发相同的行为，而不关心其他的钩子，则可以用以下代码的方式统一实现：

```
Vue.directive('color - swatch', function (el, binding) {
  el.style.backgroundColor = binding.value
})
```

7.2.4 对象字面量

如果指令需要多个值，则可以传入一个 JavaScript 对象字面量。记住，指令函数能够接受所有合法的 JavaScript 表达式，代码如下：

```
< div v - demo = "{ color: 'white', text: 'hello! ' }"></div >
Vue.directive('demo', function (el, binding) {
  console.log(binding.value.color) // = > "white"
  console.log(binding.value.text) // = > "hello! "
})
```

7.3 渲染函数与 JSX

通常情况下，Vue.js 推荐使用 template 定义视图内容(html)，然而有时使用渲染函数生成视图内容要方便得多。用 template 定义一个子组件，用 h*n* 显示插槽里面的内容，代码如下：

```
<! -- 定义 Vue.js 的视图 -- >
< div id = "app">
    < sub - component :level = "1"> Hello </sub - component >
    < sub - component :level = "2"> Hello </sub - component >
    < sub - component :level = "4"> Hello </sub - component >
</div >
< script type = "text/JavaScript">
    const SubComponent = {
        //template 定义视图
        template:`< div >
              < h1 v - if = "level === 1">< slot ></slot ></h1 >
              < h2 v - else - if = "level === 2">< slot ></slot ></h2 >
              < h3 v - else>< slot ></slot ></h3 >
          </div >`,
        props:{
          level:{
```

```
                    type: Number,
                    required: true
                }
            }
        }
        //创建 Vue.js 对象
        const vm = new Vue({
            el: "#app",
            components:{
                SubComponent
            }
        })
    </script>
```

在 template 里面，用 v-if 根据 level 的不同值，渲染不同的 h，如果代码量太大，则可改成渲染函数生成子组件的 html，代码如下：

```
<!-- 定义 Vue.js 的视图 -->
<div id = "app">
    <sub-component :level = "1">Hello</sub-component>
    <sub-component :level = "2">Hello</sub-component>
    <sub-component :level = "4">Hello</sub-component>
</div>
<script type = "text/JavaScript">
    const SubComponent = {
        //用 render()函数渲染 view 的 html
        render(createElement){
            return createElement(
                'h' + this.level,
                this.$slots.default          //获取传入插槽的值
            )
        },
        props:{
            level:{
                type: Number,
                required: true
            }
        }
    }
    //创建 Vue.js 对象
    const vm = new Vue({
        el: "#app",
        components:{
            SubComponent
        }
    })
</script>
```

这种情况下，渲染函数的使用要比 template 实现方便很多。

7.3.1　虚拟 DOM

浏览器接收到 html 页面后，会先将页面内容解析成树形结构，保存在内存中，再显示到页面上。每个元素是一个独立的节点对象，文本也是。

在上面的案例中，createElement('h'+this.level, this.$slots.default)执行的结果，也是创建一个 h 的节点对象，只是这个对象不是原汁原味的 DOM 节点，而是 Vue.js 中的节点描述对象，包含所有描述信息，以便 Vue.js 将其渲染到页面。这种节点，通常称为虚拟节点(Virtual Node)，简称为 VNode。虚拟 DOM 是在 Vue.js 里面对所有 Vue.js 组件数的总称。

7.3.2　createElement 参数

createElement 的语法是：createElement ({String | Object | function}, {Object}, {String, Array})，带 3 个参数。

第 1 个参数可以是 String|Object|Function 类型，必选，可以是 html 元素名称、组件选项对象或者分解成 String 或 Object 的任何一种异步函数。

第 2 个参数是 Object 类型，可选，同模板中的属性对象对应的数据对象。

第 3 个参数可以是 String|Array 类型，可选，可以是虚拟节点的子文本节点，或者其他子虚拟节点。

关于第 2 个参数，经常又称为数据对象，里面包含的是组件属性对象对应的数据，代码如下：

```
//与 v-bind:class 的 API 相同
//接收一个字符串、对象或字符串和对象组成的数组
'class': {
  foo: true,
  bar: false
},
//与 v-bind:style 的 API 相同
//接收一个字符串、对象,或对象组成的数组
style: {
  color: 'red',
  fontSize: '14px'
},
//普通的 HTML attribute
attrs: {
  id: 'foo'
},
//组件 prop
props: {
  myProp: 'bar'
```

```
    },
    //DOM property
    domProps: {
      innerHTML: 'baz'
    },
    //事件监听器在 on 内,
    //但不再支持如 v-on:keyup.enter 这样的修饰器
    //需要在处理函数中手动检查 keyCode
    on: {
      click: this.clickHandler
    },
    //仅用于组件,用于监听原生事件,而不是在组件内部使用
    //vm.$emit 触发的事件
    nativeOn: {
      click: this.nativeClickHandler
    },
    //自定义指令.注意,无法对 binding 中的 oldValue 赋值
    //因为 Vue.js 已经自动进行了同步
    directives: [
      {
        name: 'my-custom-directive',
        value: '2',
        expression: '1 + 1',
        arg: 'foo',
        modifiers: {
          bar: true
        }
      }
    ],
    //作用域插槽的格式为
    //{ name: props => VNode | Array<VNode> }
    scopedSlots: {
      default: props => createElement('span', props.text)
    },
    //如果组件是其他组件的子组件,则需为插槽指定名称
    slot: 'name-of-slot',
    //其他特殊顶层 property
    key: 'myKey',
    ref: 'myRef',
    //如果在渲染函数中给多个元素应用了相同的 ref 名
    //则 $refs.myRef 会变成一个数组
    refInFor: true
  }
```

特别要注意的是,在 Vue.js 组件树中,VNode 是唯一的。如下代码是不正确的,因为在渲染出来的对象里面有两个一样的 VNode 对象,代码如下:

```
render: function (createElement) {
  var myParagraphVNode = createElement('p', 'hi')
  return createElement('div', [
    //错误,重复的 VNode
    myParagraphVNode, myParagraphVNode
  ])
}
```

7.4　插件

插件通常用来为 Vue.js 添加全局功能。插件的功能一般有下面几种。

(1) 添加全局方法或者 property,如 vue-custom-element。

(2) 添加全局资源,指令/过滤器/过渡等,如 vue-touch。

(3) 通过全局混入来添加一些组件选项,如 vue-router。

(4) 添加 Vue.js 实例方法,通过把它们添加到 Vue.prototype 上实现。

(5) 一个库,提供自己的 API,同时提供上面提到的一个或多个功能,如 vue-router。

7.4.1　使用插件

通过全局方法 Vue.use()使用插件。需要在调用 new Vue()启动应用之前完成,一般在 main.js 文件中,代码如下:

```
//调用 MyPlugin.install(Vue)
Vue.use(MyPlugin)

new Vue({
  //...组件选项
})
```

也可以传入一个可选的选项对象,代码如下:

```
Vue.use(MyPlugin, { someOption: true })
```

Vue.use 会自动阻止多次注册相同的插件,届时即使多次调用也只会注册一次该插件。

Vue.js 官方提供的一些插件(例如 vue-router)在检测到 Vue.js 是可访问的全局变量时会自动调用 Vue.use(),然而在像 CommonJS 这样的模块环境中,应该始终显式地调用 Vue.use(),代码如下:

```
//用 Browserify 或 webpack 提供的 CommonJS 模块环境时
var Vue = require('vue')
var VueRouter = require('vue-router')
```

```
//不要忘了调用此方法
Vue.use(VueRouter)
```

awesome-vue 集合了大量由社区贡献的插件和库。

7.4.2 开发插件

Vue.js 的插件应该暴露一个 install 方法，这个方法的第 1 个参数是 Vue.js 构造器，第 2 个参数是一个可选的选项对象，代码如下：

```
MyPlugin.install = function (Vue, options) {
  //1. 添加全局方法或 property
  Vue.myGlobalMethod = function () {
    //逻辑...
  }

  //2. 添加全局资源
  Vue.directive('my-directive', {
    bind (el, binding, vnode, oldVnode) {
      //逻辑...
    }
    ...
  })

  //3. 注入组件选项
  Vue.mixin({
    created: function () {
      //逻辑...
    }
    ...
  })

  //4. 添加实例方法
  Vue.prototype.$myMethod = function (methodOptions) {
    //逻辑...
  }
}
```

实现一个可以设置提示位置(bottom、center、top)的 toast 插件，代码如下：

```
//第 7 章 toast 插件.html

  <!-- 定义 Vue.js 的视图 -->
    <div id="app">
```

```
    </div>
    <script type="text/JavaScript">
        var Toast = {};
        var showToast = false,                                //存储 toast 的显示状态
        showLoad = false,                                     //存储 loading 的显示状态
        toastVM = null,                                       //存储 toast 的 vm
        loadNode = null;                                      //存储 loading 的节点元素
        Toast.install = function (Vue, options) {
            let opt = {
            defaultType:'bottom',                             //默认显示位置
            duration:'2500'                                   //持续时间
        }
        for(let property in options){
            opt[property] = options[property];                //使用 options 的配置
        }
        Vue.prototype.$toast = (tips,type) => {
            var curType = type ? type : opt.defaultType;
            var wordWrap = opt.wordWrap ? 'lx-word-wrap' : '';
            var style = opt.width ? 'style="width: ' + opt.width + '"' : '';
            var tmp = '<div v-show="show" :class="type" class="lx-toast ' + wordWrap + '"
 ' + style + '>{{tip}}</div>';

            if (showToast) {
                //如果 toast 还在,则不再执行
                return;
            }
            if (!toastVM) {
                var toastTpl = Vue.extend({
                    data: function () {
                        return {
                            show: showToast,
                            tip: tips,
                            type: 'lx-toast-' + curType
                        }
                    },
                    template: tmp
                });
                toastVM = new toastTpl()
                var tpl = toastVM.$mount().$el;
                document.body.appendChild(tpl);
            }
            toastVM.type = 'lx-toast-' + curType;
            toastVM.tip = tips;
            toastVM.show = showToast = true;
```

```
            setTimeout(function () {
                toastVM.show = showToast = false;
            }, opt.duration)
        };
        ['bottom', 'center', 'top'].forEach(function (type) {
            Vue.prototype.$toast[type] = function (tips) {
                return Vue.prototype.$toast(tips, type)
            }
        });

        Vue.prototype.$loading = function (tips, type) {
            if (type == 'close') {
                loadNode.show = showLoad = false;
            } else {
                if (showLoad) {
                    //如果 loading 还在,则不再执行
                    return;
                }
                var loadTpl = Vue.extend({
                    data: function () {
                        return {
                            show: showLoad
                        }
                    },
                    template: '<div v-show="show" class="lx-load-mark"><div class="lx-load-box"><div class="lx-loading"><div class="loading_leaf loading_leaf_0"></div><div class="loading_leaf loading_leaf_1"></div><div class="loading_leaf loading_leaf_2"></div><div class="loading_leaf loading_leaf_3"></div><div class="loading_leaf loading_leaf_4"></div><div class="loading_leaf loading_leaf_5"></div><div class="loading_leaf loading_leaf_6"></div><div class="loading_leaf loading_leaf_7"></div><div class="loading_leaf loading_leaf_8"></div><div class="loading_leaf loading_leaf_9"></div><div class="loading_leaf loading_leaf_10"></div><div class="loading_leaf loading_leaf_11"></div></div><div class="lx-load-content">' + tips + '</div></div></div>'
                });
                loadNode = new loadTpl();
                var tpl = loadNode.$mount().$el;

                document.body.appendChild(tpl);
                loadNode.show = showLoad = true;
            }
        });

        ['open', 'close'].forEach(function (type) {
            Vue.prototype.$loading[type] = function (tips) {
                return Vue.prototype.$loading(tips, type)
            }
```

```
        });
        }

        Vue.use(Toast)
        //创建 Vue.js 对象
        const vm = new Vue({
            el: "#app",
            created(){
                this.$toast('你好','center')
            }
        })
    </script>
    <style>
        .lx-toast {
            position: fixed;
            bottom: 100px;
            left: 50%;
            box-sizing: border-box;
            max-width: 80%;
            height: 40px;
            line-height: 20px;
            padding: 10px 20px;
            transform: translateX(-50%);
            -webkit-transform: translateX(-50%);
            text-align: center;
            z-index: 9999;
            font-size: 14px;
            color: #fff;
            border-radius: 5px;
            background: rgba(0, 0, 0, 0.7);
            animation: show-toast .5s;
            -webkit-animation: show-toast .5s;
            overflow: hidden;
            text-overflow: ellipsis;
            white-space: nowrap;
        }

        .lx-toast.lx-word-wrap {
            width: 80%;
            white-space: inherit;
            height: auto;
        }

        .lx-toast.lx-toast-top {
            top: 50px;
            bottom: inherit;
```

```
}

.lx-toast.lx-toast-center {
    top: 50%;
    margin-top: -20px;
    bottom: inherit;
}

@keyframes show-toast {
    from {
        opacity: 0;
        transform: translate(-50%, -10px);
        -webkit-transform: translate(-50%, -10px);
    }
    to {
        opacity: 1;
        transform: translate(-50%, 0);
        -webkit-transform: translate(-50%, 0);
    }
}

.lx-load-mark {
    position: fixed;
    left: 0;
    top: 0;
    width: 100%;
    height: 100%;
    z-index: 9999;
}

.lx-load-box {
    position: fixed;
    z-index: 3;
    width: 7.6em;
    min-height: 7.6em;
    top: 180px;
    left: 50%;
    margin-left: -3.8em;
    background: rgba(0, 0, 0, 0.7);
    text-align: center;
    border-radius: 5px;
    color: #FFFFFF;
}

.lx-load-content {
    margin-top: 64%;
```

```
        font-size: 14px;
    }

    .lx-loading {
        position: absolute;
        width: 0px;
        left: 50%;
        top: 38%;
    }

    .loading_leaf {
        position: absolute;
        top: -1px;
        opacity: 0.25;
    }

    .loading_leaf:before {
        content: " ";
        position: absolute;
        width: 9.14px;
        height: 3.08px;
        background: #d1d1d5;
        box-shadow: rgba(0, 0, 0, 0.0980392) 0px 0px 1px;
        border-radius: 1px;
        -webkit-transform-origin: left 50% 0px;
        transform-origin: left 50% 0px;
    }

    .loading_leaf_0 {
        -webkit-animation: opacity-0 1.25s linear infinite;
        animation: opacity-0 1.25s linear infinite;
    }

    .loading_leaf_0:before {
        -webkit-transform: rotate(0deg) translate(7.92px, 0px);
        transform: rotate(0deg) translate(7.92px, 0px);
    }

    .loading_leaf_1 {
        -webkit-animation: opacity-1 1.25s linear infinite;
        animation: opacity-1 1.25s linear infinite;
    }

    .loading_leaf_1:before {
        -webkit-transform: rotate(30deg) translate(7.92px, 0px);
        transform: rotate(30deg) translate(7.92px, 0px);
```

```
}

.loading_leaf_2 {
    -webkit-animation: opacity-2 1.25s linear infinite;
    animation: opacity-2 1.25s linear infinite;
}

.loading_leaf_2:before {
    -webkit-transform: rotate(60deg) translate(7.92px, 0px);
    transform: rotate(60deg) translate(7.92px, 0px);
}

.loading_leaf_3 {
    -webkit-animation: opacity-3 1.25s linear infinite;
    animation: opacity-3 1.25s linear infinite;
}

.loading_leaf_3:before {
    -webkit-transform: rotate(90deg) translate(7.92px, 0px);
    transform: rotate(90deg) translate(7.92px, 0px);
}

.loading_leaf_4 {
    -webkit-animation: opacity-4 1.25s linear infinite;
    animation: opacity-4 1.25s linear infinite;
}

.loading_leaf_4:before {
    -webkit-transform: rotate(120deg) translate(7.92px, 0px);
    transform: rotate(120deg) translate(7.92px, 0px);
}

.loading_leaf_5 {
    -webkit-animation: opacity-5 1.25s linear infinite;
    animation: opacity-5 1.25s linear infinite;
}

.loading_leaf_5:before {
    -webkit-transform: rotate(150deg) translate(7.92px, 0px);
    transform: rotate(150deg) translate(7.92px, 0px);
}

.loading_leaf_6 {
    -webkit-animation: opacity-6 1.25s linear infinite;
    animation: opacity-6 1.25s linear infinite;
}
```

```
.loading_leaf_6:before {
    -webkit-transform: rotate(180deg) translate(7.92px, 0px);
    transform: rotate(180deg) translate(7.92px, 0px);
}

.loading_leaf_7 {
    -webkit-animation: opacity-7 1.25s linear infinite;
    animation: opacity-7 1.25s linear infinite;
}

.loading_leaf_7:before {
    -webkit-transform: rotate(210deg) translate(7.92px, 0px);
    transform: rotate(210deg) translate(7.92px, 0px);
}

.loading_leaf_8 {
    -webkit-animation: opacity-8 1.25s linear infinite;
    animation: opacity-8 1.25s linear infinite;
}

.loading_leaf_8:before {
    -webkit-transform: rotate(240deg) translate(7.92px, 0px);
    transform: rotate(240deg) translate(7.92px, 0px);
}

.loading_leaf_9 {
    -webkit-animation: opacity-9 1.25s linear infinite;
    animation: opacity-9 1.25s linear infinite;
}

.loading_leaf_9:before {
    -webkit-transform: rotate(270deg) translate(7.92px, 0px);
    transform: rotate(270deg) translate(7.92px, 0px);
}

.loading_leaf_10 {
    -webkit-animation: opacity-10 1.25s linear infinite;
    animation: opacity-10 1.25s linear infinite;
}

.loading_leaf_10:before {
    -webkit-transform: rotate(300deg) translate(7.92px, 0px);
    transform: rotate(300deg) translate(7.92px, 0px);
}
```

```
.loading_leaf_11 {
    -webkit-animation: opacity-11 1.25s linear infinite;
    animation: opacity-11 1.25s linear infinite;
}

.loading_leaf_11:before {
    -webkit-transform: rotate(330deg) translate(7.92px, 0px);
    transform: rotate(330deg) translate(7.92px, 0px);
}

@-webkit-keyframes opacity-0 {
    0% {
        opacity: 0.25;
    }
    0.01% {
        opacity: 0.25;
    }
    0.02% {
        opacity: 1;
    }
    60.01% {
        opacity: 0.25;
    }
    100% {
        opacity: 0.25;
    }
}

@-webkit-keyframes opacity-1 {
    0% {
        opacity: 0.25;
    }
    8.34333% {
        opacity: 0.25;
    }
    8.35333% {
        opacity: 1;
    }
    68.3433% {
        opacity: 0.25;
    }
    100% {
        opacity: 0.25;
    }
}
```

```
@-webkit-keyframes opacity-2 {
    0% {
        opacity: 0.25;
    }
    16.6767% {
        opacity: 0.25;
    }
    16.6867% {
        opacity: 1;
    }
    76.6767% {
        opacity: 0.25;
    }
    100% {
        opacity: 0.25;
    }
}

@-webkit-keyframes opacity-3 {
    0% {
        opacity: 0.25;
    }
    25.01% {
        opacity: 0.25;
    }
    25.02% {
        opacity: 1;
    }
    85.01% {
        opacity: 0.25;
    }
    100% {
        opacity: 0.25;
    }
}

@-webkit-keyframes opacity-4 {
    0% {
        opacity: 0.25;
    }
    33.3433% {
        opacity: 0.25;
    }
    33.3533% {
        opacity: 1;
    }
```

```
        93.3433% {
            opacity: 0.25;
        }
        100% {
            opacity: 0.25;
        }
    }

    @-webkit-keyframes opacity-5 {
        0% {
            opacity: 0.270958333333333;
        }
        41.6767% {
            opacity: 0.25;
        }
        41.6867% {
            opacity: 1;
        }
        1.67667% {
            opacity: 0.25;
        }
        100% {
            opacity: 0.270958333333333;
        }
    }

    @-webkit-keyframes opacity-6 {
        0% {
            opacity: 0.375125;
        }
        50.01% {
            opacity: 0.25;
        }
        50.02% {
            opacity: 1;
        }
        10.01% {
            opacity: 0.25;
        }
        100% {
            opacity: 0.375125;
        }
    }

    @-webkit-keyframes opacity-7 {
        0% {
```

```
            opacity: 0.479291666666667;
        }
        58.3433% {
            opacity: 0.25;
        }
        58.3533% {
            opacity: 1;
        }
        18.3433% {
            opacity: 0.25;
        }
        100% {
            opacity: 0.479291666666667;
        }
    }

    @-webkit-keyframes opacity-8 {
        0% {
            opacity: 0.583458333333333;
        }
        66.6767% {
            opacity: 0.25;
        }
        66.6867% {
            opacity: 1;
        }
        26.6767% {
            opacity: 0.25;
        }
        100% {
            opacity: 0.583458333333333;
        }
    }

    @-webkit-keyframes opacity-9 {
        0% {
            opacity: 0.687625;
        }
        75.01% {
            opacity: 0.25;
        }
        75.02% {
            opacity: 1;
        }
        35.01% {
            opacity: 0.25;
```

```
        }
        100% {
            opacity: 0.687625;
        }
    }

    @-webkit-keyframes opacity-10 {
        0% {
            opacity: 0.791791666666667;
        }
        83.3433% {
            opacity: 0.25;
        }
        83.3533% {
            opacity: 1;
        }
        43.3433% {
            opacity: 0.25;
        }
        100% {
            opacity: 0.791791666666667;
        }
    }

    @-webkit-keyframes opacity-11 {
        0% {
            opacity: 0.895958333333333;
        }
        91.6767% {
            opacity: 0.25;
        }
        91.6867% {
            opacity: 1;
        }
        51.6767% {
            opacity: 0.25;
        }
        100% {
            opacity: 0.895958333333333;
        }
    }
</style>
```

7.5 过滤器

Vue.js中提供了过滤器机制，开发人员可以利用过滤器机制，实现常用文本的格式化。过滤器可以用在双花括号插值和v-bind表达式里。过滤器应该被添加到JavaScript表达式的尾部，由管道符指示，代码如下：

```
<!-- 在双花括号中 -->
{{ message | capitalize }}

<!-- 在 `v-bind` 中 -->
<div v-bind:id="rawId | formatId"></div>
```

可以定义两种过滤器。一种是在组件内部，用于定义一个本地过滤器，只能在组件内部使用，代码如下：

```
filters: {
  capitalize: function (value) {
    if (!value) return ''
    value = value.toString()
    return value.charAt(0).toUpperCase() + value.slice(1)
  }
}
```

另一种在创建Vue.js实例之前，定义全局过滤器，代码如下：

```
Vue.filter('capitalize', function (value) {
  if (!value) return ''
  value = value.toString()
  return value.charAt(0).toUpperCase() + value.slice(1)
})

new Vue({
  //对象属性
})
```

分别定义和使用一个局部和全局过滤器。在SubComponent组件中，定义capitalize局部过滤器和globalCapitalize全局过滤器，代码如下：

```
//第7章/过滤器.html

    <!-- 定义 Vue.js 的视图 -->
    <div id="app">
```

```
        <input type="text" v-model="value"/>
        <sub-component :value="value"></sub-component>
        <input type='text' v-bind:value="value|globalCapitalize">
    </div>
    <script type="text/JavaScript">
        const SubComponent = {
            template: `<div>{{value|capitalize}}</div>`,
            props:['value'],
            filters:{
                capitalize(value){                    //局部过滤器
                    if(!value){
                        return
                    }
                    value = value.toString()
                    let values = value.split(' ');
                    values.forEach(function(element,index, values(){
                        values[index] = element.charAt(0).toUpperCase() + element.slice(1)
                    })
                    return values.join(' ')
                }
            }
        }
        //全局过滤器
        Vue.filter('globalCapitalize',
            function(value){
                if(!value){
                        return
                }
                value = value.toString()
                let values = value.split(' ');
                values.forEach(function(element,index, values){
                    values[index] = element.charAt(0).toUpperCase() + element.slice(1)
                })
                return values.join(' ')
            }
        )
        //创建 Vue.js 对象
        const vm = new Vue({
            el: "#app",
            data(){
                return {
                    value:'default value'
                }
            },
            components:{
                SubComponent
            }
        })
    </script>
```

第8章 路由基础

在前端项目中，当某个事件发生时，需要显示 Vue.js 实例下的另外一个组件。在 Vue.js 中，需要使用 Vue Router 进行导向和控制。本章介绍 Vue Router 的基础知识。

8.1 Vue Router 简介

Vue Router 是 Vue.js 官方的路由管理器。它和 Vue.js 的核心深度集成，方便前端开发人员简单快捷地构建单页面应用。Vue Router 主要有以下几个功能。

(1) 嵌套的路由/视图表。

(2) 模块化的、基于组件的路由配置。

(3) 路由参数、查询、通配符。

(4) 基于 Vue.js 过渡系统的视图过渡效果。

(5) 细粒度的导航控制。

(6) 带有自动激活的 CSS class 的链接。

(7) HTML5 历史模式或 hash 模式，在 IE 9 中自动降级。

(8) 自定义的滚动条行为。

8.2 安装 Vue Router

在使用 Vue Router 之前，需先安装 Vue Router。有 3 种方式可以安装 Vue Router。

1. 直接下载/CDN

unpkg.com 提供了基于 NPM 的 CDN 链接。https://unpkg.com/vue-router/dist/vue-router.js 链接会一直指向在 NPM 发布的最新版本。开发者也可以像 https://unpkg.com/vue-router@2.0.0/dist/vue-router.js 这样指定版本号或者 Tag。开发人员可以使用 script 元素，引用 Vue.js 和 vue-router.js 安装使用 router，代码如下：

```
<script src="…Vue.js"></script>
<script src="…router.js"></script>
```

当然也可以通过访问 https://unpkg.com/vue-router/dist/vue-router.js 链接，将里面的内容保存到本地的一个 JS 文件中，再在页面中基于 script 元素引入 JS 文件。

2. NPM 安装

在模块化工程中，可以使用 npm install vue-router 命令安装 Vue Router，然后用 Vue.use()注册全局路由，一般在模块化工程的 main.js 文件中注册，代码如下：

```
import Vue from 'vue'
import VueRouter from 'vue-router'

Vue.use(VueRouter)
```

3. Vue CLI 脚手架安装

在使用 Vue CLI 脚手架的项目时，可以执行 vue add router 命令，以项目插件的形式添加 Vue Router。CLI 可以生成上述代码及两个示例路由。它也会覆盖 App.vue，因此需要确保在项目中运行 vue add router 命令之前备份 App.vue 文件。

4. 构建开发版

如果开发人员想使用最新的开发版，则可以从 GitHub 上直接克隆，然后按如下的命令构建一个最新的 Vue-Router。

```
git clone https://github.com/vuejs/vue-router.git node_modules/vue-router
cd node_modules/vue-router
npm install
npm run build
```

8.3 第 1 个路由

用 Vue.js + Vue Router 创建单页应用非常简单。通过前面的介绍，开发人员可以组合组件来组成应用程序，如果要把 Vue Router 添加进来，则开发人员需要做的是将组件(Components)映射到路由(Routes)，然后告诉 Vue Router 在哪里渲染它们。如下案例介绍 Vue.js+Vue Router 的简单使用。

创建 html，使用< router-view >元素制定组件的渲染位置，使用< router-link >元素制定组件的导航，代码如下：

```
//第 8 章/第 1 个路由.html
...
    <script src="../static/js/Vue.js" type="text/JavaScript"></script>
<script src="../static/js/vue-router.js" type="text/JavaScript"></script>
...
    <div id="app">
```

```
        <h1>Hello App!</h1>
          <p>
            <!-- 使用 router-link 组件来导航. -->
            <!-- 通过传入 `to` 属性指定链接. -->
            <!-- <router-link> 默认会被渲染成一个 `<a>` 标签 -->
            <router-link to="/foo">GotoFoo</router-link>
            <router-link to="/bar">GotoBar</router-link>
          </p>
          <!-- 路由出口 -->
          <!-- 由路由匹配到的组件将渲染在这里 -->
          <router-view></router-view>
      </div>
...
```

在第1个路由.html文件中，添加JavaScript脚本，在Vue.js对象中引入Router对象，基于Router实现Vue.js中组件的路由导航，代码如下：

```
//第1个路由.html
...
   <script>
        //1. 定义 (路由) 组件
        //可以从其他文件 import 进来
        const Foo = { template: '<div>foo</div>' }
        const Bar = { template: '<div>bar</div>' }

        //2. 定义路由
        //每个路由应该映射一个组件.其中"component" 可以是
        //通过 Vue.extend() 创建的组件构造器
        //或者,只是一个组件配置对象
        //我们以后再讨论嵌套模式路由
        const routes = [
          { path: '/foo', component: Foo },
          { path: '/bar', component: Bar }
        ]

        //3. 创建 router 实例,然后传 'routes' 配置
        //还可以传别的配置参数, 不过先采用这种简单方式
        const router = new VueRouter({
          routes //(缩写) 相当于 routes: routes
        })

        //4. 创建和挂载根实例
        //记得要通过 router 配置的参数注入路由
        //从而让整个应用都有路由功能
        const vm = new Vue({
```

```
            el:"#app",
            router,//通过注入路由器,可以在任何组件内通过 `this.$router` 访问路由器
            computed: {
                username() {
                  //很快就会看到 `params` 是什么
                  return this.$route.params.username
                }
            },
            methods: {
                goBack() {
                  window.history.length > 1 ? this.$router.go(-1) : this.$router.push('/')
                }
            }
        })
    </script>
...
```

通过注入路由器,开发人员可以在任何组件内通过 this. $router 访问路由器,也可以通过 this. $router 访问当前路由。其实 this. $router 和 router 使用起来完全一样。本书中使用 this. $router 的原因是不想在每个需要独立封装路由的组件中都导入路由。

需要注意,当< router-link >对应的路由匹配成功时,将自动设置 class 属性值.router-link-active。

8.4 路由种类

根据路由的应用场景、特点和实现方式,可以将路由分为三类,分别是动态路由、嵌套模式路由和编程式路由。接下来逐个介绍这些路由。

8.4.1 动态路由

动态路由,也叫作参数路由,即在路由路径中添加动态的参数,以便于一个组件可以匹配渲染一类路由。例如在开发过程中,需要用 User 组件,渲染所有 ID 用户的路由,此种情况就可以使用动态路由。在 Vue-Router 的路由路径中使用“动态路径参数”实现所有用户都用同一个组件渲染的效果,代码如下:

```
const User = {
  template: '<div>User</div>'
}

const router = new VueRouter({
  routes: [
    //动态路径参数 以冒号开头
```

```
    { path: '/user/:id', component: User }
  ]
})
```

这样，/user/foot 和/user/bar 都被映射到相同的路由上（User 组件渲染）。

一个"路径参数"使用冒号"："标记。当匹配到一个路由时，参数值会被设置到"this. $route. params"，可以在每个组件内使用。于是，开发人员可以更新 User 的模板，输出当前用户的 ID，代码如下：

```
const User = {
  template: '<div>User {{ $route.params.id }}</div>'
}
```

完整样例的代码如下：

```
<div id="app">
    <h1>Hello App!</h1>
      <p>
      <!-- 使用 router-link 组件来导航. -->
      <!-- 通过传入 `to` 属性指定链接. -->
      <!-- <router-link> 默认会被渲染成一个 `<a>` 标签 -->
          <router-link to="/user/foo">GotoFoo</router-link>
          <router-link to="/user/bar">GotoBar</router-link>
      </p>
      <!-- 路由出口 -->
      <!-- 由路由匹配到的组件将渲染在这里 -->
      <router-view></router-view>
</div>
<script>
    const User = {
      template: '<div>User {{ $route.params.id }}</div>',
      //切换路由时不会重复调用(组件创建后缓存了)
      created:function){
        console.log('create..... User')
       },
      watch:{
        //to 和 from 都是路由对象 $route
        $route(to,from){
          console.log(to.params.id)
          console.log(from)
        }
      }
    }

    const router = new VueRouter({
```

```
        routes: [
          //动态路径参数，以冒号开头
          { path: '/user/:id', component: User }
        ]
      })
      const vm = new Vue({
          el:"#app",
          router,

      })
    </script>
```

我们可以在一个路由中设置多段“路径参数”，对应的值都会设置到 $route. params 中，如表 8-1 所示。

表 8-1 路径参数及对应的值

模　　式	匹配路径	$route. params
/user/：username	/user/evan	{ username：'evan' }
/user/：username/post/：post_id	/user/evan/post/123	{ username：'evan'，post_id：'123' }

除了 $route. params 外，$route 对象还提供了其他有用的信息，例如，$route. query（如果 URL 中有查询参数）、$route. hash 等，这些以后再逐个介绍。接下来介绍动态路由在使用过程中的特殊应用。

1. 监视路由参数的变化

当使用路由参数时，例如从/user/foo 导航到/user/bar，原来的组件实例会被复用。因为两个路由都渲染同一个组件，比起销毁再创建，复用则显得更加高效。不过，这也意味着组件的生命周期钩子不会再被调用。当复用组件时，如果想对路由参数的变化做出响应，开发人员则可以简单地 watch（监测变化）$route 对象，代码如下：

```
const User = {
  template: '...',
  watch: {
    $route(to, from) {
      //对路由变化做出响应
    }
  }
}
```

或者使用后面将介绍的导航守卫，代码如下：

```
const User = {
  template: '...',
```

```
    beforeRouteUpdate (to, from, next) {
      //react to route changes...
      //don't forget to call next
    }
  }
```

有时候使用 watch 不能及时地检测到 $route 对象的路径参数的变化，可以将 immediate 设置成 true 解决这种问题，代码如下：

```
watch:{
    $route:{
        immediate: true,               //立即监听
        handler(from, to){             //处理监听方法
            this.userName = this.$route.params.userName
            this.age = this.$route.params.age
        }
    }
}
```

2. 捕获所有路由或 404 路由

常规参数只会匹配被“/”分隔的 URL 片段中的字符。如果想匹配任意路径，则可以使用通配符(＊)，代码如下：

```
{
  //会匹配所有路径
  path: '*'
}
{
  //会匹配以 `/user-` 开头的任意路径
  path: '/user-*'
}
```

当使用通配符路由时，需要确保路由的顺序是正确的，也就是说含有通配符的路由应该放在最后。路由 { path: '*' } 通常用于客户端 404 错误。如果使用了 History 模式，则需要确保正确地配置了服务器。

当使用一个通配符时，$route.params 内会自动添加一个名为 pathMatch 的参数。它包含了 URL 通过通配符被匹配的部分，代码如下：

```
//给出一个路由 { path: '/user-*' }
this.$router.push('/user-admin')
this.$route.params.pathMatch //'admin'
//给出一个路由 { path: '*' }
this.$router.push('/non-existing')
this.$route.params.pathMatch //'/non-existing'
```

3. 高级匹配模式

vue-router 使用 path-to-regexp 作为路径匹配引擎，所以支持很多高级的匹配模式，例如：可选的动态路径参数、匹配零个或多个、一个或多个，甚至是自定义正则匹配。详细内容可以参考 https://github.com/pillarjs/path-to-regexp/tree/v1.7.0，这里就不做进一步阐述了。

4. 匹配优先级

有时候，同一个路径可以匹配多个路由，此时，匹配的优先级就按照路由的定义顺序决定，谁先定义，谁的优先级就最高。第 1 个路由的优先级要高于第 2 个路由的优先级，代码如下：

```
const router = new VueRouter({
    routes: [
    {path:'/user/ * ', component:NotFound},           //定义在前面,优先级最高
//后面正确的路径无法匹配
    { path: '/user/:id', component: User }
    ]
})
```

8.4.2 嵌套模式路由

实际生活中的应用界面，通常由多层嵌套的组件组合而成。同样地，URL 中各段动态路径也按某种结构对应嵌套的各层组件，如图 8.1 所示。

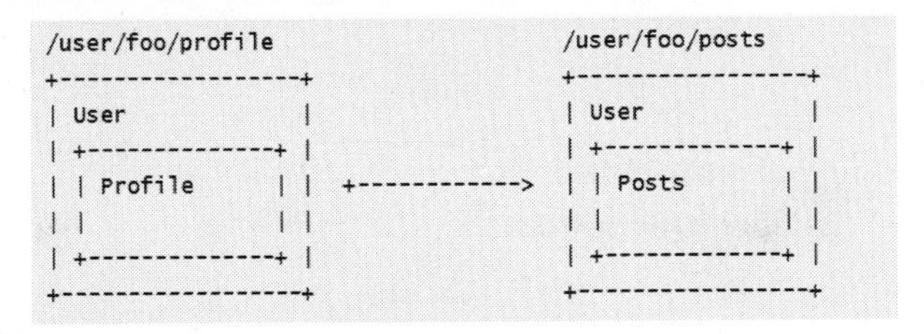

图 8.1 组件结构

借助 Vue-Router，使用嵌套模式路由配置，就可以很简单地表达这种关系。

接着上节创建的应用代码如下：

```
<div id = "app">
  <router - view></router - view>
</div>
const User = {
  template: '<div> User {{  $route.params.id }}</div>'
}

const router = new VueRouter({
```

```
  routes: [
    { path: '/user/:id', component: User }
  ]
})
```

这里的<router-view>是最顶层的出口，渲染最高级路由匹配到的组件。同样地，一个被渲染的组件同样可以包含自己的嵌套<router-view>。例如，在User组件的模板添加一个<router-view>，代码如下：

```
const User = {
  template: `
    <div class = "user">
      <h2>User {{ $route.params.id }}</h2>
      <router-view></router-view>
    </div>
  `
}
```

要在嵌套的出口中渲染组件，需要在Vue Router的参数中使用children配置，代码如下：

```
const router = new VueRouter({
  routes: [
{
path: '/user/:id',
component: User,
      children: [
        {
          //当 /user/:id/profile 匹配成功
          //UserProfile 会被渲染在 User 的 <router-view> 中
          path: 'profile',
          component: UserProfile
        },
        {
          //当 /user/:id/posts 匹配成功
          //UserPosts 会被渲染在 User 的 <router-view> 中
          path: 'posts',
          component: UserPosts
        }
      ]
    }
  ]
})
```

注意：以"/"开头的嵌套模式路径会被当作根路径。这样做可以充分地使用嵌套组件而无须设置嵌套的路径。

children 配置就是像 routes 配置一样的路由配置数组，所以开发人员可以嵌套多层路由。

此时，基于上面的配置，当访问/user/foo 时，User 的出口是不会渲染任何子路由的，这是因为没有匹配到合适的子路由。如果想要渲染点什么，则可以提供一个空的子路由，代码如下：

```
const router = new VueRouter({
  routes: [
    {
      path: '/user/:id', component: User,
      children: [
        //当 /user/:id 匹配成功
        //UserHome 会被渲染在 User 的 <router-view> 中
        { path: '', component: UserHome },

        //...其他子路由
      ]
    }
  ]
})
```

完整可运行的样例代码如下：

```
<div id="app">
    <h1>Hello App!</h1>
      <p>
        <!-- 使用 router-link 组件来导航. -->
        <!-- 通过传入 `to` 属性指定链接. -->
        <!-- <router-link> 默认会被渲染成一个 `<a>` 标签 -->
                    <router-link to="/user/foo/profile">GotoFoo</router-link>
                    <router-link to="/user/bar/posts">GotoBar</router-link>
      </p>
      <!-- 路由出口 -->
      <!-- 由路由匹配到的组件将渲染在这里 -->
      <router-view></router-view>
</div>
<script>
    const UserProfile = {
        template:`<span>User Profile</span>`
    }
    const UserPosts = {
        template:`<span>User Posts</span>`
    }
    const User = {
```

```
    template: `
      <div class = "user">
        <h2 > User {{ $route.params.id }}</h2 >
        <router - view></router - view >
      </div >
    `
  }
  const router = new VueRouter({
    routes: [
            { path: '/user/:id', component: User,
              children: [
                {
                  //当 /user/:id/profile 匹配成功
                  //UserProfile 会被渲染在 User 的 < router - view > 中
                  path: 'profile',
                  component: UserProfile
            },
            {
              //当 /user/:id/posts 匹配成功
              //UserPosts 会被渲染在 User 的 < router - view > 中
              path: 'posts',
                component: UserPosts
            }
        ]
      }
    ]
  })
  const vm = new Vue({
      el:"#app",
      router
  })
</script >
```

8.4.3 编程式路由

router 对象提供了 push、replace 和 go 共 3 种方法，开发人员可以在 JavaScript 中调用这 3 种方法实现组件路由。

除了使用< router-link >创建 a 标签来定义导航链接，开发人员还可以借助 router 的实例方法，通过编写代码实现。

1. router.push(location,onComplete?,onAbort?)

在 Vue.js 实例内部，开发人员可以通过 $router 访问路由实例，因此可以调用 this.$router.push()方法。

如果想要导航到不同的 URL，则可使用 router.push()方法。这种方法会向 history 栈添加一个新的记录，所以，当用户单击浏览器后退按钮时，会回到之前的 URL。

当用户单击< router-link >时，这种方法会在内部调用，所以，单击< router-link：to="...">等同于调用 router. push(...)。

该方法的参数可以是一个字符串路径，或者一个描述地址的对象，代码如下：

```
//字符串
router.push('home')

//对象
router.push({ path: 'home' })

//命名的路由
router.push({ name: 'user', params: { userId: '123' }})

//带查询参数,变成 /register?plan = private
router.push({ path: 'register', query: { plan: 'private' }})
```

如果提供了 path，则 params 会被忽略，上述例子中的 query 并不属于这种情况。取而代之的是下面例子的做法，需要提供路由的 name 或手写完整的带有参数的 path，代码如下：

```
const userId = '123'
router.push({ name: 'user', params: { userId }}) // -> /user/123
router.push({ path: `/user/ ${userId}` }) // -> /user/123
//这里的 params 不生效
router.push({ path: '/user', params: { userId }}) // -> /user
```

同样的规则也适用于 router-link 组件的 to 属性。

在 2.2.0+版本中，可选地在 router. push()或 router. replace()方法中提供 onComplete 和 onAbort 回调作为第 2 个和第 3 个参数。这些回调将会在导航成功完成(在所有的异步钩子被解析之后)或终止(导航到相同的路由或在当前导航完成之前导航到另一个不同的路由)的时候进行相应的调用。在 3.1.0+版本中，可以省略第 2 个和第 3 个参数，此时如果支持 Promise，则 router. push()或 router. replace()方法将返回一个 Promise。

注意：如果目的地和当前路由相同，只有参数发生了改变(例如从一个用户资料到另一个/users/1->/users/2)，则需要使用 beforeRouteUpdate 响应路由参数的变化来监视捕获这个变化(例如抓取用户信息)。

2. router.replace(location,onComplete?,onAborts?)

跟 router. push()方法很像，唯一的不同就是，它不会向 history 添加新记录，而是跟它的方法名一样——替换掉当前的 history 记录。

3. router.go(n)

这种方法的参数是一个整数，意思是在 history 记录中向前或者向后退多少步，类似

window.history.go(n)方法，代码如下：

```
//在浏览器记录中前进一步，等同于 history.forward()
router.go(1)

//后退一步记录，等同于 history.back()
router.go(-1)

//前进 3 步记录
router.go(3)

//如果 history 记录不够用，就只能默默地失败了
router.go(-100)
router.go(100)
```

router.push()、router.replace()和 router.go()跟 window.history.pushState()、window.history.replaceStat()和 window.history.go()很像，实际上它们确实效仿了 window.history API。

还有值得提及的是 Vue Router 的导航方法(push、replace、go)在各类路由模式(history、hash 和 abstract)下表现一致。

8.5 路由视图

前面介绍过，router-view 用来指定路由渲染的地方。根据使用的方式和应用场景的不同，可以分为命名视图和嵌套命名视图两种。

8.5.1 命名视图

有时候，在同一级需要展现多个路由视图，每个路由视图中要渲染不同的组件，这样就需要在一个 Vue View 中定义多个 router-view，以便将对应的组件渲染进去。定义 3 个路由视图，分别用来渲染导航按钮和导航页面的详细内容，代码如下：

```
//第 8 章/命名视图路由.html

<div id="app">
    <!-- 第 1 个路由视图，显示导航按钮 -->
    <router-view></router-view>
    <div style="display: flex;">
        <!-- 第 2 个路由视图，显示左侧导航 -->
        <router-view name="leftView"></router-view>
        <!-- 第 3 个路由视图，显示导航页面的详细操作内容 -->
        <router-view name="rightView"></router-view>
    </div>
</div>
```

3个router-view分别要渲染不同的组件。这时候可以给这些视图起个名称(用name属性指定)。路由里面,通过该名称,指定将哪个组件渲染到哪个router-view中,代码如下:

```
// 第8章/命名路由视图.html

<script>
    //定义按钮导航组件
    const NavButtons = {
            template:`
            <div>
            <button @click = "to('root')"> Home </button>
            <button @click = "to('email')"> SettingEmail </button>
            <button @click = "to('profile')"> SettingProfile </button></div>
    `,
      methods:{
        to(name){
          this. $router. push({name:name})
        }
      }
    }
    //定义左侧导航组件
    const NavLeft = {
            template:
                <div><ul>
                <li><router - link to = '/email'> E - mail </router - link></li>
                <li><router - link to = '/profile'> Profile </router - link></li>
                </ul></div>

    }
    //定义欢迎组件
    const WelcomeSetting = {
        template:`<div>欢迎进入 setting 页面</div>`
    }
    //定义设置E-mail组件
    const E - mailSetting = {
        template:`<div> Setting E - mail </div>`
    }
    const ProfileSetting = {
        template:`<div> Setting Profile </div>`
    }

    const router = new VueRouter({
        routes:[
            {
                name:'root',
                path:'/',            //进入页面,自动路由到根目录
```

```
            components:{
                default:NavButtons,
                leftView:NavLeft,
                rightView:WelcomeSetting
            }
    },
    {
            name:'email',
            path:'/email',
            components:{
                default:NavButtons,
                leftView:NavLeft,
                rightView:EmailSetting
        }
    },
    {
        name:'profile',
        path:'/profile',
        components:{
            default:NavButtons,
            leftView:NavLeft,
            rightView:ProfileSetting
        }
      }
    ]
  })

  const vm = new Vue({
        el:"#app",
        router
  })
</script>
```

8.5.2 嵌套命名视图

如上面的命名视图案例，Vue.js 里面包含了 3 个 router-view，名称分别是 default、leftView 和 rightView，渲染的组件分别是 NavButtons、NavLeft 和 WelcomeSetting、EmailSetting、ProfileSetting。

在 WelcomeSetting、EmailSetting 和 ProfileSetting 中都是简单的元素，如果在这些组件中继续包含 router-view，以便于渲染第三方组件，这就形成了嵌套命名视图（在 router-view 中渲染一个带有 router-view 的另外一个组件）。

将上面案例中的 ProfileSetting 修改成包含两个 router-view 的组件，名称分别是 profileBase 和 profileDetail，同时增加 ProfileBase 和 ProfileDetail 两个对象，代码如下：

```
    //原 ProfileSetting
    const ProfileSetting = {
            template:`<div>Setting Profile</div>`
    }
    //改成
    const ProfileSetting = {
            template:`
            <div>Setting Profile
              <router-view name="profileBase"></router-view>
              <router-view name="profileDetail"></router-view>
</div>`
    }
    //增加
    const ProfileBase = {
        template:`<div>Profile Base</div>`
    }
    const ProfileDetail = {
        template:`<div>Profile Detail</div>`
    }
```

在 router 对象的路由表中，改造 profile 路由对象，添加 children 嵌套模式路由，以便支持嵌套模式路由视图，代码如下：

```
const router = new VueRouter({
    routes:[
      ...
    {
      name:'profile',
      path:'/profile',
      components:{
        default:NavButtons,
        leftView:NavLeft,
        rightView:ProfileSetting
    },
    //增加 children 嵌套模式路由
    children:[
      {                    //指定 ProfileSetting 中 router-view 对应的组件
          name:'profileExt',
          path:'/profileExt',
          components:{
            profileBase:ProfileBase,
            profileDetail:ProfileDetail
          }
        }
      ]
    }
    ]
})
```

将NavButtons中的SettingProfile按钮的导航，改成导航到profileExt，以便于能显示嵌套命名视图，代码如下：

```
//原代码
<button @click="to('profile')">SettingProfile</button></div>
//改成
<button @click="to('profileExt')">SettingProfile</button></div>
```

完整代码如下：

```
//第8章/嵌套模式路由视图.html

<div id="app">
    <router-view></router-view>
    <div style="display: flex;">
            <router-view name="leftView"></router-view>
            <router-view name="rightView"></router-view>
    </div>
</div>
<script>
    const NavButtons = {
            template:`
              <div>
              <button @click="to('root')">Home</button>
              <button @click="to('email')">SettingEmail</button>
              <button @click="to('profileExt')">SettingProfile</button></div>

            methods:{
              to(name){
                this.$router.push({name:name})
              }
            }
    }
    const NavLeft = {
            template:
              <div><ul>
              <li><router-link to='/email'>E-mail</router-link></li>
              <li><router-link to='/profile'>Profile</router-link></li>
              </ul></div>

    }
    const WelcomeSetting = {
        template:`<div>欢迎进入setting页面</div>`
    }
    const EmailSetting = {
        template:`<div>Setting Email</div>`
```

```
}
const ProfileSetting = {
        template:
        < div > Setting Profile
          < router - view name = "profileBase"></router - view >
          < router - view name = "profileDetail"></router - view >
        </div >
}
const ProfileBase = {
        template:`< div > Profile Base </div >`
}
const ProfileDetail = {
        template:`< div > Profile Detail </div >`
}

const router = new VueRouter({
routes:[
  {
    name:'root',
    path:'/',          //进入页面,自动路由到根目录
    components:{
      default:NavButtons,
      leftView:NavLeft,
      rightView:WelcomeSetting
    }
  },
  {
    name:'email',
    path:'/email',
    components:{
      default:NavButtons,
      leftView:NavLeft,
      rightView:EmailSetting
    }
  },
  {
    name:'profile',
    path:'/profile',
    components:{
      default:NavButtons,
      leftView:NavLeft,
      rightView:ProfileSetting
    },
    children:[
    {
      name:'profileExt',
```

```
            path:'/profileExt',
            components:{
              profileBase:ProfileBase,
              profileDetail:ProfileDetail
              }
            }
          ]
        }
      ]
    })
    const vm = new Vue({
            el:"#app",
            router
    })
</script>
```

单击 Profile 和 SettingProfile 的效果如图 8.2 所示。

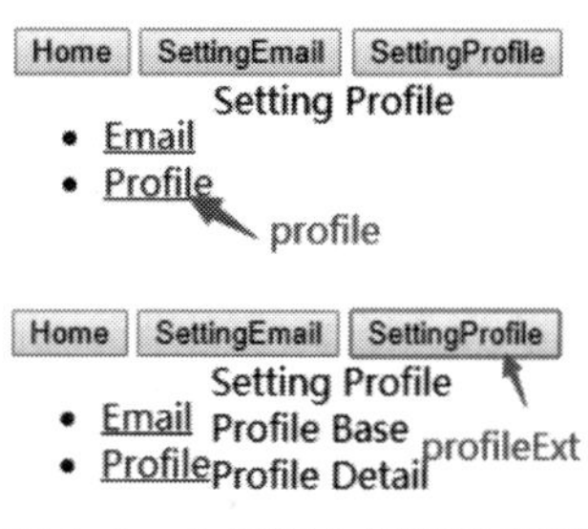

图 8.2　内嵌路由视图效果图

profile 只显示了 ProfileSetting 组件，profileExt 将 ProfileSetting 中的两个 router-view 都渲染出来了。

8.6　别名和重定向

在实际应用中，有时候需要将多个路径路由到同一个组件，这时候就可以在路由表中将多个路径重定向同一个路由路径。在路由表中用 redirect 属性指定要重定向的目标。指定目标有目标路径、目标对象或箭头函数 3 种方式，代码如下：

```
const router = new VueRouter({
  routes:[
    {
      name:'a',
      path:'/a',
      component:CompA
    },
```

```
        {
          name:'b',
          path:'/b',
          component:CompB
        },
        {
          name:'d',
          path:'/d',
          //redirect:'/b'                //重定向路径
          //redirect:{name:'b'}          //重定向对象
          redirect:to =>{               //重定向箭头函数
            //console.log(to)
            //return '/b'
            return {name:'b'}
          }
        }
      ]
  })
```

在路由表中，除了可以给一个路由指定重定向路径，一样可以给路由起个别名，通过别名来路由，代码如下：

```
  const router = new VueRouter({
    routes:[
      {
        name:'c',
        path:'/c',
        alias:'/e',                //声明一个别名
        component:CompC
      }
    ]
  })
```

重定向和别名的对比：

“重定向”的意思是，当用户访问/a时，URL将会被替换成/b，然后将路由匹配为/b，那么“别名”又是什么呢？

/a的别名是/b，意味着，当用户访问/b时，URL会保持为/b，但是路由匹配则为/a，就像用户访问/a一样。

重定向和别名的完整案例代码如下：

```
//第8章/别名和重定向.html
<div id="app">
    <router-link :to="{name:'a'}" tag="button">/a</router-link>
    <router-link :to="{name:'b'}" tag="button">/b</router-link>
```

```
        <router-link :to="{name:'c'}" tag="button">/c</router-link>
        <!-- <router-link :to="{name:'d'}" tag="button">/d</router-link> -->
        <!-- <router-link :to="'/e'" tag="button">/e</router-link> -->
        <button @click="to('/d')">重定向 d-b</button>
        <button @click="to('/e')">别名 e-c</button>
        <router-view></router-view>
    </div>
    <script>
        const CompA = {
            template:`<div>Component A</div>`
        }
        const CompB = {
            template:`<div>Component B</div>`
        }
        const CompC = {
            template:`<div>Component C</div>`
        }
        const router = new VueRouter({
            routes:[
                {
                    name:'a',
                    path:'/a',
                    component:CompA
                },
                {
                    name:'b',
                    path:'/b',
                    component:CompB
                },
                {
                    name:'d',
                    path:'/d',
                    //redirect:'/b'
                    //redirect:{name:'b'}
                    redirect:to =>{
                      //console.log(to)
                      //return '/b'
                      return {name:'b'}
                }
        },
        {
                name:'c',
                path:'/c',
                alias:'/e',
                component:CompC
              }
```

```
        ]
    })
    const vm = new Vue({
            el:"#app",
            router,
            methods:{
              to(path){
                this.$router.push(path)
                ///d 重定向到/b,当前 path 为/b
                //访问别名/e,当前 path 为/e
                console.log(this.$route.fullPath)
              }
            }
    })
</script>
```

8.7 给路由组件传值

在 router 中,路由组件中可以使用 this.$route 对象的 params 和 query 属性,获取路由传入的参数值(params 用来接收参数值,query 用来接收查询参数值),代码如下:

```
<router-link :to="'/a/3'" tag="button">测试/a/3</router-link>
<router-link :to="{name:'a_id',params:{id:4,name:'zhangsan'}}" tag="button">测试 params
</router-link>
<router-link :to="{path:'/a/5',query:{name:'wangwu'}}" tag="button">测试 query</router-link>

<!-- 在接收路由参数的组件里面直接使用 this.$route 获取参数值 -->
const ReceiveCompA = {
        template:`
          <div>
                <p>param-path-id:{{this.$route.params.id}}</p>
                <p>param-obj-name:{{this.$route.params.name}}</p>
                <p>query-name:{{this.$route.query.name}}</p>
          </div>
`
}
```

如上方式,虽然可以在组件中获取路由传入的值,但在组件中要明确使用 this.$route 对象,这样会造成组件同 this.$route 有严密的耦合。

开发人员可以使用 props 属性,实现如上那种耦合,步骤如下:

(1) 在组件上定义 props 属性。

(2) 在路由对象中,将 props 声明为 true 或给 props 对象赋值。

具体有 3 种方式：①将 props 设置为 true，自动地将路由的参数获取到 props 的属性中；②给 props 赋予一个静态对象；③基于函数，返回 props 对象。

代码如下：

```
<div id = "app">
    <router - link :to = "'/a/3'" tag = "button">测试/a/3 </router - link >
    <router - link :to = "{name:'a_id',params:{id:4,name:'zhangsan'}}" tag = "button">测试
params </router - link >
    <router - link :to = "{path:'/a/5',query:{name:'wangwu'}}" tag = "button">测试 query
</router - link ><br/>
    <router - link :to = "'/b1/3'" tag = "button">测试/b1/3 </router - link >
    <router - link :to = "{name:'b1_id',params:{id:4,name:'zhangsan'}}" tag = "button">测试
params </router - link >
    <router - link :to = "{path:'/b1/5',query:{name:'wangwu'}}" tag = "button">测试 query
</router - link >
    <router - view ></router - view >
</div >
<script >
    <! -- 在接收路由参数的组件里面直接使用 this. $route 获取参数值 -->
    const ReceiveCompA = {
      template:`
        <div >
          <p > param - path - id:{{this. $route. params. id}}</p >
          <p > param - obj - name:{{this. $route. params. name}}</p >
          <p > query - name:{{this. $route. query. name}}</p >
        </div >
      `
    }
    <! -- 以 props 的形式接收参数,同 this. $route 解耦 -->
    const ReceiveCompB1 = {
      template:`
        <div >
          <p > param - path - id:{{id}}</p >
          <p > param - obj - name:{{name}}</p >
          <p > query - name:{{name}}</p >
        </div >
      `,
      props:['id','name']
    }
    const router = new VueRouter({
      routes:[
        {
          name:'a_id',
```

```
            path:'/a/:id',
            component:ReceiveCompA
        },
        //{ //方式 1:将 props 声明为 true
        //    name:'b1_id',
        //    path:'/b1/:id',
        //    component:ReceiveCompB1,
        //    props:true
        //},
        //{   //方式 2:给 props 赋予静态对象
        //    name:'b1_id',
        //    path:'/b1/:id',
        //    component:ReceiveCompB1,
        //    props:{id:5,name:'赵六'}
        //},
        {   //方式 3:使用箭头函数,返回 props 对象
            name:'b1_id',
            path:'/b1/:id',
            component:ReceiveCompB1,
            props:(route1) =>{return {id:route1.params.id, name:route1.params.name + '--' +
route1.query.name}}
        },
        ]
    })
    const vm = new Vue({
            el:"#app",
            router
    })
</script>
```

8.8 路由的请求模式

Router 有 hash、history 和 abstract 共 3 种模式，默认为 hash 模式。在浏览器环境下使用 hash 和 history 模式，而 abstract 使用在 Node.js 环境下。

在浏览器下使用 hash 模式，也是默认模式，会在浏览器的网址栏中包含 #，看起来不是很舒服，如 http://yoursite.com/user/#/b1/3。

为了让它不显示得那么“丑”(http://yoursite.com/user/b1/3)，可以使用 router 的 mode 属性，指定使用 history 模式，代码如下：

```
const router = new VueRouter({
  mode: 'history',
```

```
  routes: [...]
})
```

不过这种模式要处理好，还需要后台配置支持。因为应用是个单页客户端应用，如果后台没有正确地配置，当用户在浏览器直接访问 http://oursite.com/user/id 时就会返回 404，这对用户来讲就不友好了。为了避免这种不友好的情况，开发人员要在服务器端增加一个覆盖所有情况的候选资源：如果 URL 匹配不到任何静态资源，则应该返回同一个 index.html 页面，这个页面就是 App 应用的实际页面。

第三篇　Vue.js高级应用

第9章 高级 Vue Router

第8章主要介绍了 Vue Router 的基础，本章进一步介绍 Vue Router 的高级部分，包含导航守卫、路由元数据、获取相应数据和路由懒加载等。

9.1 导航守卫

导航守卫是 Vue-Router 中提供的一种机制，通过该机制，开发人员可以在路由的导航过程中插入更多的业务逻辑，从而控制组件的跳转或取消等功能。根据守卫的范围，导航路由可以分为全局守卫、独享守卫和组件内路由导航守卫。

注意：当路由的参数或查询数据发生改变时，不会触发路由守卫(进入路由/离开路由)。

9.1.1 全局守卫

全局路由守卫是指给所有的路由都添加守卫逻辑。可以通过 Router 对象的 beforeEach()、beforeResolve()和 afterEach()方法，添加全局的进入和离开路由业务。

beforeEach()方法是全局前置守卫方法。当一个导航触发时，全局前置守卫将按顺序进行调用。特别注意，守卫是异步解析执行，此时导航在所有守卫解析完之前一直处于等待中。

每个守卫方法接收3个参数。

(1) to：Route，即将要进入的目标[路由对象](https://router.vuejs.org/zh/api/#路由对象)。

(2) from：Route，当前导航正要离开的路由。

(3) next：Function，一定要调用该方法来解析这个钩子。执行效果依赖 next()方法的调用参数。

next()：执行管道中的下一个钩子。如果全部钩子执行完了，则导航的状态就是 confirmed(确认的)。

next(false)：中断当前的导航。如果浏览器的 URL 改变了(可能由用户手动或者浏览器后退按钮导致)，则 URL 地址会重置到 from 路由对应的地址。

next('/')或者 next({ path: '/' })：跳转到一个不同的地址。当前的导航被中断，然后进行一个新的导航。开发人员可以向 next 传递任意位置对象，并且允许设置诸如 replace：true、name：'home' 之类的选项及任何用在 < router-link >的 to prop（https://router.vuejs.org/zh/api/#to）或 $router.push（https://router.vuejs.org/zh/api/#router-push）中的选项。

next(error)：（2.4.0+版本）如果传入 next 的参数是一个 Error 实例，则导航会被终止且该错误会被传递给 [router.onError()](https://router.vuejs.org/zh/api/#router-onerror)注册过的回调。

确保 next()方法在任何给定的导航守卫中都被严格调用一次。它可以出现多次，但是只能在所有的逻辑路径都不重叠的情况下，否则钩子永远都不会被解析或报错。

基于 beforeEarch()方法实现全局前置守卫，代码如下：

```
//第 9 章/全局前置守卫.html

<div id = "app">
    <router - link tag = "button" to = "a"> to A </router - link>
    <router - link tag = "button" to = "b"> to B </router - link>
    <router - link tag = "button" to = "/c"> to C </router - link>
    <router - link tag = "button" to = "e"> to E </router - link>
    <router - view></router - view>
</div>
<script>
    const CompA = {
            template:`<div> Component A </div>`
    }
    const CompB = {
            template:`<div> Component B </div>`
    }
    const CompError = {
            template:`<div> Component Error </div>`
    }
    const router = new VueRouter({
            routes:[
              {
                            name:'a',
                            path:'/a',
                            component:CompA
              },
              {
                            name:'b',
                            path:'/b',
                            component:CompB
              },
```

```
                {
                                name:'c',
                                path:'/c',
                                component:CompB
                },
                {
                                name:'e',
                                path:'/e',
                                component:CompError
                },
        ],
    };
     //进入路由守卫
    router.beforeEach(to, from, next) => {
      //console.log(to)
      console.log(from)
      if(to.path == '/b'){
                console.log('不能请求/b')
                next('/e')
      }else if(to.path == '/c'){
                next(error)
      }else{
                next()
      }
    })
    const vm = new Vue({
            el:"#app",
            router
    })
</script>
```

在2.5.0+版本中，Vue-Router在router对象中，提供了beforeResolve()方法，此方法用于注册全局解析守卫，和router.beforeEach()方法类似，其区别是在导航被确认之前，同时在所有组件内守卫和异步路由组件被解析之后，解析守卫会被调用。

当然，开发人员也可以基于router.afterEach()方法注册全局后置守卫，然而和守卫不同的是，这个钩子不会接受next()方法，也不会改变导航本身，代码如下：

```
router.afterEach(to, from) => {
  ...
}
```

9.1.2 路由独享守卫

除了可以给所有路由注册全局守卫外，开发人员也可以在路由配置上直接定义

beforeEnter()方法，注册当前路由的独享守卫，代码如下：

```
const router = new VueRouter({
  routes: [
    {
      path: '/foo',
      component: Foo,
      beforeEnter: (to, from, next) => {
        //...
      }
    }
  ]
})
```

这些守卫与全局前置守卫的方法参数是一样的。

9.1.3 组件内的路由导航守卫

最后，开发人员还可以在路由组件内直接定义 beforeRouteEnter()、beforeRouteUpdate()（2.2 新增版）和 beforeRouteLeave()路由导航守卫，代码如下：

```
const Foo = {
  template: `...`,
  beforeRouteEnter(to, from, next) {
    //在渲染该组件的对应路由被确认前调用
    //不能获取组件实例 `this`
    //因为当守卫执行前，组件实例还没被创建
  },
  beforeRouteUpdate(to, from, next) {
    //在当前路由改变，但是在该组件被复用时调用
    //举例来讲，对于一个带有动态参数的路径 /foo/:id，在 /foo/1 和 /foo/2 之间跳转的时候
    //由于会渲染同样的 Foo 组件，因此组件实例会被复用，而这个钩子就会在这个情况下被调用
    //可以访问组件实例 `this`
  },
  beforeRouteLeave(to, from, next) {
    //当导航离开该组件的对应路由时调用
    //可以访问组件实例 `this`
  }
}
```

beforeRouteEnter 守卫 不能 访问 this，因为守卫在导航确认前被调用，这时即将登场的新组件还没被创建。

不过开发人员可以通过传一个回调给 next，以便访问组件实例。在导航被确认的时候执行回调，并且把组件实例作为回调方法的参数，代码如下：

```
beforeRouteEnter (to, from, next) {
  next(vm => {
    //通过 vm 访问组件实例
  })
}
```

beforeRouteEnter 是支持给 next 传递回调的唯一守卫。对于 beforeRouteUpdate 和 beforeRouteLeave 来讲，this 已经可用了，所以不支持传递回调，因为没有必要了，代码如下：

```
beforeRouteUpdate (to, from, next) {
  //just use `this`
  this.name = to.params.name
  next()
}
```

beforeRouteleave 守卫通常用来禁止用户在还未保存修改前突然离开。该导航可以通过 next(false)来取消，代码如下：

```
beforeRouteLeave (to, from, next) {
  const answer = window.confirm('Do you really want to leave? you have unsaved changes!')
  if (answer) {
    next()
  } else {
    next(false)
  }
}
```

9.1.4 完整的路由解析流程

完整的路由解析流程如下：

(1) 导航被触发。

(2) 在失活的组件里调用 beforeRouteLeave 守卫。

(3) 调用全局的 beforeEach 守卫。

(4) 在重用的组件里调用 beforeRouteUpdate 守卫(2.2+版本)。

(5) 在路由配置里调用 beforeEnter。

(6) 解析异步路由组件。

(7) 在被激活的组件里调用 beforeRouteEnter。

(8) 调用全局的 beforeResolve 守卫(2.5+版本)。

(9) 导航被确认。

(10) 调用全局的 afterEach 钩子。

(11) 触发 DOM 更新。

(12) 调用 beforeRouteEnter 守卫中传给 next 的回调函数,创建好的组件实例会作为回调函数的参数传入。

完整的路由守卫代码如下:

```
//第 9 章/路由守卫执行流程.html

  <div id = "app">
  <router - link to = "/"> first </router - link >  
  <router - link to = "/second/1"> second </router - link >  
  <router - link to = "/second/2"> second </router - link >

  <router - view ></router - view >
  </div >
  <script type = "text/JavaScript">
  const first = {
      name: 'first',
      template: '< div > this is first component </div >',
      beforeRouteEnter(to, from , next){
              console.log('first:beforeRouteEnter')
              next()
      },
      beforeRouteUpdate(to, from , next){
              console.log('first:beforeRouteUpdate')
              next()
      },
      beforeRouteLeave(to, from, next){
              console.log('first:beforeRouteLeave')
              next()
      }
    }

  const second = {
      name: 'second',
      template: '< div > this is second component </div >',
      beforeRouteEnter(to, from , next){
              console.log('second:beforeRouteEnter')
              next()
      },
      beforeRouteUpdate(to, from , next){
              console.log('second:beforeRouteUpdate')
              next()
      },
      beforeRouteLeave(to, from, next){
              console.log('second:beforeRouteLeave')
```

```
            next()
        }
    }

const router = new VueRouter({
    routes: [
        {
            name: 'first',
            path: '/',
            component: first
        },
        {
            name: 'second',
            path: `/second/:id`,
            component: second,
            beforeEnter: (to, from ,next) => {
                console.log('独享:router.beforeEnter')
                next()
            }
        }
    ]
}
//注册全局路由守卫
router.beforeEach(to, from , next) => {
    console.log("全局:router.beforeEach ")
    next()
})
router.beforeResolve(to, from, next) => {
    console.log('全局:router.beforeResolve')
    next()
}
router.afterEach(to, from) => {
    console.log('全局:router.afterEach')
}
const vm = new Vue({
    el: "#app",
    router
})
</script>
```

9.2 路由元信息

开发人员可以在路由表中的每个路由对象（路由记录）中用 meta 属性配置若干元信息，以便在路由过程中根据这些信息进行业务处理。在 meta 中配置的信息就是路由元信息。在路由过程中，可以通过 route. matched 对象获取匹配的路由表的全部信息，代码如下：

```
const router = new VueRouter({
routes: [
  {
    path: '/foo',
    component: Foo,
    children: [
      {
        path: 'bar',
        component: Bar,
        //a meta field
        meta: { requiresAuth: true }
      }
    ]
  }
]
})

router.beforeEach(to, from, next) => {
if (to.matched.some(record => record.meta.requiresAuth) {
  //this route requires auth, check if logged in
  //if not, redirect to login page.
  if (!auth.loggedIn) {
    next({
      path: '/login',
      query: { redirect: to.fullPath }
    })
  } else {
    next()
  }
} else {
  next()                    //确保一定要调用 next()
}
})
```

备注：

Array. some()方法

some()方法用来检测数组的每个元素是否满足对应函数指定的条件，如果有任何一个条件满足，则返回值为 true，否则返回值为 false。在执行过程中，不会检测数组是否为空，也不会改变原来的数组。

语法：

Array. some(function(currValue,index,arr),thisValue)

currValue：遍历的每个元素。

index：元素下标。

arr：原数组对象。

thisValue：可选。对象作为该执行回调时使用，传递给函数，用作 this 的值。如果省略了 thisValue，则 this 的值为"undefined"。

检测数组中是否存在大于 30 的值，代码如下：

```
const ages = [21,23,22,30,42,53,12]
const age = 30
const result = ages.some((currValue,index,arr) =>{if(currValue > age) return true;})
```

9.3 获取响应数据

有时候，进入某个路由后，需要从服务器获取数据。例如，在渲染用户信息时，需要从服务器获取用户的数据。可以通过以下两种方式实现。

(1) 导航完成之后获取：先完成导航，然后在接下来的组件生命周期钩子中获取数据。在数据获取期间显示"加载中"之类的指示。

(2) 导航完成之前获取：导航完成前，在路由进入的守卫中获取数据，在数据获取成功后执行导航。

从技术角度讲，这两种方式都不错，主要看向用户提供哪种体验。

9.3.1 导航完成后获取响应数据

当开发人员使用这种方式时，Vue-Router 会马上导航和渲染组件，然后在组件的 created 钩子中获取数据。这让开发人员有机会在数据获取期间展示一个 loading 状态，还可以在不同视图间展示不同的 loading 状态。

假设有一个 Post 组件，需要基于 $route.params.id 获取文章数据，代码如下：

```
//第 9 章/导航完成后获取相应数据.html

<div id = "app">
    <router-link to = "/"> Home </router-link>
    <router-link to = "/other/3"> Others </router-link>
    <router-view></router-view>
</div>
<script>
    const CompA = {
        template:`<div> Home </div>`
    }
    const CompB = {
```

```
        template:`
            <div>
            <div class = "post">
                <div v - if = "loading" class = "loading">
                Loading...
                </div>

                <div v - if = "error" class = "error">
                {{ error }}
                </div>

                <div v - if = "post" class = "content">
                <h2>{{ post.title }}</h2>
                <p>{{ post.body }}</p>
                </div>
            </div>
            </div>
        `,
            data () {
            return {
                loading: false,
                post: null,
                error: null
            }
            },
            created () {
            //组件创建完后获取数据
            //此时 data 已经被观测到了
            this.fetchData()
            },
            watch: {
            //如果路由有变化,会再次执行该方法
            '$route': 'fetchData'
            },
            methods: {
            fetchData () {
                this.error = this.post = null
                this.loading = true
                result = getPost(this.$route.params.id)
                this.loading = false;
                this.error = result.error;
                this.post = result.post;
            }
            }
    }
    const router = new VueRouter({
```

```
        routes:[
            {
                path:'/',
                component:CompA
            },
            {
                path:'/other/:id',
                component:CompB
            }
        ]
    })
    const vm = new Vue({
        el:"#app",
        router
    })


    function getPost(id){
        return {
            res:"res from " + id,
            error:false,
            post:{title:'hello',body:'how are you' + Math.random()}
        }
    }
</script>
```

9.3.2 导航完成前获取响应数据

通过这种方式,开发人员可以实现在导航转入新的路由前获取数据。可以在组件的 beforeRouteEnter 守卫中获取数据,当数据获取成功后只需调用 next()方法,代码如下:

```
//第 9 章/导航完成前获取相应数据.html

<div id="app">
    <router-link to="/">Home</router-link>
    <router-link to="/other/3">Others</router-link>
    <router-view></router-view>
</div>
<script>
    const CompA = {
        template:`<div>Home</div>`
    }
    const CompB = {
        template:`
            <div>
```

```
        <div class = "post">
            <div v - if = "loading" class = "loading">
            Loading...
            </div>

            <div v - if = "error" class = "error">
            {{ error }}
            </div>

            <div v - if = "post" class = "content">
            <h2>{{ post.title }}</h2>
            <p>{{ post.body }}</p>
            </div>
        </div>
        </div>
    `,
        data () {
        return {
            loading: false,
            post: null,
            error: null
        }
        },
        beforeRouteEnter (to, from, next) {
            let result = getPost(to.params.id);
            next(vm) => vm.setData(result.error, result.post)
        },
        //路由改变前,组件就已经渲染完了
        //逻辑稍稍不同
        beforeRouteUpdate (to, from, next) {
            this.post = null
            let result = getPost(to.params.id);
            this.setData(err, post)
            next()
        },
        methods: {
            setData (err, post) {
                this.error = err
                this.post = post
            }
        }
}
const router = new VueRouter({
    routes:[
        {
            path:'/',
```

```
                component:CompA
            },
            {
                path:'/other/:id',
                component:CompB
            }
        ]
    })
    const vm = new Vue({
        el:"#app",
        router
    })


    function getPost(id){
        return {
            res:"res from " + id,
            error:false,
            post:{title:'hello',body:'how are you' + Math.random()}
        }
    }
</script>
```

9.4　路由懒加载

当打包构建应用时，JavaScript 包会变得非常大，影响页面加载。如果能把不同路由对应的组件分割成不同的代码块，然后当路由被访问的时候加载对应组件，这样就更加高效了。

结合 Vue.js 的异步组件和 webpack 的代码分割功能，可轻松实现路由组件的懒加载。

首先，可以将异步组件定义为返回一个 Promise 的工厂函数(该函数返回的 Promise 应该 resolve 组件本身)，代码如下：

```
const Foo = () => Promise.resolve({ /* 组件定义对象 */ })
```

其次，在 webpack 2 中，开发人员可以使用动态 import 语法来定义代码分块点 (Split Point)，代码如下：

```
import('./Foo.vue')        //返回 Promise
```

如果使用的是 Babel，则需要添加 syntax-dynamic-import 插件，这样才能使 Babel 可以正确地解析语法。

结合这两者，便可定义一个能够被 webpack 自动代码分割的异步组件。

```
const Foo = () => import('./Foo.vue')
```

在路由配置中什么都不需要改变，只需像往常一样使用 Foo，代码如下：

```
const router = new VueRouter({
routes: [
  { path: '/foo', component: Foo }
]
})
```

有时候开发人员想把某个路由下的所有组件都打包在同个异步块(chunk)中，此时只需使用命名 chunk，即以一个特殊的注释语法来提供 Chunk Name(webpack 的版本号需要大于 2.4)，代码如下：

```
const Foo = () => import(/* webpackChunkName: "group-foo" */ './Foo.vue')
const Bar = () => import(/* webpackChunkName: "group-foo" */ './Bar.vue')
const Baz = () => import(/* webpackChunkName: "group-foo" */ './Baz.vue')
```

webpack 会将任何一个异步模块与相同的块名称组合到相同的异步块中。

第10章

Promise对象

Promise是ECMAScript 6提供的一个原生对象，为开发人员基于JavaScript实现异步编程提供了一种解决方法。通过Promise对象，可以获取异步操作的消息。从本意上说，它给开发人员提供一个承诺，承诺一段时间后，会给出一个结果。本章将专门介绍Promise的特点和使用。

10.1 Promise对象基础

开发人员可以直接基于Promise的构造器创建对象，语法如下：

```
const promise = new Promise(function(resolve, reject){...})
```

构造器接收一个名为executor的函数，该函数接收两个参数，即resolve和reject，这两个参数也是函数。以便在executor函数中执行异步代码，当异步代码执行成功的时候，执行resolve函数，否则执行reject函数，代码如下：

```
//第10章/创建Promise对象.html

<script type = "text/JavaScript">
    const promise = new Promise(function(resolve,reject){
        setTimeout(function(){
            let num = Math.random();
            if(num < 0.5){
                resolve('小于0.5');
            }else{
                reject('大于或等于0.5')
            }
        })
    });
    console.log(promise)
    promise.then(function(v){                    //v是resolve方法传入的值
        console.log(v + "then:" )
```

```
        console.log(promise)
    }).catch(function(v){                          //v是reject方法传入的值
        console.log(v + "catch:" )
        console.log(promise)
    }).finally(function(v){
        console.log(v + "finally:" )
        console.log(promise)
    })
</script>
```

Promise对象代表一个异步操作，共有3种状态，分别如下。

（1）pending状态：初始状态，也叫作进行状态，既没有被兑现，也没有被拒绝。

（2）fulfilled状态：已兑现状态，意味着操作成功，也叫作成功状态。

（3）rejected状态：已拒绝状态，意味着操作失败，也叫作失败状态。

Promise对象的状态，不受外界的影响，只有异步操作的结果，可以决定当前是哪种状态，任何其他操作都无法改变它的状态。这也是Promise这个名字的由来，它的英语意思就是承诺，表示其他手段无法改变。

Promise对象的状态一旦改变，就不会再变，任何时候都可以得到这个结果。Promise对象的状态改变，只有两种可能：从Pending变为Resolved和从Pending变为Rejected。只要这两种情况发生，状态就凝固了，即不会再变了，会一直保持这个结果。就算改变已经发生了，开发人员再对Promise对象添加回调函数，也会立即得到这个结果。这与事件(Event)完全不同，事件的特点是，如果错过了它，则再去监听，是得不到结果的。

有了Promise对象，就可以将异步操作以同步操作的流程表达出来，避免了层层嵌套的回调函数。此外，Promise对象提供了统一的接口，使控制异步操作更加容易。

当然Promise也有一些缺点。首先，无法取消Promise，一旦新建它就会立即执行，无法中途取消。其次，如果不设置回调函数，则由Promise内部抛出的错误不会反映到外部。最后，当处于Pending状态时，无法得知目前进展到哪个阶段(刚刚开始还是即将完成)。

10.2 Promise对象的方法

Promise对象支持很多方法，如then、catch、finally、all、any、reject、resolve等方法，开发人员可以基于这些方法实现自己的业务逻辑。

10.2.1 原型方法

Promise对象中有3个原型方法，它们分别是then(onFulfilled, OnRejected)、catch(onRejected)和finally(onFinally)。如果Promise对象执行成功(fulfilled)，则自动调用then()方法，如果执行失败(rejected)，则自动调用catch()方法，执行完后，不管成功与失败，都会自动调用finally()方法。Promise对象的原型方法的执行流程如图10.1所示。

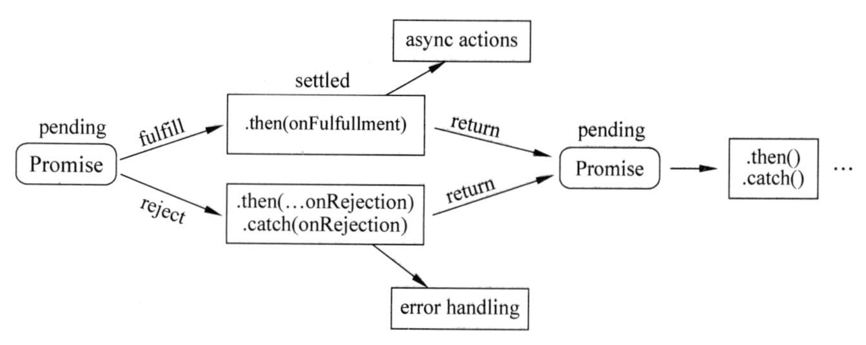

图 10.1　Promise 对象的原型方法的执行流程

关于这 3 种方法的使用案例，可以参考 10.1 节里面的创建 Promise 对象.html 的样例代码，里面介绍过这 3 种方法的使用。

10.2.2　静态方法

Promise 除了前面介绍的 then、catch 和 finally 原型方法外，还有若干静态方法。

1. Promise.all(iterable)方法

Promise.all()方法的作用是批量执行用 iterable 方式传入的 Promise 对象，并返回一个新的 Promise 对象，新 Promise 对象执行的结果是数组，里面包含的是以 iterable 形式传入的每个 Promise 执行后的结果。任何一个传入的 Promise 会被 reject 抛出异常，返回的 Promise 对象立马 reject，并返回第 1 个被 reject 的信息，代码如下：

```
//第 10 章/Promise 的 all 方法.html
<script>
    const promise1 = Promise.resolve(1)
    const promise2 = 2
    const promise3 = new Promise( (resolve, reject) => {
        setTimeout(resolve, 100, '3')
    })

    const promise = Promise.all([promise1, promise2, promise3])
    console.log('完成所有的 Promise 对象')
    promise.then( msg => {
        //没有 reject 的 promise,输出[1,2,3]
        console.log('success:' + msg)
    }).catch( msg => {
        console.log('fail:' + msg)
    }).finally( _ => {
        console.log('finally')
    })

    const promise4 = Promise.reject('e4')
```

```
    const promise5 = Promise.reject('e5')
    Promise.all([promise1, promise4, promise3, promise5])
    .then( msg => {
        console.log('success:' + msg)
    }).catch( msg => {
        //reject,返回 promise4 的 reject reason
        console.log('fail:' + msg)
    })
</script>
```

2. Promise.allSettled(iterable)方法

Promise.allSettled()方法的作用同 all()方法的作用一样,批量执行传入的 Promise 对象,并返回一个数组,数组里面包含的是所有 Promise 对象执行后的结果,不管是正常执行还是拒绝,代码如下:

```
//第 10 章/allSettled 方法.html

const promise1 = Promise.resolve(3);
const promise2 = new Promise((resolve, reject) => setTimeout(reject, 100, 'foo'));
const promises = [promise1, promise2];

Promise.allSettled(promises).
  then((results) => results.forEach(
(result) => console.log(result.status)));
```

最后输出 fulfilled 和 rejected。

3. Promise.any(iterable)方法

Promise.any()方法用于接收一个 Promise 可迭代对象,只要其中的一个 Promise 成功,就返回那个已经成功的 Promise。如果可迭代对象中没有一个 Promise 成功(所有的 Promises 都失败/拒绝),就返回一个失败的 Promise 和 AggregateError 类型的实例,它是 Error 的一个子类,用于把单一的错误集合在一起。本质上,这种方法和 Promise.all()方法是相反的,返回 promise1 对象,代码如下:

```
//第 10 章/any 方法.html

const promise0 = Promise.reject(0)
const promise1 = Promise.resolve(1)
const promise2 = 2
const promise3 = new Promise( (resolve, reject) => {
    setTimeout(resolve, 100, '3')
})

const result = Promise.any([promise0, promise1, promise2, promise3])
result.then( v => console.log("s:" + v)).catch( v => console.log("f:" + v))
```

注意：Promise.any()方法依然是实验性的，尚未被所有的浏览器完全支持。

4. Promise.race(iterable)方法

Promise.race(iterable) 方法用于返回一个 Promise，一旦迭代器中的某个 Promise 被解决或拒绝，返回的 Promise 就会被解决或拒绝，代码如下：

```
//第 10 章/race 方法.html

const promise1 = new Promise((resolve, reject) => {
  setTimeout(resolve, 500, 'one');
});

const promise2 = new Promise((resolve, reject) => {
  setTimeout(resolve, 100, 'two');
});

Promise.race([promise1, promise2]).then((value) => {
  console.log(value);
  //都解决了,但是 promise2 要快些,所以返回的是 promise2
});
```

5. Promise.reject(reason)方法

Promise.reject()方法用于返回一个带有拒绝原因的 Promise 对象，代码如下：

```
function resolved(result) {
  console.log('Resolved');
}

function rejected(result) {
  console.error(result);
}
//输出 fail
Promise.reject(new Error('fail').then(resolved, rejected);
```

6. Promise.resolve(reason)方法

Promise.resolve(value)方法用于返回一个以给定值解析后的 Promise 对象。如果这个值是一个 Promise，则将返回这个 Promise；如果这个值是 thenable(带有"then"方法)，则返回的 Promise 会“跟随”这个 thenable 的对象，采用它的最终状态；否则返回的 Promise 将以此值完成。此函数将类 Promise 对象的多层嵌套展平，代码如下：

```
const promise1 = Promise.resolve(123);

promise1.then((value) => {
  console.log(value);
  //输出 123
});
```

注意：不要在解析为自身的 thenable 上调用 Promise. resolve()方法，这将导致无限递归。

10.3 Promise 对象的使用经验

Promise 对象在开发前端应用的时候用得比较多，如下总结了几个优先使用 Promise 对象的场景和经验，开发人员在实际项目中可以作为参考。

(1) 调用包含异步或阻塞带的方法时，使用 Promise。

(2) 为了代码的可读性，resolve()方法对应 then()方法，reject()方法对应 catch()方法。

(3) 确保同时写入 catch()和 then()方法实训所有的 Promise。

(4) 如果不管成功或失败都要执行，则请求使用 finally()方法。

(5) 我们只有一次改变每个 Promise 的机会。

(6) Promise 对象中所有方法的返回类型，都是 Promise。

(7) 在 Promise. all()方法中，无论哪个 Promise 首先被完成，Promise 的顺序都被保存在值变量中。

第 11 章 axios

11.1 axios 简介

axios 是一个基于 Promise 的网络请求库，作用于 Node.js 和浏览器中，它是 isomorphic 的(同一套代码可以运行在浏览器和 Node.js 中)。在服务器端它使用原生 Node.js 中的 http 模块，而在客户端(浏览器)则使用 XMLHttpRequests。axios 有以下几个方面的特性：

(1) 从浏览器中创建 XMLHttpRequests 对象。

(2) 从 Node.js 中创建 XMLHttpRequests 对象。

(3) 支持 PromiseAPI。

(4) 支持请求和响应拦截。

(5) 支持取消请求。

(6) 自动转换 JSON 数据。

(7) 客户端支持防御 XSRF。

视情况不同，可以通过 3 种方式安装 axios。

使用 npm 安装 axios，代码如下：

```
npm install axios
```

使用 bower 安装 axios，代码如下：

```
bower install axios
```

使用 cds 安装 axios，代码如下：

```
<script src = "https://unpkg.com/axios/dist/axios.min.js"></script>
```

以下是一个简单的 axios 使用示例：

```
<!DOCTYPE html>
<html lang = "en">
```

```
<head>
    <meta charset="UTF-8">
    <meta http-equiv="X-UA-Compatible" content="IE=edge">
    <meta name="viewport" content="width=device-width, initial-scale=1.0">
    <title>Document</title>
    <script type="text/JavaScript" src="../static/js/axios.js"></script>
</head>
<body>
<script>
    //注意,正确请求,需要有后台支持
        axios.get('/user/23').then( res => {
            console.log(res)
        }).catch( err => {
            console.log(err)
        })
    </script>
</body>
</html>
```

11.2 axios API

axios 作为一个基于 Promise 的网络请求库,给开发人员提供了很多方法,帮助开发人员快速编码,发送请求。接下来逐个介绍 axios 的每个请求方法及使用。

11.2.1 基本方法

axios 可以通过向 axios 传递相关配置来创建请求,语法如下:

```
axios(config)
```

或

```
axios(url[,config])
```

代码如下:

```
//发送 POST 请求
axios({
    method: 'post',
    url: '/user/12345',
    data: {
        firstName: 'Fred',
        lastName: 'Flintstone'
    }
})
```

```
//发送 GET 请求(默认的方法)
axios('/user/12345');
```

11.2.2 请求别名

为了方便开发人员编码，axios 还为各种请求提供了别名方法。

1. axios.request(config)

基于自定义请求属性发送请求，代码如下：

```
//原始的 Axios 请求方式
axios({
  method: 'post',
  url: '/user/12345',
  data: {
    firstName: 'Fred',
    lastName: 'Flintstone'
  },
  timeout: 1000,
  ...//其他相关配置
})
```

2. axios.get(url [,config])

基于指定的 url 发送 get 请求，config 可选，添加请求属性，代码如下：

```
axios.get('demo/url', {
    params: {
        id: 123,
        name: 'Henry',
    },
   timeout: 1000,
  ...          //其他相关配置
})
```

3. axios.delete(url [,config])

基于指定的 url 发送 delete 请求，config 可选，添加请求属性，代码如下：

```
//如果服务器端将参数作为 Java 对象来封装接受
axios.delete('demo/url', {
    data: {
        id: 123,
        name: 'Henry',
    },
```

```
    timeout: 1000,
    ...//其他相关配置
})
//如果服务器端将参数作为 url 参数来接受,则请求的 url 为:www.demo/url?a = 1&b = 2 形式
axios.delete('demo/url', {
    params: {
        id: 123,
        name: 'Henry',
    },
    timeout: 1000,
    ...//其他相关配置
})
```

4. axios.head(url [,config])

基于指定的 url,发送 head 请求,config 可选,添加请求属性,代码如下:

```
axios.head('axios/head',{
    headers:{
        'userName':'zhangsan'
    }
}).then(res => console.log(res))
  .catch(err => console.log(err))
```

5. axios.options(url [,config])

基于指定的 url,发送 options 请求,config 可选,添加请求属性,代码如下:

```
axios.options('axios/head',{
    headers:{
        'userName':'zhangsan'
    }
}).then(res => console.log(res))
  .catch(err => console.log(err))
```

6. axios.post(url [,data[,config]])

基于指定的 url 发送 post 请求,data 和 config 可选。data 用于指定请求数据,config 用于添加其他属性,代码如下:

```
axios.post('demo/url', {
    id: 123,
    name: 'Henry',
},{
   timeout: 1000,
    ...            //其他相关配置
})
```

7. axios.put(url [,data[,config]])

基于指定的 url 发送 put 请求，data 和 config 可选。data 用于指定请求数据，config 用于添加其他属性，代码如下：

```
axios.put('demo/url', {
    id: 123,
    name: 'Henry',
},{
   timeout: 1000,
    ...             //其他相关配置
})
```

8. axios.patch(url [,data[,config]])

基于指定的 url 发送 patch 请求，data 和 config 可选。data 用于指定请求数据，config 用于添加其他属性，代码如下：

```
axios.patch('demo/url', {
    id: 123,
    name: 'Henry',
},{
   timeout: 1000,
    ...           //其他相关配置
})
```

通过以上案例可以看出，get、delete、head、options 请求方式中，第 1 个参数为请求的 url 网址，第 2 个参数为请求的一些配置项，需要传递给后端的参数包含在配置项的 data 或者 params 属性中，而 post、put、patch 请求的第 1 个参数为 url 网址，第 2 个参数是需要入参的 JSON 数据，第 3 个参数是入参以外的其他配置项。同时也需要注意，当使用 get、delete、put、post 等别名方法的时候，url、method 和 data 等属性不用在 config 属性中重复声明。

11.2.3 并发方法

前面的别名方法，调用一次需发送一个请求，每个请求的响应需单独处理。axios 提供了 all()方法，能一次发送多个请求，同时还提供了 spread()方法，用于处理多个请求的响应，代码如下：

```
//执行多个请求
function getUserAccount() {
return axios.get('user',{params:{id:11}});
}
```

```
function getUserPermissions() {
    return axios.post('user',{firstName:'zhang',lastName:'san'});
}

axios.all([getUserAccount(), getUserPermissions()])
    .then(axios.spread(function (res1, res2) {          //两个请求的响应
            console.log('-------------')
            console.log(res1),
            console.log(res2)
    }));
```

如果需要同时调用多个请求，为了避免多并发问题，可以使用这两种方法。

11.3 axios 实例

在实际项目中，开发人员除了可以使用 axios 的方法发送处理请求外，也可以先创建 axios 的实例，再调用同一个实例的不同方法发送多个处理请求。这样就可以在一个实例中配置公共参数，以便重复使用。

11.3.1 创建实例

开发人员可以使用 axios.create([config]) 方法创建一个公共的实例 instance，通过 config 自定义公共的配置属性，创建一个实例，配置 baseUrl、timeout 和 X-Custom-Header 头等 3 个属性，代码如下：

```
const instance = axios.create({
  baseURL: 'https://some-domain.com/api/',
  timeout: 1000,
  headers: {'X-Custom-Header': 'foobar'}
});
```

接下来就可以通过 instance 调用 axios 的别名方法，发送和处理对应的请求。语法同 11.2 节中直接调用 axios 的别名方法一样。

11.3.2 请求配置和响应结构

在基于 axios 发送请求的时候，可以使用 config 配置很多属性，其中只有 url 属性是必需的，其他属性都可以不配置。其中如果没有指定 method 属性，则缺省值是 get，具体属性代码如下：

```
{
  //url 是用于请求的服务器 URL
  url: '/user',
```

```
//method 是创建请求时使用的方法
method: 'get', //default

//baseURL 将自动加在 url 前面,除非 url 是一个绝对 URL
//它可以通过设置一个 baseURL 便于为 axios 实例的方法传递相对 URL
baseURL: 'https://some-domain.com/api/',

//transformRequest 允许在向服务器发送前,修改请求数据
//只能用在 PUT、POST 和 PATCH 这几个请求方法
//后面数组中的函数必须返回一个字符串或 ArrayBuffer 或 Stream
transformRequest: [function (data, headers) {
  //对 data 进行任意转换处理
  return data;
}],

//transformResponse 在传递给 then/catch 前,允许修改响应数据
transformResponse: [function (data) {
  //对 data 进行任意转换处理
  return data;
}],

//headers 是即将被发送的自定义请求头
headers: {'X-Requested-With': 'XMLHttpRequest'},

//params 是即将与请求一起发送的 URL 参数
//必须是一个无格式对象(plain object)或 URLSearchParams 对象
params: {
  ID: 12345
},

//paramsSerializer 是一个负责 params 序列化的函数
//(e.g. https://www.npmjs.com/package/qs,
//http://api.jquery.com/jquery.param/)
paramsSerializer: function(params) {
  return Qs.stringify(params, {arrayFormat: 'brackets'})
},

//data 是作为请求主体被发送的数据
//只适用于这些请求方法 PUT、POST 和 PATCH
//在没有设置 transformRequest 时,必须是以下类型之一:
//- string, plain object, ArrayBuffer, ArrayBufferView, URLSearchParams
//- 浏览器专属:FormData、File、Blob
//- Node 专属: Stream
data: {
  firstName: 'Fred'
```

```
  },

  //timeout 指定请求超时的毫秒数(0 表示无超时时间)
  //如果请求花费了超过 timeout 的时间,请求将被中断
  timeout: 1000,

   //withCredentials 表示跨域请求时是否需要使用凭证
  withCredentials: false,              //default

  //adapter 允许自定义处理请求,以使测试更轻松
  //返回一个 promise 并应用一个有效的响应
  adapter: function (config) {
    /* ... */
  },

//auth 表示应该使用 HTTP 基础验证,并提供凭据
//这将设置一个 Authorization 头,覆写掉现有的任意使用 headers 设置的自定义 Authorization 头
  auth: {
    username: 'janedoe',
    password: 's00pers3cret'
  },

  //responseType 表示服务器响应的数据类型,可以是 arraybuffer、blob, document、json、
  //text、stream
  responseType: 'json', //default

  //responseEncoding indicates encoding to use for decoding responses
  //Note: Ignored for responseType of stream or client - side requests
  responseEncoding: 'utf8', //default

   //xsrfCookieName 是用作 xsrf token 的值的 Cookie 的名称
  xsrfCookieName: 'XSRF - TOKEN', //default

  //xsrfHeaderName is the name of the http header that carries the xsrf token value
  xsrfHeaderName: 'X - XSRF - TOKEN', //default

   //onUploadProgress 允许为上传处理进度事件
  onUploadProgress: function (progressEvent) {
  //Do whatever you want with the native progress event
  },

    //onDownloadProgress 允许为下载处理进度事件
  onDownloadProgress: function (progressEvent) {
    //对原生进度事件的处理
```

```
  },

  //maxContentLength 定义允许的响应内容的最大长度
  maxContentLength: 2000,

  //validateStatus 定义对于给定的 HTTP 响应状态码是 resolve 或 reject promise。
  //如果 validateStatus 返回 true(或者设置为 null 或 undefined),
  //则 promise 将被 resolve; 否则,promise 将被 reject
  validateStatus: function (status) {
    return status >= 200 && status < 300; //default
  },

  //maxRedirects 定义在 Node.js 中,表示最大的重定向数目
  //如果设置为 0,则将不会跟随任何重定向
  maxRedirects: 5, //default

  //socketPath defines a UNIX Socket to be used in Node.js
  //e.g. '/var/run/docker.sock' to send requests to the docker daemon
  //Only either socketPath or proxy can be specified
  //If both are specified, socketPath is used.
  socketPath: null, //default

  //httpAgent 和 httpsAgent 分别在 Node.js 中,用于定义在执行 http 和 https
  //时使用的自定义代理。允许像这样配置选项
  //keepAlive 默认没有启用
  httpAgent: new http.Agent({ keepAlive: true }),
  httpsAgent: new https.Agent({ keepAlive: true }),

  //proxy 用于定义代理服务器的主机名称和端口
  //auth 表示 HTTP 基础验证应当用于连接代理,并提供凭据
  //这将会设置一个 Proxy-Authorization 头,覆写掉已有的通过 header 设置
  //的自定义 Proxy-Authorization 头
  proxy: {
    host: '127.0.0.1',
    port: 9000,
    auth: {
      username: 'mikeymike',
      password: 'rapunz3l'
    }
  },

  //cancelToken 指定用于取消请求的 cancel token
  //查看后面的 Cancellation,以便了解更多
  cancelToken: new CancelToken(function (cancel) {
  })
}
```

同样，axios 接收的响应也有固定结构，开发人员需要掌握这些属性和意义，以便正确地进行响应处理，代码如下：

```
{
  //data 是由服务器提供的响应
  data: {},

  //status 是来自服务器响应的 HTTP 状态码
  status: 200,

  //statusText 是来自服务器响应的 HTTP 状态信息
  statusText: 'OK',

  //headers 是服务器响应的头
  headers: {},

   //config 是为请求提供的配置信息
  config: {},

  //request
  //request is the request that generated this response
  //It is the last ClientRequest instance in Node.js (in redirects)
  //and an XMLHttpRequest instance the browser
  request: {}
}
```

11.4 默认配置

在基于 axios 发送请求前，可以提前给 axios 设置属性的默认配置。根据默认配置的作用范围，分为全局默认配置和局部默认配置。

全局默认配置是指所有的 axios 请求都起效果，跨所有的 axios 实例，代码如下：

```
//配置全局 baseUrl
axios.defaults.baseURL = 'https://api.example.com';
//配置全局自定义 Authorization 头
axios.defaults.headers.common['Authorization'] = AUTH_TOKEN;
//配置全局对应协议(post)Content-Type 头
axios.defaults.headers.post['Content-Type'] = 'application/x-www-form-urlencoded';
```

局部默认配置是在某个 axios 实例中配置默认属性，它们的作用范围是当前的 axios 实例，代码如下：

```
//Set config defaults when creating the instance
//Alter defaults after instance has been created
```

```
const instance = axios.create({ baseURL: 'https://api.example.com'});
//设置 instance 的自定义头 Authorization
instance.defaults.headers.common['Authorization'] = AUTH_TOKEN;
```

axios 最后会将所有属性进行合并，再向后台发送请求。合并的顺序是先加载 lib/default.js 中的配置属性，接着加载全局配置属性，再加载局部配置属性，最后加载方法中通过 config 配置的属性。合并规则是后加载的配置属性值会覆盖前面加载的配置属性值，也就是说后者的优先级高于前者的优先级。最后请求的 timeout 属性的值是 5000，而不是 0 或 2500，代码如下：

```
//使用由库提供的配置的默认值来创建实例
//此时超时配置的默认值是 0
var instance = axios.create();

//覆写库的超时默认值
//现在，在超时前，所有请求都会等待 2.5s
instance.defaults.timeout = 2500;

//为已知需要花费很长时间的请求覆写超时设置
instance.get('/longRequest', { timeout: 5000});
```

11.5 拦截器

axios 内置了拦截机制的实现，开发人员可以通过给 axios 的请求和响应部分添加拦截器，从而在发送请求前和接收的响应被 then 或 cath 处理前，对请求和响应进行拦截处理，代码如下：

```
<script type = "text/JavaScript">
    const axiosInstance = axios.create({
      baseUrl:'http://localhost:8088'
    });
    //添加请求拦截器
    axiosInstance.interceptors.request.use(
      config = >{
          //在发送请求前执行
          console.log('发送请求前执行')
          return config;
      },
      error = >{
          //在请求处理错误时执行
          console.log('请求错误……')
          return Promise.reject(error;
```

```
        }
      )
      //添加响应拦截器
      axiosInstance.interceptors.response.use(
        res =>{
            //处理响应数据
            console.log('处理响应数据…….')
            return res;
        },
        error =>{
            //处理响应错误
            return Promise.reject(error)
        }
      )
      //在发送get请求过程中,自动执行请求和响应拦截器
      axiosInstance.get('/user',{params:{id:13}})
        .then(res =>{
          console.log(res)
        })
        .catch(error =>{
          console.log(error)
        })
    </script>
```

第 12 章

模板模式开发 Vue.js 应用

前面介绍的技术和编写的案例都是在一个页面文件中实现的代码，但是在实际项目中，不应把所有代码写在一个文件中，需要在一个统一的开发环境下开发、编译、打包和运行整个工程文件。接下来介绍如何搭建集成开发环境，并且基于这个环境模块式地开发 Vue.js 应用。

12.1 Node.js

Node.js 是一个开源且跨平台的 JavaScript 运行时环境。它是一个可用于几乎任何项目的流行工具。Node.js 在浏览器外运行 V8JavaScript 引擎（Google Chrome 的内核）。这使 Node.js 表现得非常出色。Node.js 应用程序运行于单个进程中，无须为每个请求创建新的线程。

Node.js 在其标准库中提供了一组异步的 I/O 原生功能（用以防止 JavaScript 代码被阻塞），并且 Node.js 中的库通常是用非阻塞的范式编写的（从而使阻塞行为成为例外而不是规范）。当 Node.js 执行 I/O 操作时（例如从网络读取、访问数据库或文件系统），Node.js 会在响应返回时恢复操作，而不是阻塞线程并浪费 CPU 循环等待。这使 Node.js 可以在一台服务器上处理数千个并发连接，而无须引入管理线程并发的负担（这可能是重大 Bug 的来源）。

Node.js 具有独特的优势，因为为浏览器编写 JavaScript 的数百万前端开发者现在除了客户端代码之外还可以编写服务器端代码，而无须学习完全不同的语言。

在 Node.js 中，可以毫无问题地使用新的 ECMAScript 标准，因为不必等待所有用户更新其浏览器，可以通过更改 Node.js 版本来决定要使用的 ECMAScript 版本，并且还可以通过运行带有标志的 Node.js 来启用特定的实验中的特性。

基于 Node.js 启动一个 Web 服务，返回 Hello World 响应，代码如下：

```
//第 12 章/第 1 个 Node.js
const http = require('http')


const hostname = '127.0.0.1'
```

```
const port = 3000

const server = http.createServer(req, res) => {
  res.statusCode = 200
  res.setHeader('Content-Type', 'text/plain;charset=utf-8')
  res.end('你好世界\n')
})

server.listen((port, hostname, ) => {
  console.log(`服务器运行在 http://${hostname}:${port}/`)
})
```

此代码首先引入了 Node.js 中的 http 模块。Node.js 具有出色的标准库，包括对网络的一流支持。

http 的 createServer()方法会创建新的 HTTP 服务器并返回。

服务器被设置为监听指定的端口和主机名。当服务器就绪后，回调函数会被调用，在此示例中会通知我们服务器正在运行。

每当接收到新的请求时，request 事件会被调用，并提供两个对象：一个请求（http.IncomingMessage 对象）和一个响应（http.ServerResponse 对象）。

这两个对象对于处理 HTTP 调用至关重要。第 1 个对象提供了请求的详细信息。在这个简单的示例中没有使用它，但是可以访问请求头和请求数据。第 2 个对象用于将数据返回调用方。

在以下示例中：

```
res.statusCode = 200
```

将 statusCode 属性设置为 200，以表明响应成功。

设置 Content-Type 响应头：

```
res.setHeader('Content-Type', 'text/plain;charset=utf-8')
```

关闭响应，添加内容作为 end()方法的参数：

```
res.end('你好世界\n')
```

在安装 Node.js 的过程中，会顺便安装 npm。npm 是一个包管理工具，因为 npm 的结构比较简单，方便 npm 仓库托管公共的第三方开源库包，目前有超过 1 000 000 开源库包被托管到 npm 仓库，有效地构建了 Node.js 的生态。比较常用的库清单如下。

（1）AdonisJS：一个全栈框架，高度专注于开发者的效率、稳定和信任。Adonis 是最快的 Node.js Web 框架之一。

（2）Express：提供了创建 Web 服务器的最简单但功能最强大的方法之一。它的极简主义方法专注于服务器的核心功能，是其成功的关键。

(3) Fastify：一个 Web 框架，高度专注于提供最佳的开发者体验(最少的开销和强大的插件架构)。Fastify 是最快的 Node. js Web 框架之一。

(4) Gatsby：一个基于 React、由 GraphQL 驱动的静态网站生成器，具有非常丰富的插件和启动器生态系统。

(5) hapi：一个富框架，用于构建应用程序和服务，使开发者可以专注于编写可重用的应用程序逻辑，而不必花费时间来搭建基础架构。

(6) koa：由 Express 背后的同一个团队构建，旨在变得更简单、更轻巧。新项目的诞生是为了创建不兼容的更改而又不破坏现有社区。

(7) Loopback. io：使构建需要复杂集成的现代应用程序变得容易。

(8) Meteor：一个强大的全栈框架，以同构的方式使用 JavaScript 构建应用(在客户端和服务器上共享代码)。曾经是提供了所有功能的现成工具，现在可以与前端库 React、Vue. js 和 Angular 集成，也可以用于创建移动应用。

(9) Micro：提供了一个非常轻量级的服务器，用于创建异步的 HTTP 微服务。

(10) NestJS：一个基于 TypeScript 的渐进式 Node. js 框架，用于构建企业级的高效、可靠和可扩展的服务器端应用程序。

(11) Next. js：一个 React 框架，可为开发者提供生产所需的所有功能的最佳体验，如混合静态和服务器渲染、TypeScript 支持、智能捆绑、路由预取等。

(12) Nx：使用 NestJS、Express、React、Angular 等进行全栈开发的工具包。Nx 有助于将开发工作从一个团队(构建一个应用程序)扩展到多个团队(在多个应用程序上进行协作)。

(13) Sapper：Sapper 是一个用于构建各种规模的 Web 应用程序的框架，具有出色的开发体验和灵活的、基于文件系统的路由，还提供 SSR 等。

(14) Socket. io：一个实时通信引擎，用于构建网络应用程序。

(15) Strapi：Strapi 是一个灵活的开源 Headless CMS，让开发者可以自由选择自己喜欢的工具和框架，同时还允许编辑人员轻松地管理和分发其内容。通过管理面板和 API 可以将插件系统进行扩展，Strapi 使大公司能够加速内容交付，同时构建优美的数字体验。

12.1.1　下载并安装 Node. js

进入 https://nodejs. org/zh-cn/网页，可以下载 Windows 系统下最新版本的 Node. js 安装包，也可以进入 https://nodejs. org/zh-cn/download 下载其他版本的安装包。本书所用的 Node. js 是基于 node-v14. 16. 0-x64. msi 安装包安装的。

在 Windows 系统下，直接双击安装文件，一直单击“下一步”按钮，就可以完成 Node. js 的安装。安装完后，打开 cmd 窗口，输入 node-v 命令，如果能显示 Node. js 的版本号，如图 12. 1 所示，就表示 Node. js 安装成功了。

```
C:\Users\Administrator>node -v
v14.16.0
```

图 12.1　查看 Node.js 版本

12.1.2　npm 的使用

npm 是随 Node.js 一起安装的包管理工具，能解决 Node.js 代码部署中的很多问题，常见的使用场景有以下几种。

(1) 允许用户从 npm 服务器下载别人编写的第三方包到本地使用。

(2) 允许用户从 npm 服务器下载并安装别人编写的命令行程序到本地使用。

(3) 允许用户将自己编写的包或命令行程序上传到 npm 服务器供别人使用。

由于新版的 Node.js 已经集成了 npm，所以 npm 也一并安装好了。同样可以通过输入 npm-v 命令来测试是否成功安装。如图 12.2 所示，表示 npm 安装成功。

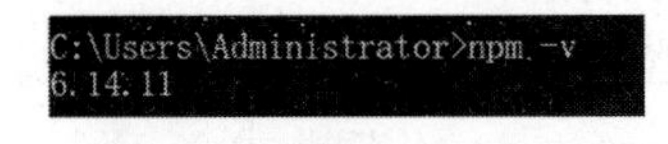

图 12.2　查看 npm 版本

如果安装的是旧版本，也可以使用 npm install npm-g 命令升级，还可以使用淘宝镜像命令，代码如下：

```
npm install npm -g
//淘宝镜像命令
npm install -g cnpm --registry=https://registry.npm.taobao.org
```

使用 npm 可以很方便地安装模板。安装模块的语法如下：

```
npm install <模块名称>
```

安装 Web 框架模块的 express，代码如下：

```
npm install express
```

使用 npm 安装模块的方式有两种，分别是全局安装和本地安装。全局安装的方式是将模块安装在 Node.js 的安装目录，可以直接在命令行中使用。本地安装是将模块安装到工程的./node_modules 目录下，如果当前工程目录下没有 node_modules 目录，则 npm 会自动创建一个 node_modules 目录。在项目代码中，可以通过 require 来引入本地安装包。

全局安装的语法如下：

```
npm install <模块名称> -g
```

本地安装的语法如下：

```
npm install <模块名称>
```

开发人员可以使用 npm list-g 命令查看全局安装的所有模块和版本信息，也可以使用 npm list <模块名称>-g 命令查看全局安装的指定模块的版本信息。如图 12.3 所示，通过 npm list-g 命令查看全局安装的模块和版本信息。如图 12.4 所示，通过 npm list express-g 命令查看全局安装的 express 模块的版本信息。

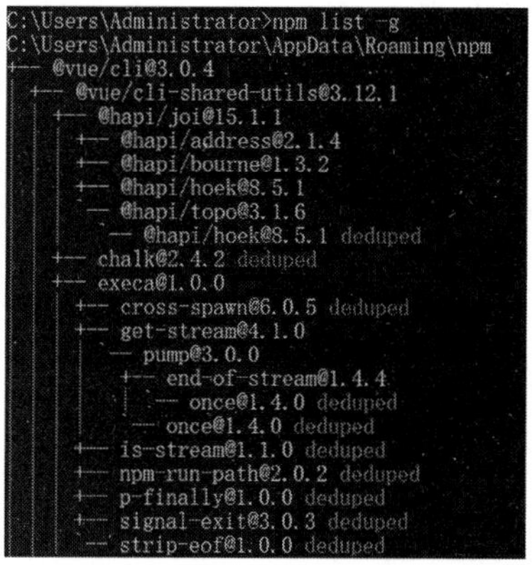

```
C:\Users\Administrator>npm list -g
C:\Users\Administrator\AppData\Roaming\npm
+-- @vue/cli@3.0.4
  +-- @vue/cli-shared-utils@3.12.1
    +-- @hapi/joi@15.1.1
      +-- @hapi/address@2.1.4
      +-- @hapi/bourne@1.3.2
      +-- @hapi/hoek@8.5.1
      -- @hapi/topo@3.1.6
        -- @hapi/hoek@8.5.1 deduped
    +-- chalk@2.4.2 deduped
    +-- execa@1.0.0
      +-- cross-spawn@6.0.5 deduped
      +-- get-stream@4.1.0
        -- pump@3.0.0
          +-- end-of-stream@1.4.4
          | -- once@1.4.0 deduped
          -- once@1.4.0 deduped
      +-- is-stream@1.1.0 deduped
      +-- npm-run-path@2.0.2 deduped
      +-- p-finally@1.0.0 deduped
      +-- signal-exit@3.0.3 deduped
      -- strip-eof@1.0.0 deduped
```

图 12.3　查看全局安装的模块和版本信息

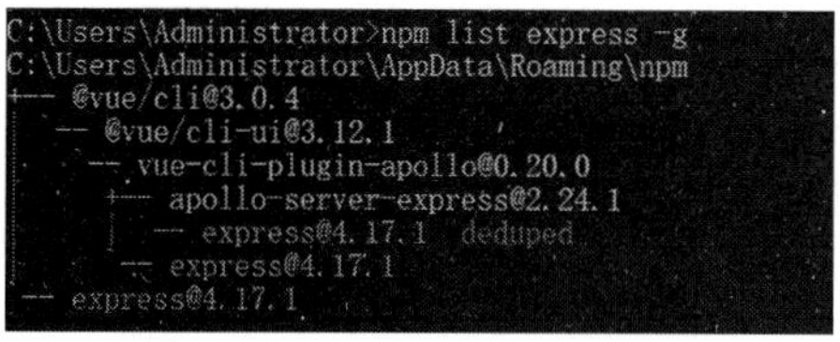

```
C:\Users\Administrator>npm list express -g
C:\Users\Administrator\AppData\Roaming\npm
+-- @vue/cli@3.0.4
  -- @vue/cli-ui@3.12.1
    -- vue-cli-plugin-apollo@0.20.0
      +-- apollo-server-express@2.24.1
      | -- express@4.17.1 deduped
      -- express@4.17.1
-- express@4.17.1
```

图 12.4　查看全局安装的 express 模块的版本信息

如果要查看本地安装的模块和版本信息，则可以查看工程目录下的 package.json 文件，里面描述了所有依赖模块和对应的版本信息。

package.json 的样例代码如下：

```
{
  ...
  "dependencies": {
    "axios": "0.18.1",
    "clipboard": "2.0.4",
    "codemirror": "5.45.0",
    "core-js": "3.6.5",
    ...
  },
  "devDependencies": {
    "@vue/cli-plugin-babel": "4.4.4",
    "@vue/cli-plugin-eslint": "4.4.4",
    "@vue/cli-plugin-unit-jest": "4.4.4",
    ...
  },
  ...
}
```

同样，也可以使用 npm 卸载已经安装的依赖。

卸载依赖的语法如下：

```
npm uninstall <模块名称>
```

12.1.3 切换镜像站点

npm 默认的镜像站点是 https://registry.npmjs.org，这个站点是国外站点，下载速度可能较慢，开发人员可以自己设置镜像站点，以提高下载速度。

切换镜像站点有两种方式。

1. 通过更改 npm 配置文件切换站点

使用 npm config set registry <url> 命令，修改 npm 的配置文件进行切换，如：

```
npm config set registry https://registry.npm.taobao.com
```

使用 npm config get registry 或 npm config list 命令，获取当前的镜像站点，如：

```
npm config get registry
```

或

```
npm config list
```

使用 npm config rm registry 命令删除镜像，代码如下：

```
npm config rm registry
```

2. 通过 nrm 进行镜像管理

1）下载并安装 nrm

使用 npm install nrm -g 命令，安装 nrm 镜像管理器，如图 12.5 所示。

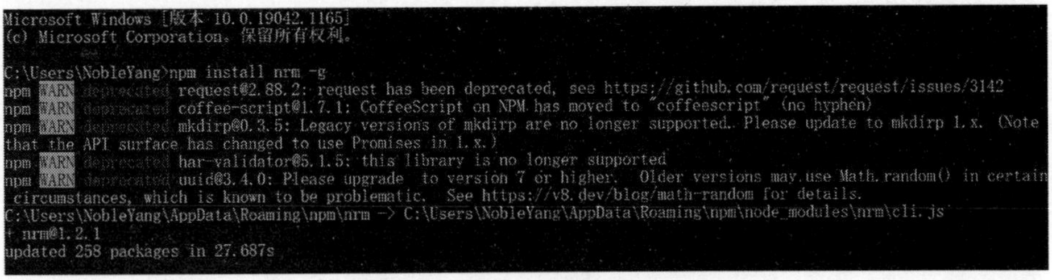

```
Microsoft Windows [版本 10.0.19042.1165]
(c) Microsoft Corporation。保留所有权利。

C:\Users\NobleYang>npm install nrm -g
npm WARN deprecated request@2.88.2: request has been deprecated, see https://github.com/request/request/issues/3142
npm WARN deprecated coffee-script@1.7.1: CoffeeScript on NPM has moved to "coffeescript" (no hyphen)
npm WARN deprecated mkdirp@0.3.5: Legacy versions of mkdirp are no longer supported. Please update to mkdirp 1.x. (Note that the API surface has changed to use Promises in 1.x.)
npm WARN deprecated har-validator@5.1.5: this library is no longer supported
npm WARN deprecated uuid@3.4.0: Please upgrade  to version 7 or higher.  Older versions may use Math.random() in certain circumstances, which is known to be problematic.  See https://v8.dev/blog/math-random for details.
C:\Users\NobleYang\AppData\Roaming\npm\nrm -> C:\Users\NobleYang\AppData\Roaming\npm\node_modules\nrm\cli.js
+ nrm@1.2.1
updated 258 packages in 27.687s
```

图 12.5 安装 nrm 镜像管理器

2）查看镜像源

使用 nrm ls 命令，可以查看镜像源。本书安装的 Node.js 的版本是 v14.16.0，安装的 nrm 的版本是 1.2.1，使用 nrm ls 命令的时候系统会抛出 internal/validators.js：124 异常，如图 12.6 所示。

```
C:\Users\NobleYang>nrm ls
internal/validators.js:124
    throw new ERR_INVALID_ARG_TYPE(name, 'string', value);

[TypeError [ERR_INVALID_ARG_TYPE]: The "path" argument must be of type string. Received undefined
  at validateString (internal/validators.js:124:11)
  at Object.join (path.js:375:7)
  at Object.<anonymous> (C:\Users\NobleYang\AppData\Roaming\npm\node_modules\nrm\cli.js:17:20)
  at Module._compile (internal/modules/cjs/loader.js:1063:30)
  at Object.Module._extensions..js (internal/modules/cjs/loader.js:1092:10)
  at Module.load (internal/modules/cjs/loader.js:928:32)
  at Function.Module._load (internal/modules/cjs/loader.js:769:14)
  at Function.executeUserEntryPoint [as runMain] (internal/modules/run_main.js:72:12)
  at internal/main/run_main_module.js:17:47
] {
  code: 'ERR_INVALID_ARG_TYPE'
}
```

图 12.6　nrm ls 命令异常

根据异常提示，找到 C：\Users\NobleYang\AppData\Roaming\npm\node_modules\nrm\cli. js 文件中的第 17 行代码，注释掉此行代码，添加“const NRMRC = path. join(process. env[(process. platform == 'win32')？'USERPROFILE'：'HOME']，'. nrmrc')；”，代码如下：

```
//const NRMRC = path.join(process.env.HOME, '.nrmrc'); (将原代码注释掉)
const NRMRC = path.join(process.env[(process.platform == 'win32')? 'USERPROFILE' : 'HOME'],
'.nrmrc');
```

再执行 nrm ls 命令，查看镜像源结果，如图 12.7 所示。

```
C:\Users\NobleYang>nrm ls

* npm -------- https://registry.npmjs.org/
  yarn ------- https://registry.yarnpkg.com/
  cnpm ------- http://r.cnpmjs.org/
  taobao ----- https://registry.npm.taobao.org/
  nj --------- https://registry.nodejitsu.com/
  npmMirror -- https://skimdb.npmjs.com/registry/
  edunpm ----- http://registry.enpmjs.org/
```

图 12.7　使用 nrm ls 命令查看镜像源

其中 npm 前面有 * 标记，表示是当前使用的镜像源。npm 镜像源是默认镜像源。

3）切换镜像源

使用 nrm use <镜像名>命令，切换镜像源。如图 12.8 所示，使用 nrm use taobao 命令，

```
C:\Users\NobleYang>nrm use taobao


  Registry has been set to: https://registry.npm.taobao.org/


C:\Users\NobleYang>nrm ls

  npm -------- https://registry.npmjs.org/
  yarn ------- https://registry.yarnpkg.com/
  cnpm ------- http://r.cnpmjs.org/
* taobao ----- https://registry.npm.taobao.org/
  nj --------- https://registry.nodejitsu.com/
  npmMirror -- https://skimdb.npmjs.com/registry/
  edunpm ----- http://registry.enpmjs.org/
```

图 12.8　使用 nrm 命令切换镜像源

切换到淘宝镜像源，再用 nrm ls 查看镜像源，* 标记在 taobao 前面，说明镜像源切换成功。

12.2 webpack 工具

在模板模式开发的前端项目中，会按模块编写很多独立的 JavaScript、CSS 等文件，运行前，需要将这些文件按照它们之间的相互依赖关系打包整合到一起，这时就需要使用工具帮开发人员完成相关的工作。

webpack 是一个现代 JavaScript 应用程序的静态模块打包器(Module Bundler)。当 webpack 处理应用程序时，它会递归地构建一个依赖关系图(Dependency Graph)，其中包含应用程序需要的每个模块，然后将所有模块打包成一个或多个 bundle。

webpack 工具可以实现以下几个功能。

(1) 将许多小文件打包成一个整体，减少单页面内的衍生请求次数，从而提高网站效率。

(2) 将 ES6 的高级语法进行转换编译，以兼容旧版本的浏览器。

(3) 将代码打包的同时进行混淆，以此提高代码的安全性。

webpack 的工作原理如图 12.9 所示。

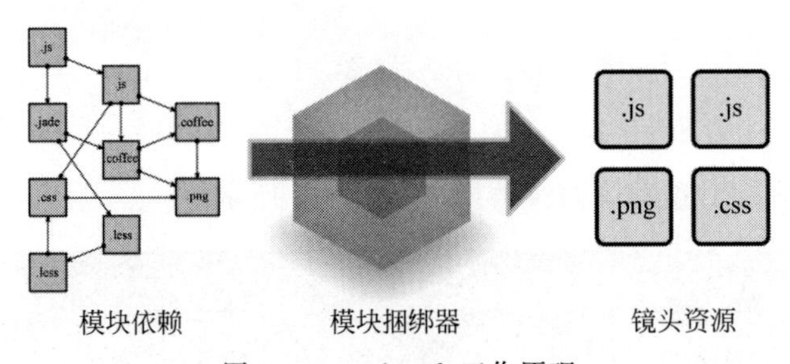

图 12.9 webpack 工作原理

webpack 的详细介绍可以参考 webpack 的官网 https://webpackjs.com。

12.2.1 安装 webpack

webpack 支持全局安装和本地安装，官方推荐的安装方法是本地安装，命令如下：

```
npm install --save-dev webpack -g
```

或

```
npm install webpack
```

对于 webpack 4+的版本，除了需要安装 webpack 外，还需要安装 webpack-cli，命令如下：

```
npm install --save-dev webpack webpack-cli -g
```

可以使用 webpack -v 命令查看安装后的 webpack 版本，如图 12.10 所示，当前 webpack 的版本是 5.50.0，webpack-cli 的版本是 4.8.0。

图 12.10　查看 webpack 的版本

12.2.2　手动体验 webpack

接下来基于案例，手动体验使用 webpack 工具打包。首先创建一个项目目录(webpack)，然后在 webpack 目录中创建 src 和 dist 两个目录，再创建 webpack/src/js/module01.js 文件，代码如下：

```
//第 12 章/js/module01.js
let user = {
    name:'zhangsan',
    age:12,
    sum:function(a,b){
  return a + b;
    }
}

let name = '李四'
let sub = function(a,b){
    return a - b;
}
//导出 user 用户
export default user
export {name};
export {sub};
```

创建 webpack/src/js/main01.js 文件，代码如下：

```
//第 12 章/js/main01.js
//const user = require('./module01.js')
//默认导入
import user from './module01.js'
import {name, sub} from './module01.js'
console.log(user)
console.log(name)
console.log(sub(6,2));
```

创建 webpack/src/index.html 文件，代码如下：

```
//第 12 章/index.html
<!DOCTYPE html>
<html>
    <head>
        <meta charset="utf-8">
  <title></title>
  <!--引入 dist/main.js,该文件通过 webpack 打包后会自动生成-->
        <script type="text/JavaScript" src="../dist/js/main.js"></script>
    </head>
    <body>
        Hello webpack
    </body>
</html>
```

在 webpack 目录下，执行 webpack 命令，将 module01.js 和 main01.js 文件打包到 dist/js/main.js 文件中。webpack 的打包命令如下：

```
webpack --mode development --entry ./src/js/main01.js -o ./dist/js
```

打包结果如图 12.11 所示。

```
E:\work\code\vscodeWorkspace\vuebook\webpack>webpack --mode development --entry ./src/js/main01.js -o ./dist/js
asset main.js 4.8 KiB [emitted] (name: main)
runtime modules 670 bytes 3 modules
cacheable modules 502 bytes
  ./src/js/main01.js 263 bytes [built] [code generated]
  ./src/js/module01.js 239 bytes [built] [code generated]
webpack 5.50.0 compiled successfully in 127 ms
```

图 12.11 webpack 打包

打开浏览器的开发工具，查看 console，运行 index.html 文件后可以在 console 看到 JavaScript 的输出内容，如图 12.12 所示。

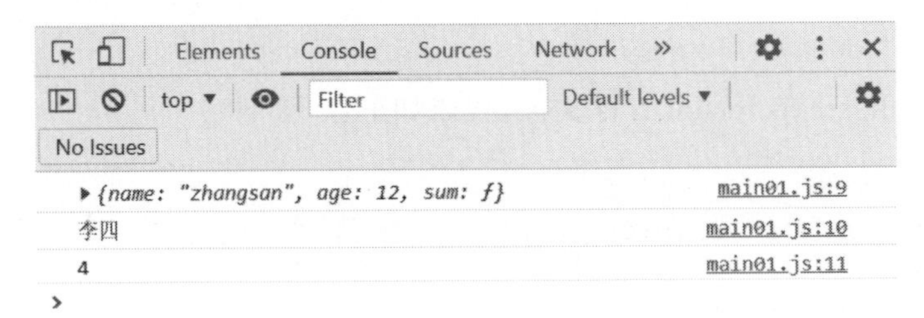

图 12.12 index.html 文件在控制台的输出

12.2.3 基于配置体验 webpack 打包

通过前面的案例，开发人员可以体验到 webpack 的打包过程，但是还不够智能，接下来介绍基于配置文件的 webpack 打包。

在 12.2.2 节的 webpack 目录下，执行 npm init -y 命令，如图 12.13 所示。

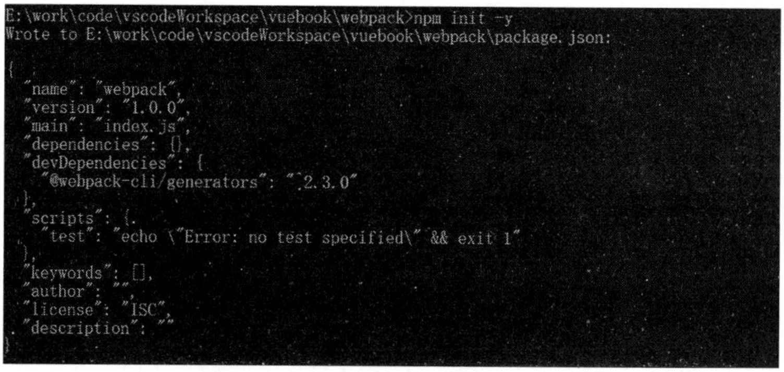

```
E:\work\code\vscodeWorkspace\vuebook\webpack>npm init -y
Wrote to E:\work\code\vscodeWorkspace\vuebook\webpack\package.json:

{
  "name": "webpack",
  "version": "1.0.0",
  "main": "index.js",
  "dependencies": {},
  "devDependencies": {
    "@webpack-cli/generators": "^2.3.0"
  },
  "scripts": {
    "test": "echo \"Error: no test specified\" && exit 1"
  },
  "keywords": [],
  "author": "",
  "license": "ISC",
  "description": ""
}
```

图 12.13　初始化 package.json 文件

在 webpack 目录下，执行 webpack init 命令，初始化 webpack 工程，如图 12.14 所示。

```
E:\work\code\vscodeWorkspace\vuebook\webpack>webpack init
[webpack-cli] For using this command you need to install: '@webpack-cli/generators' package.
[webpack-cli] Would you like to install '@webpack-cli/generators' package? (That will run 'npm install -D @webpack-cli/g
enerators') (Y/n) Y
npm WARN deprecated resolve-url@0.2.1: https://github.com/lydell/resolve-url#deprecated
npm WARN deprecated urix@0.1.0: Please see https://github.com/lydell/urix#deprecated
npm WARN @webpack-cli/generators@2.3.0 requires a peer of webpack@4.x.x || 5.x.x but none is installed. You must install
 peer dependencies yourself.
npm WARN @webpack-cli/generators@2.3.0 requires a peer of webpack-cli@4.x.x but none is installed. You must install peer
 dependencies yourself.
npm WARN webpack@1.0.0 No description
npm WARN webpack@1.0.0 No repository field.

+ @webpack-cli/generators@2.3.0
updated 1 package in 13.186s
[webpack-cli] No commands found to run
```

图 12.14　初始化 webpack 工程

在 webpack 目录下，创建 webpack.config.js 文件，加入要打包的源文件和目标文件的描述信息，代码如下：

```
//第 12 章/webpack.config.js

const path = require('path')                                  //导入 path 模块

module.exports = {
    entry:{
      fileName:path.resolve(__dirname, 'src/js/main01.js')
    },//'./src/main.js',                                      //输入文件
    output:{                                                  //输出信息
    path:path.resolve(__dirname, './dist/js'),                //拼接
//__dirname(webpack.config.js 所在目录)和'./dist'
    filename:'main.js'                                        //目标文件
    }
}
```

在 webpack 目录中，执行 webpack --mode development 命令，完成 webpack 目录下的 JS 文件的打包。运行 index.html 文件，效果同手动完成的效果一样，代码如下：

```
webpack  -- mode development
```

在 package. json 文件中，添加 url-loader 依赖和 build 脚本，代码如下：

```
//第 12 章/package. json
{
  "name": "webpack",
  "version": "1.0.0",
  "main": "index. js",
  "dependencies": {},
  "devDependencies": {
    "@webpack - cli/generators": "^2.3.0",
    "url - loader": "^4.1.1" //url - loader 依赖
  },
  "scripts": {
    "test": "echo \"Error: no test specified\" && exit 1",
    "build": "webpack  -- mode development" //webpack 打包 script
  },
  ...
}
```

在 webpack 目录下，运行 npm run build 命令，效果同此前运行 webpack- -mode development 命令一样。

12.3 基于 Vue-CLI 脚手架创建项目开发

Vue-CLI 是一个基于 Vue. js 进行快速开发的完整系统，它提供了以下几方面的功能。

(1) 通过@vue/cli 实现交互式的项目脚手架。

(2) 通过@vue/cli+@vue/cli-service-global 实现零配置原型开发。

(3) 基于一个运行时依赖(@vue/cli-service)，该依赖具备可以升级、基于 webpack 构建并带有合理的默认配置、可以通过项目内的配置文件进行配置、可以通过插件进行扩展等特征。

(4) 它是一个丰富的官方插件集合，集成了前端生态中最好的工具。

(5) 它是一套完全图形化的创建和管理 Vue. js 项目的用户界面。

Vue-CLI 致力于将 Vue. js 生态中的工具基础标准化。它确保了各种构建工具能够基于智能的默认配置即可平稳衔接，这样开发人员可以专注在撰写应用上，而不必花几天时间去纠结配置的问题。与此同时，它也为每个工具提供了调整配置的灵活性。

接下来介绍如何基于 Vue-CLI 脚手架创建 Vue. js 项目。

使用 npm install 命令，安装@vue/cli，代码如下：

```
npm install  - g @vue/cli@3.0.4                //全局安装
```

使用 vue --version 命令查看版本，代码如下：

```
vue -- version
```

在命令行窗口中输入 vue create vue02 命令，创建 vue02 项目，会出现如图 12.15 所示的提示。项目名称不能是大写字母。

```
Vue CLI v3.0.4
? Please pick a preset: (Use arrow keys)
> default (babel, eslint)
  Manually select features
```

图 12.15　选择 preset

default(babel,eslint)：默认设置(直接按 Enter 键)非常适合快速创建一个新项目的原型，没有带任何辅助功能的 npm 包。

Manually select features：自定义配置(按方向键↓)是我们所需要的面向生产的项目，提供可选功能的 npm 包。

通过方向键，选择 Manually select features，按 Enter 键，会出现如图 12.16 所示的界面。

```
Vue CLI v3.0.4
? Please pick a preset: Manually select features
? Check the features needed for your project: (Press <space> to select, <a> to toggle all, <i> to invert selection)
>(*) Babel
 ( ) TypeScript
 ( ) Progressive Web App (PWA) Support
 ( ) Router
 ( ) Vuex
 ( ) CSS Pre-processors
 (*) Linter / Formatter
 ( ) Unit Testing
 ( ) E2E Testing
```

图 12.16　自定义配置

(1) Babel：转码器，可以将 ES6 代码转换为 ES5 代码，从而在现有环境执行。

(2) TypeScript：TypeScript 是一个 JavaScript(后缀为.js)的超集(后缀为.ts)，包含并扩展了 JavaScript 的语法，需要被编译并输出为 JavaScript 才可在浏览器运行，目前使用的人数较少。

(3) Progressive Web App(PWA)Support：渐进式 Web 应用程序。

(4) Router：vue-router(Vue.js 路由)。

(5) Vuex：vuex(Vue.js 的状态管理模式)。

(6) CSS Pre-processors：CSS 预处理器(如 less、sass)。

(7) Linter/Formatter：代码风格检查和格式化(如 ESlint)。

(8) Unit Testing：单元测试(Unit Tests)。

(9) E2E Testing：E2E(End to End)测试。

通过方向键移动，使用空格键进行选择或取消，选中需要的选项。建议选中 Babel、Router、Vuex、CSS Pre-processors 等选项。

通过方向键和空格键，选中合适的选项，按 Enter 键，会出现如图 12.17 所示的界面，询问是否采用 history 路由模式。输入 Y 或 n 表示使用或不使用。

```
Vue CLI v3.0.4
? Please pick a preset: Manually select features
? Check the features needed for your project: Babel, Router, Vuex, CSS Pre-processors
? Use history mode for router? (Requires proper server setup for index fallback in production) (Y/n)
```

图 12.17　选择路由模式

输入 Y 后按 Enter 键，会出现如图 12.18 所示的界面，提示选择 CSS pre-processor 样式预处理器，可以使用方向键选择合适的预处理器。

```
Vue CLI v3.0.4
? Please pick a preset: Manually select features
? Check the features needed for your project: Babel, Router, Vuex, CSS Pre-processors
? Use history mode for router? (Requires proper server setup for index fallback in production) Yes
? Pick a CSS pre-processor (PostCSS, Autoprefixer and CSS Modules are supported by default): (Use arrow keys)
  Sass/SCSS
  Less
  Stylus
```

图 12.18　选择 CSS 预处理器

用方向键选中 CSS 预处理器后按 Enter 键，进入如图 12.19 所示的界面，选择配置文件的保存方式。

```
Vue CLI v3.0.4
? Please pick a preset: Manually select features
? Check the features needed for your project: Babel, Router, Vuex, CSS Pre-processors
? Use history mode for router? (Requires proper server setup for index fallback in production) Yes
? Pick a CSS pre-processor (PostCSS, Autoprefixer and CSS Modules are supported by default): Sass/SCSS
? Where do you prefer placing config for Babel, PostCSS, ESLint, etc.? (Use arrow keys)
  In dedicated config files
  In package.json
```

图 12.19　配置文件的保存方式

(1) In dedicated config files：单独保存在各自的配置文件中。

(2) In package. json：保存到 package. json 文件中。

选中配置文件的保存方式后按 Enter 键，进入如图 12.20 所示的界面，询问是否保存当前的选择信息，如果输入 y，则提示输入保存的文件名称，如果输入 N，则表示不保存。

```
Vue CLI v3.0.4
? Please pick a preset: Manually select features
? Check the features needed for your project: Babel, Router, Vuex, CSS Pre-processors
? Use history mode for router? (Requires proper server setup for index fallback in production) Yes
? Pick a CSS pre-processor (PostCSS, Autoprefixer and CSS Modules are supported by default): Sass/SCSS
? Where do you prefer placing config for Babel, PostCSS, ESLint, etc.? In package.json
? Save this as a preset for future projects? (y/N)
```

图 12.20　是否保存配置选项

等待 Vue-CLI 完成项目初始化，结果如图 12.21 所示。

```
□ Running completion hooks...

� Generating README.md...

� Successfully created project vue02.
� Get started with the following commands:

$ cd vue02
$ npm run serve
```

图 12.21　完成项目初始化

根据提示，进入 vue02 目录，执行 npm run serve 命令，启动项目，用浏览器访问指定的链接，会出现如图 12.22 所示的预览效果。

图 12.22　预览测试页面

（1）dist：用于存放使用 npm run build 命令打包的项目文件。

（2）node_modules：用于存放项目的各种依赖。

（3）public：用于存放静态资源。

（4）public/index.html：是一个模板文件，其作用是生成项目的入口文件，这样浏览器访问项目的时候就会默认打开生成的 index.html 文件。

（5）src：是存放各种 Vue.js 文件的地方。

（6）src/assets：用于存放各种静态文件，例如图片。

（7）src/components：用于存放公共组件，例如 header、footer 等。

（8）src/router/index.js：vue-router 路由文件。需要引入 src/views 文件夹下的.vue，以及配置 path、name、component。

（9）src/store/index.js：是 Vuex 的文件，主要用于项目里的一些状态保存，例如 state、mutations、actions、getters、modules。

（10）src/views：用于存放已写好的各种页面，例如 Login.vue 和 Home.vue。

（11）src/App.vue：是主 Vue.js 模块，主要使用 router-link 引入其他模块，App.vue 是项目的主组件，所有的页面都是在 App.vue 下切换的。

（12）src/main.js：入口文件，其主要作用是初始化 Vue.js 实例，同时可以在此文件中引用某些组件库或者全局挂载一些变量。

（13）.gitignore：配置通过 git 上传时想要忽略的文件格式。

（14）babel.config.js：是一个工具链，主要用于在当前和版本较低的浏览器或环境中将 ES6 的代码转换为向后兼容（兼容低版本 ES）。

（15）package.json：模块基本信息项目开发所需要的模块、版本、项目名称。

（16）package-lock.json：是在执行 npm install 命令的时候生成的文件，用于记录当前状态下实际安装的各个 npm package 的具体来源和版本号。

实测发现，在基于 Vue.js 脚手架创建项目的时候，往往依赖 Python 2，因此需要安装 Python 2 版本。使用 npm config 设置 Python，命令如下：

```
npm config set python <python 的安装路径>
```

有时还可能需要强制清除缓存，命令如下：

```
npm cache clean --force
```

1. 自定义环境变量和启动模式

除了可以使用 Vue-CLI 脚手架创建目录和文件外，开发人员根据自己的需求，可以在项目根路径下创建文件，以及自定义环境变量，以便在项目中使用。

文件名的格式代码如下：

```
.env                        # 在所有的环境中被载入
.env.local                  # 在所有的环境中被载入，但会被 git 忽略
.env.[mode]                 # 只在指定的模式中被载入
.env.[mode].local           # 只在指定的模式中被载入，但会被 git 忽略
```

文件内容，代码如下：

```
NODE_ENV = "dev"
VUE_APP_URL = "http://localhost:8088/api"
VUE_APP_* = 其他

在组件代码中获取值的方式
console.log(process.env.NODE_ENV)
console.log(process.env.VUE_APP_URL)
console.log(process.env.VUE_APP_*)
```

注意：自定义变量的 key 一定要以 VUE_APP_开头，否则无效。

在 package.json 文件的 scripts 的启动脚本中使用 --mode 指定使用哪个环境变量文件，代码如下：

```
...
  "scripts": {
    "serve": "Vue-CLI-service serve",
    "build": "Vue-CLI-service build",
    "local": "Vue-CLI-service serve --mode local"//.env.local 文件
  },
...
```

2. 删除创建项目时的配置选项

当开发人员在图 12.20 上选择保存当前配置的时候，Vue-CLI 会将当前的选项以配置

文件的方式保存下来,以便以后重复使用,不再需要重新选择。这些配置信息保存在当前登录用户目录的.vuerc文件中(笔者的当前用户目录是C:\Users\NobleYang)。.vuerc文件的内容如下:

```
{
  "presets": {
    "mydefault": {
      "useConfigFiles": false,
      "plugins": {
        "@vue/cli-plugin-babel": {}
      },
      "router": true,
      "routerHistoryMode": true,
      "vuex": true,
      "cssPreprocessor": "sass"
    }
  },
  "useTaobaoRegistry": true
}
```

presetts中保存的是配置选项,如在上面的代码中,就保存了一个名称为mydefault的历史选项。如果想删除某个选项,则删除对应名称和后面的内容即可。

第 13 章

Vuex 状态管理

在按模块化方式开发的前端项目中会有很多不同的组件，而且组件之间经常会存在数据通信问题，也就是状态的共享，甚至在共享状态的同时，还会存在必要的业务处理需求。本章介绍的 Vuex 可以有针对性地解析这些项目需求。

13.1 Vuex 简介

Vuex 是一个专为 Vue.js 应用程序开发的状态管理模式，它采用集中式存储管理应用的所有组件的状态，并以相应的规则保证状态以一种可预测的方式发生变化。Vuex 已被集成到 Vue.js 的官方调试工具 devtools extension(opens new window)，提供了诸如零配置的 time-travel 调试、状态快照导入/导出等高级调试功能。

13.1.1 状态管理模式

在一个 Vue.js 组件中状态管理一般分为三部分：state、view 和 actions。state 是数据源，一般定义在 Vue.js 组件的 data 属性中；view 以声明方式映射了 state 的视图；actions 是针对 view 状态变化的响应，一般是定义在 methods 中的方法。如具有状态管理功能的计算器代码如下：

```
new Vue({
  //state
  data () {
    return {
      count: 0
    }
  },
  //view
  template: `
    <div>{{ count }}</div>
  `,
  //actions
```

```
  methods: {
    increment() {
      this.count++
    }
  }
})
```

这样的数据流是单向的，如图13.1所示。

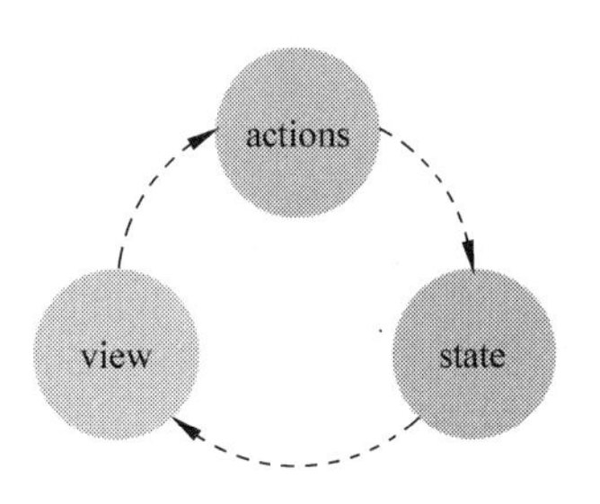

图13.1 单向数据流

这种单向数据流的状态管理比较简洁，在一个组件内部使用起来很方便，但是如果状态要实现多组件共享，则立刻会面对多个视图依赖于同一种状态和不同的视图行为，因此存在需要变更同一种状态的问题。

对于多个视图依赖同一种状态的问题，在父子组件之间，可以通过方法参数进行传递，但是会遇到烦琐的多层嵌套问题，而且还不能很好地解决兄弟组件之间的状态传递。对于多个视图会存在不同行为的问题，可以采用父子组件直接引用或者通过事件来变更和同步状态的多个副本，但是这样实现会造成代码很难维护。

有一种新的状态管理模式，将组件的共享状态抽取出来，以一个全局单例模式管理。在这种模式下，组件树构成了一个巨大的"视图"，不管在树的哪个位置，任何组件都能获取状态或者触发行为，并且通过定义和隔离状态管理中的各种概念以及通过强制规则维持视图和状态间的独立性，使代码变得更结构化且易维护。

Vuex就是基于这种新模式，专门为Vue.js设计的状态管理库，以利用Vue.js的细粒度数据响应机制进行高效的状态更新。Vuex状态管理模型如图13.2所示。

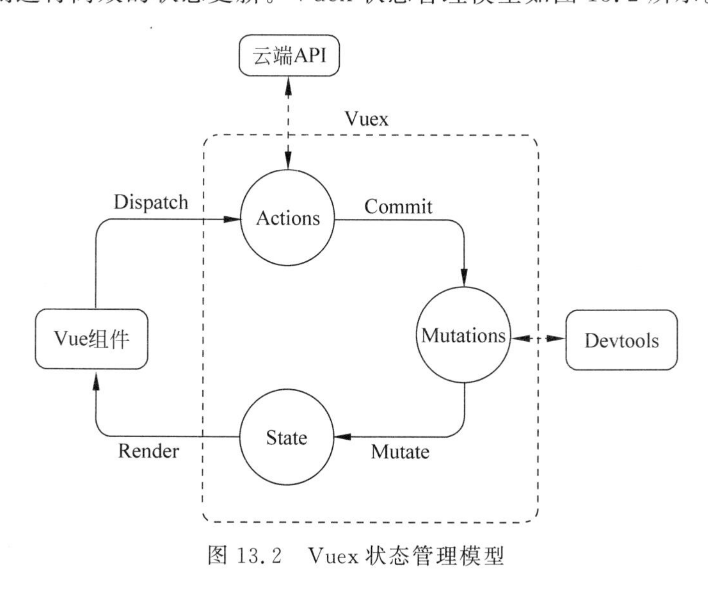

图13.2 Vuex状态管理模型

13.1.2 安装 Vuex

有两种方式可以安装 Vuex，一种是在页面中，使用 CDN 连接完成当前页面的 Vuex 的安装，另一种是在模块开发环境下，使用 npm 或 yarn 命令安装 Vuex 依赖。具体的样例代码如下：

```
//使用 CDN 连接引入安装
<script src = "/path/to/Vue.js"></script>
<script src = "/path/to/vuex.js"></script>

//npm 安装
npm install vuex -- save

//yarn 安装
yarn add vuex
```

在模块化开发环境的项目中，用 npm 或 yarn 命令安装好 Vuex 后，需要通过 Vue.use()方法注册 Vuex，这样才能在组件中使用 Vuex，代码如下：

```
import Vue from 'vue'
import Vuex from 'vuex'

Vue.use(Vuex)
```

需要注意的是，Vuex 依赖 Promise。如果浏览器没有实现 Promise，例如 IE，则需要使用 polyfill(填料)库，例如 es6-promise。可以使用 npm 或 yarn 命令安装 es6-promise 依赖，同时在使用 Vuex 的地方，使用 import 命令引入 es6-promise，代码如下：

```
//npm 安装
npm install es6 - promise -- save
//yarn 安装
yarn add es6 - promise

import 'es6 - promise'
import Vuex from 'vuex'
```

13.1.3 第 1 个案例

为体验 Vuex 的状态管理，接下来基于 Vuex 实现一个计算器，具体实现步骤如下。

1. 创建工程

基于 Vue-CLI 脚手架创建一个带有 Vuex 依赖的工程，详细的创建过程可以参考 12.3 节的内容。对应的样例工程是 vuebook/store。

在此工程里面，自动创建了 src/store.js 文件，文件里面包含了对 Vuex 的引用，并且 export 了基本的 Vuex.Store 对象。同时在 src/main.js 文件中，import 了 store 对象，并且全局注册到 Vue 中，代码如下：

```
//第 13 章/src/store.js

import Vue from 'vue'
import Vuex from 'vuex'

Vue.use(Vuex)

export default new Vuex.Store({
  state: {

  },
  mutations: {

  },
  actions: {

  }
})

//第 13 章 src/main.js
import Vue from 'vue'
import App from './App.vue'
import router from './router'
import store from './store'                    //导入 store 对象

Vue.config.productionTip = false

new Vue({
  router,
  store,                                        //注册 store 对象
  render: h => h(App)
}).$mount('#app')
```

2. 在 Vuex 中添加状态和更新状态的 action

在 src/store.js 文件中，添加记录数量的 count 属性和状态递增的 increment 行为，代码如下：

```
//第 13 章/src/store.js

import Vue from 'vue'
import Vuex from 'vuex'
```

```
Vue.use(Vuex)

export default new Vuex.Store({
  state: {
    count:0                              //计算状态
  },
  mutations: {
    increment(state){                    //更新状态的 action
      state.count++
    }
  },
  actions: {

  }
})
```

3. 添加 Counter 组件

在 src/components 目录下，创建 Counter.vue 组件，显示和计算 count 的值，同时添加事件，修改 count 的状态，代码如下：

```
//第 13 章/src/components/Counter.vue

<template>
  <div>
      <!-- 给 button 绑定事件,更新状态,同时显示 count 的值 -->
      <button @click="toIncrement">我被单击{{this.$store.state.count}}次</button>
  </div>
</template>

<script>
export default {
  methods: {
      toIncrement(){
          //提交 increment,更新 store 中的 count 状态
          this.$store.commit('increment');
      }
  }
}
</script>

<style>
</style>
```

4. 在应用中使用 Counter 组件

修改 src/App.vue 文件，在里面注册并且使用 Counter 组件，代码如下：

```
//第 13 章/src/App.vue

<template>
  <div id="app">
    <counter/>
  </div>
</template>

<script>
import Counter from '@/components/Counter.vue'

export default {
  name:'app',
  components: {
    Counter
  }
}
</script>
```

5. 启动应用测试

执行 npm run serve 命令，启动应用，如图 13.3 所示。

打开浏览器，访问应用，单击按钮，这样就可以查看单击次数的变化，如图 13.4 所示。

图 13.3　启动 store 样例应用

我被单击1次

图 13.4　状态页面

13.2 Vuex 核心概念

Vuex 有 5 个核心对象，分别是 state、getter、mutation、actions 和 module。state 对象用来定义和保存状态；getters 用来优化获取状态值的方式；mutation 和 action 用来更新状态，mutations 支持同步更新，actions 支持异步更新；module 用来将复杂的状态分割成不同的模块，每个模块有自己的 state、getter、mutation 和 action，方便状态的管理。接下来逐个介绍这些核心对象的特征和使用。

13.2.1 state

state 是一个单一状态树，用一个对象包含应用级别的全部状态，而且在整个应用中只有一个，并且是唯一的一个对象。Vuex 内部解决了单一状态树对象同多个模块可能出现的冲突。

因为 Vuex 的状态存储是响应式的，从 state 实例对象中获取状态的最简单方式就是在组件中定义计算属性，返回状态的值，代码如下：

```
//创建一个 Counter 组件
const Counter = {
  template: `<div>{{ count }}</div>`,
  computed: {
    count () {
      return store.state.count
    }
  }
}
```

每当 store. state. count 变化的时候，都会重新求取计算属性，并且触发与更新相关联的 DOM。

然而，这种模式导致组件依赖全局状态单一。在模块化的构建系统中，在每个需要使用 state 的组件中需要频繁地导入，并且在测试组件时需要模拟状态。Vuex 通过 store 选项提供了一种机制，可以将状态从根组件“注入”到每个子组件中（需调用 Vue. use（Vuex）），代码如下：

```
const app = new Vue({
  el: '#app',
  //把 store 对象提供给 "store" 选项，可以把 store 的实例注入所有的子组件
  store,
  components: { Counter },
  template: `
    <div class="app">
      <counter></counter>
    </div>
  `
})
```

通过在根实例中注册 store 选项，该 store 实例会注入根组件下的所有子组件中，并且子组件能通过 this. $store 访问。更新后的 Counter 组件的代码如下：

```
const Counter = {
  template: `<div>{{ count }}</div>`,
  computed: {
    count () {
      return this.$store.state.count
    }
  }
}
```

当一个组件需要获取多种状态的时候，将这些状态都声明为计算属性会有些重复和冗余。为了解决这个问题，开发人员可以使用 mapState 辅助函数生成计算属性。以下例子用于说明具体的使用。

在 state 对象中添加多种状态（本例中是 count 和 age 两种状态），代码如下：

```
import Vue from 'vue'
import Vuex from 'vuex'

Vue.use(Vuex)

export default new Vuex.Store({
  state: {
    count:0,                //第 1 种状态
    age:10                  //第 2 种状态
  },
  mutations: {
    increment(state){
      state.count++
    }
  },
  actions: {

  }
})
```

创建 MapState.vue 组件，使用 mapState 函数生成多个计算属性并且使用，代码如下：

```
<template>
  <div>
      count:{{count}}<br/>
      age:{{age}}<br/>
      localCount:{{localCount}}<br>
      computedCount:{{computedCount}}
  </div>
</template>

<script>
//引入 mapState 函数
import {mapState} from 'vuex'
export default {
  name:'MapState',
  data(){
    return{
        localCount:10
    }
```

```
    },
    computed:{
        //基于 mapState 生成 count、age 和 computedCount 共 3 个计算属性
        ...mapState({
            count: state => state.count,
            age: 'age',            //等同于 state => state.age
            computedCount(state){
                return this.localCount + state.count;
            }
        })
    }
}
</script>

<style>

</style>
```

在 App. vue 中注册使用 MapState 组件,运行测试效果如图 13.5 所示。

需要注意的是,虽然可以使用 Vuex 存放状态,也可以定义 action 更新状态,但这并不意味着将所有组件的所有状态都用 Vuex 管理。在理论上是可以实现的,但会造成代码变得冗余,并且不容易理解和维护。在实际项目中,开发人员需要根据实际情况,判断状态是属于单个组件的,还是由多个组件共享的,只需将多个组件共享的状态交给 Vuex 管理。

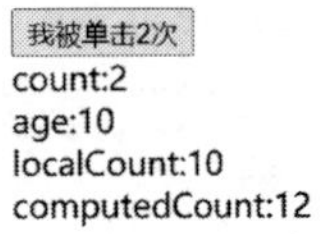

我被单击2次
count:2
age:10
localCount:10
computedCount:12

图 13.5 MapState 组件效果

13.2.2 getter

有时候,组件需要过去 state 中的派生属性,例如,需要获取 state 中的 count 和 age 属性的和。这时候,开发人员可以在每个组件中定义相同的函数进行计算,或者抽取一个共享函数,在需要用到的地方导入。不管用哪种方式,都不是很理想。这时候,可以在 Vuex 的 Store 对象中,定义一个 getter 属性,完成状态的计算后,返回计算结果,这样就可以在所有需要这个派生属性的地方,直接使用 getter 属性。定义在 Store 对象中的 getter 属性同组件中的计算属性一样,只有当依赖的属性发生变化时,才会重新计算,如果没有变化,则会直接返回缓存中的数据。

在 Store 中定义一个名为 countAge 的 getter 属性,返回 count 和 age 属性的和,代码如下:

```
//第 13 章/src/store.js
...
export default new Vuex.Store({
  state: {
    count:0,
```

```
    age:10
  },
  //定义 getters 属性
  getters: {
    countAge: state => state.count + state.age
  },
  ...
})
```

定义在 Store 中的 getter 会暴露在 store.getters 对象中，开发人员可以以属性的方式访问这些值。定义一个 getter 组件，里面的 countage 计算属性和 countAge 方法都通过 store.getters 属性获取 countAge 的 getter 属性，代码如下：

```
<template>
  <div>
      CountAge:{{countAge()}}->{{countage}}
  </div>
</template>

<script>
export default {
  name: 'Getter',
  computed:{
    //在计算属性中获取 getter
    countage(){
        return this.$store.getters.countAge
    }
  },
  methods: {
      //在方法中获取 getter
      countAge(){
          return this.$store.getters.countAge
      }
  }
}
</script>
```

在 Store 中定义 getter 属性时，可以用第 1 个参数接收 state 对象，也可以用第 2 个参数接收 getters 对象，定义一个名为 countAgeAndNamesLength 的 getter 属性，第 2 个参数传入的是 getters 对象，可以从 getters 对象中获取名为 namesLength 的 getter 的属性值，代码如下：

```
//第 13 章/src/store.js

export default new Vuex.Store({
```

```
...
  getters: {
    countAge: state => state.count + state.age,
    namesLength : state => state.names.length,
    //第 2 个参数传入 getters 对象
    countAgeAndNamesLength(state, getters){
      //返回 namesLength getter 的值
      return getters.namesLength;
    },
...
})
```

getter 除了可以返回普通数据外，也可以返回一个函数，这样就可以实现给一个 getter 传参数。定义一个名为 nameIndex 的 getter，返回一个带 name 参数的函数，在函数中，找出 name 在 state 中 names 数组的下标，代码如下：

```
...
export default new Vuex.Store({
  ...
  getters: {
    countAge: state => state.count + state.age,
    namesLength : state => state.names.length,
    countAgeAndNamesLength(state, getters){
      return getters.namesLength;
  },
  //返回带 name 参数的函数
    nameIndex: (state) =>(name) =>{
      for(let i = 0;i < state.names.length;i++){
        if(state.names[i] === name){
          return i
        }
      }
      return -1
    }
  },
  ...
})
```

在组件中，可以通过调用 getter 方法的方式，调用 getter 返回的函数，代码如下：

```
this.$store.getters.nameIndex('zhangsan')
```

注意：getter 在通过方法调用时，每次都会进行调用，不会使用缓存。

同 mapState 函数类似，Vuex 也提供了 mapGetters 辅助函数，能方便地将 store 中的 getters 映射成局部计算属性。使用 mapGetters 辅助函数，将 namesLength 和 nameIndex

两个 getter 分别映射成 namesLength 和 myNameIndex 计算属性，代码如下：

```
…
<script>
import { mapGetters } from 'vuex'
export default {
  name: 'Getter',
  …
  computed:{
    …
    ...mapGetters([
      'namesLength',

    ]),
    ...mapGetters({
        myNameIndex:'nameIndex'
    })
  },
  …
}
</script>
```

13.2.3 mutation

更改 Vuex 的 store 中的状态的唯一方法是提交 mutation。Vuex 中的 mutation 同事件类似：每个 mutation 都有一个字符串的事件类型（type）和一个回调函数（handler）。这个回调函数就是实际进行状态更改的地方，并且它会接受 state 作为第 1 个参数。在 store 对象的 mutations 中，定义一个 increment handler，代码如下：

```
const store = new Vuex.Store({
  state: {
    count: 1
  },
  mutations: {
    increment (state) {
      //变更状态
      state.count++
    }
  }
})
```

开发人员不能在组件中直接调用 increment handler，而是要同事件一样，调用 store.commit()函数，传入 increment 字符串，表示触发定义在 store 中的 Increment Mutation，代码如下：

```
store.commit('increment')
```

在基于 store.commit 触发字符串定义的 mutation 的同时，也可以通过第 2 个参数传入额外的参数，即 mutation 的载荷(payload)。在 store 定义一个能接收载荷的 increment，在组件中，基于 store.commint 触发 increment 的同时传入 10 载荷，代码如下：

```
//在 store 中定义带有载荷的 increment
mutations: {
  increment (state, n) {
    state.count += n
  }
}
//在组件中，提交一个带载荷的 increment
store.commit('increment', 10)
```

在大多数情况下，载荷应该是一个对象，这样可以包含多个字段并且记录的 mutation 会更易读，代码如下：

```
...
mutations: {
  increment (state, payload) {
    state.count += payload.amount
  }
}
store.commit('increment', {
  amount: 10
})
```

store.commit 也支持对象风格的提交方式，也就是在 commit 里面提交对象，包含 type 属性的对象和载荷，代码如下：

```
store.commit({
  type: 'increment',
  amount: 10
})
```

当使用对象风格的提交方式时，整个对象都作为载荷传给 mutation 函数，因此 handler 保持不变，代码如下：

```
mutations: {
  increment (state, payload) {
    state.count += payload.amount
  }
}
```

既然 Vuex 的 store 中的状态是响应式的，那么当变更状态时，监视状态的 Vue.js 组件也会自动更新。这也意味着 Vuex 中的 mutation 也需要与使用 Vue.js 一样，遵守以下注意事项。

(1) 最好提前在 store 中初始化所有所需属性。

(2) 当需要在对象上添加新属性时，应该使用 Vue.set(obj, 'newProp', 123)，或者以新对象替换老对象。例如以下代码使用了对象展开运算符：

```
state.obj = { ...state.obj, newProp: 123 }
```

使用常量替代 mutation 事件类型在各种 Flux 实现中是很常见的模式。这样可以使 linter 之类的工具发挥作用，同时把这些常量放在单独的文件中可以让代码合作者对整个 App 包含的 mutation 一目了然，示例代码如下：

```
//mutation-types.js
export const SOME_MUTATION = 'SOME_MUTATION'

//store.js
import Vuex from 'vuex'
import { SOME_MUTATION } from './mutation-types'

const store = new Vuex.Store({
  state: { ... },
  mutations: {
    //可以使用 ES2015 风格的计算属性命名功能来使用一个常量作为函数名
    [SOME_MUTATION] (state) {
      //mutate state
    }
  }
})
```

用不用常量取决于开发团队和开发人员——在需要多人协作的大型项目中，这会很有帮助，但如果开发团队成员不喜欢，则完全可以不这样做。

注意：mutation 必须是同步函数。

同 mapStates 和 mapGetters 类似，Vuex 也提供了 mapMutations 辅助函数，方便开发人员在组件中将 methods 映射成 store.commit 的调用，代码如下：

```
import { mapMutations } from 'vuex'

export default {
  ...
  methods: {
    ...mapMutations([
```

```
    'increment',         //将 `this.increment()` 映射为
                         //`this.$store.commit('increment')`

    //mapMutations 也支持载荷
    'incrementBy'        //将 `this.incrementBy(amount)` 映射为
                         //`this.$store.commit('incrementBy', amount)`
  ]),
  ...mapMutations({
    add: 'increment'     //将 `this.add()` 映射为
                         //`this.$store.commit('increment')`
  })
 }
}
```

13.2.4 action

action 同 mutation 一样,用来更新 store 对象中的 state,不同的是:

(1) action 提交的是 mutation,而不是直接变更状态。

(2) action 可以包含任意异步操作。

在 Store 对象中,注册一个 increment action,代码如下:

```
const store = new Vuex.Store({
  state: {
    count: 0
  },
  mutations: {
    increment (state) {
      state.count++
    }
  },
  actions: {
    increment (context) {
      context.commit('increment')
    }
  }
})
```

action 函数接收一个与 store 实例具有相同方法和属性的 context 对象,因此在 increment action 中调用 context.commit 提交一个 mutation,或者通过 context.state 和 context.getters 获取 state 和 getters。在这里需要特别注意的是,context 对象不是 store 实例。

实践中,经常用 ES2015 的参数解构来简化代码(特别是需要很多次调用 commit 的时候),代码如下:

```
actions: {
  increment ({ commit }) {
    commit('increment')
  }
}
```

在组件中，action 通过 store.dispatch()方法触发，代码如下：

```
store.dispatch('increment')
```

看上去感觉多此一举，直接分发 mutation 岂不更方便？实际上并非如此，mutation 必须同步执行，action 却不受约束。基于这点，可以在 action 内部执行异步操作，代码如下：

```
actions: {
  incrementAsync ({ commit }) {
    setTimeout(() => {
      commit('increment')
    }, 1000)
  }
}
```

action 支持用同样的载荷方式和对象方式进行分发，代码如下：

```
//以载荷形式分发
store.dispatch('incrementAsync', {
  amount: 10
})

//以对象形式分发
store.dispatch({
  type: 'incrementAsync',
  amount: 10
})
```

实际的购物车实现涉及调用异步 API 和分发多重 mutation，代码如下：

```
actions: {
  checkout ({ commit, state }, products) {
    //把当前购物车的物品备份起来
    const savedCartItems = [...state.cart.added]
    //发出结账请求，然后乐观地清空购物车
    commit(types.CHECKOUT_REQUEST)
    //购物 API 接收一个成功回调和一个失败回调
    shop.buyProducts(
      products,
```

```
        //成功操作
        () => commit(types.CHECKOUT_SUCCESS),
        //失败操作
        () => commit(types.CHECKOUT_FAILURE, savedCartItems)
      )
    }
  }
```

注意上面的代码中正在进行一系列的异步操作，并且通过提交 mutation 来记录 action 产生的副作用(状态变更)。

在组件中使用 this. $store. dispatch('xxx')分发 action，或者使用 mapActions 辅助函数将组件的 methods 映射为 store. dispatch 调用(需要先在根节点注入 store)，代码如下：

```
import { mapActions } from 'vuex'

export default {
  ...
  methods: {
    ...mapActions([
      'increment',          //将 `this.increment()` 映射为
//`this. $store.dispatch('increment')`

      //mapActions 也支持载荷
      'incrementBy'         //将 `this.incrementBy(amount)` 映射为
//`this. $store.dispatch('incrementBy', amount)`
    ]),
    ...mapActions({
      add: 'increment'      //将 `this.add()` 映射为
//`this. $store.dispatch('increment')`
    })
  }
}
```

action 通常是异步的，那么如何知道 action 什么时候结束呢？更重要的是，如何才能组合多个 action，以处理更加复杂的异步流程？

store. dispatch 可以处理被触发的 action 处理函数返回的 Promise，并且 store. dispatch 仍旧会返回 Promise，代码如下：

```
actions: {
  actionA ({ commit }) {
    return new Promise(resolve, reject) => {
      setTimeout(() => {
        commit('someMutation')
        resolve()
```

```
        }, 1000)
      }
    }
  }
```

这样，在组件方法中，可以用如下代码方式：

```
store.dispatch('actionA').then(() => {
  ...
})
```

同样，在另外一个 action 中也可以异步调用，代码如下：

```
actions: {
  ...
  actionB ({ dispatch, commit }) {
    return dispatch('actionA').then(() => {
      commit('someOtherMutation')
    })
  }
}
```

实战中，可以利用 async/await 组合 action，代码如下：

```
//假设 getData() 和 getOtherData() 返回的是 Promise

actions: {
  async actionA ({ commit }) {
    commit('gotData', await getData())
  },
  async actionB ({ dispatch, commit }) {
    await dispatch('actionA')          //等待 actionA 完成
    commit('gotOtherData', await getOtherData())
  }
}
```

注意：一个 store.dispatch 在不同模块中可以触发多个 action 函数。在这种情况下，只有当所有触发函数完成后，返回的 Promise 才会执行。

13.2.5 module

由于使用单一状态树，应用的所有状态会集中到一个比较大的对象。当应用变得非常复杂时，store 对象就有可能变得相当臃肿。

为了解决以上问题，Vuex 提供了一个机制，允许将 store 分割成模块(module)。每个模块拥有自己的 state、mutation、action、getter，甚至是嵌套子模块——从上至下进行同样方式的分割。在一个 Store 对象中，划分出了 A、B 两个模块，代码如下：

```
const moduleA = {
  state: () => ({ ... }),
  mutations: { ... },
  actions: { ... },
  getters: { ... }
}

const moduleB = {
  state: () => ({ ... }),
  mutations: { ... },
  actions: { ... }
}

const store = new Vuex.Store({
  modules: {
    a: moduleA,
    b: moduleB
  }
})

store.state.a // -> moduleA 的状态
store.state.b // -> moduleB 的状态
```

对于模块内部的 mutation 和 getter，接收的第 1 个参数是模块的局部状态对象，代码如下：

```
const moduleA = {
  state: () => ({
    count: 0
  }),
  mutations: {
    increment (state) {
      //这里的 state 对象是模块的局部状态
      state.count++
    }
  },

  getters: {
    doubleCount (state) {
      return state.count * 2
    }
  }
}
```

同样，对于模块内部的 action，局部状态通过 context. state 暴露出来，根节点状态则为 context. rootState，代码如下：

```
const moduleA = {
  ...
  actions: {
    incrementIfOddOnRootSum ({ state, commit, rootState }) {
      if (state.count + rootState.count) % 2 === 1) {
        commit('increment')
      }
    }
  }
}
```

对于模块内部的 getter，根节点状态会作为第 3 个参数暴露出来，代码如下：

```
const moduleA = {
  ...
  getters: {
    sumWithRootCount (state, getters, rootState) {
      return state.count + rootState.count
    }
  }
}
```

默认情况下，模块内部的 action、mutation 和 getter 注册在全局命名空间——这样使多个模块能够对同一 mutation 或 action 做出响应。

如果希望模块具有更高的封装度和复用性，则可以通过添加 namespaced: true 的方式使其成为带命名空间的模块。当模块被注册后，它的所有 getter、action 及 mutation 都会自动地根据模块注册的路径调整命名，代码如下：

```
const store = new Vuex.Store({
  modules: {
    account: {
      namespaced: true,

      //模块内容(module assets)
      state: () => ({ ... }),        //模块内的状态已经是嵌套的了
      //使用 namespaced 属性不会对其产生影响
      getters: {
        isAdmin() { ... }            //-> getters['account/isAdmin']
      },
      actions: {
        login() { ... }              //-> dispatch('account/login')
      },
```

```
        mutations: {
          login() { ... }              // -> commit('account/login')
        },

        //嵌套模块
        modules: {
          //继承父模块的命名空间
          myPage: {
            state:() => ({ ... }),
            getters: {
              profile() { ... }       // -> getters['account/profile']
            }
          },

          //进一步嵌套命名空间
          posts: {
            namespaced: true,

            state:() => ({ ... }),
            getters: {
              popular() { ... }       // -> getters['account/posts/popular']
            }
          }
        }
      }
    }
  })
```

启用了命名空间后的 getter 和 action 会收到局部化的 getter、dispatch 和 commit。换言之，在使用模块内容(Module Assets)时不需要在同一模块内额外添加空间名前缀，更改 namespaced 属性后不需要修改模块内的代码。

如果希望在 getter 中使用全局的 state 和 getter，则 rootState 和 rootGetters 可以通过 getter 的第 3 个和第 4 个参数传入；如果希望在 action 中使用全局的 state 和 getter，则 context 也应分别以 rootState 和 rootGetters 属性传入。

若需要在全局命名空间内分发 action 或提交 mutation，则可将{root: true}作为第 3 个参数传给 dispatch 或 commit，代码如下：

```
modules: {
  foo: {
    namespaced: true,

    getters: {
      //在这个模块的 getter 中,getters 被局部化了
      //可以使用 getter 的第 4 个参数来调用 rootGetters
```

```
      someGetter (state, getters, rootState, rootGetters) {
        getters.someOtherGetter // -> 'foo/someOtherGetter'
        rootGetters.someOtherGetter // -> 'someOtherGetter'
      },
      someOtherGetter: state => { ... }
    },

    actions: {
      //在这个模块中,dispatch 和 commit 也被局部化了
      //它们可以接收 root 属性,以访问根 dispatch 或 commit
      someAction ({ dispatch, commit, getters, rootGetters }) {
        getters.someGetter // -> 'foo/someGetter'
        rootGetters.someGetter // -> 'someGetter'

        dispatch('someOtherAction') // -> 'foo/someOtherAction'
        dispatch('someOtherAction', null, { root: true }) // -> 'someOtherAction'

        commit('someMutation') // -> 'foo/someMutation'
        commit('someMutation', null, { root: true }) // -> 'someMutation'
      },
      someOtherAction (ctx, payload) { ... }
    }
  }
}
```

若需要在带命名空间的模块中注册全局 action,则可添加 root: true,并将这个 action 的定义放在函数 handler 中,代码如下:

```
{
  actions: {
    someOtherAction ({dispatch}) {
      dispatch('someAction')
    }
  },
  modules: {
    foo: {
      namespaced: true,

      actions: {
        someAction: {
          root: true,
          handler (namespacedContext, payload) { ... } // -> 'someAction'
        }
      }
    }
  }
}
```

当使用 mapState、mapGetters、mapActions 和 mapMutations 这些函数来绑定带命名空间的模块时，写起来可能比较烦琐，代码如下：

```
computed: {
  ...mapState({
    a: state => state.some.nested.module.a,
    b: state => state.some.nested.module.b
  })
},
methods: {
  ...mapActions([
    'some/nested/module/foo', //-> this['some/nested/module/foo']
    'some/nested/module/bar' //-> this['some/nested/module/bar']
  ])
}
```

对于这种情况，可以将模块的空间名称字符串作为第 1 个参数传递给上述函数，这样所有绑定都会自动将该模块作为上下文，于是上面的例子可以简化为如下代码：

```
computed: {
  ...mapState('some/nested/module', {
    a: state => state.a,
    b: state => state.b
  })
},
methods: {
  ...mapActions('some/nested/module', [
    'foo', //-> this.foo)
    'bar' //-> this.bar)
  ])
}
```

也可以通过 createNamespacedHelpers 创建基于某个命名空间的辅助函数，它返回一个对象，对象里有新的绑定在给定命名空间值上的组件绑定辅助函数，代码如下：

```
import { createNamespacedHelpers } from 'vuex'

const { mapState, mapActions } = createNamespacedHelpers('some/nested/module')

export default {
  computed: {
    //在 some/nested/module 中查找
    ...mapState({
      a: state => state.a,
      b: state => state.b
```

```
    })
  },
  methods: {
    //在 some/nested/module 中查找
    ...mapActions([
      'foo',
      'bar'
    ])
  }
}
```

13.3　Vuex 进阶

前面对 Vuex 的基本知识点进行了逐个介绍，接下来介绍在项目中使用 Vuex 的通常策略和方式。

13.3.1　项目结构

Vuex 没有对项目代码结构做限制，但是，它规定了以下需要遵守的规则。

(1) 应用层级的状态应该集中到单个 store 对象中。

(2) 提交 mutation 是更改状态的唯一方法，并且这个过程是同步的。

(3) 异步逻辑都应该封装到 action 里面。

只要遵守以上规则，如何组织代码可随项目需求确定，但是如果 store 文件太大，则建议将 action、mutation 和 getter 分别保存到单独的文件中。

对于大型应用，往往会把与 Vuex 相关的代码分别放到模块中。下面是参考项目结构示例：

```
├── index.html
├── main.js
├── api
│   └──...                          # 抽取出 API 请求
├── components
│   ├── App.vue
│   └──...
└──store
    ├── index.js                    # 组装模块并导出 store
    ├── actions.js                  # 根级别的 action
    ├── mutations.js                # 根级别的 mutation
    └──modules
        ├── cart.js                 # 购物车模块
        └──products.js              # 产品模块
```

13.3.2 严格模式

在严格模式下，无论何时发生了状态变更且该变更不是由 mutation 函数引起的，系统都会抛出错误，这能保证所有的状态变更都能被调试工具跟踪到。开启严格模式很简单，仅需要在创建 store 的时候传入 strict：true，代码如下：

```
const store = new Vuex.Store({
  ...
  strict: true
})
```

一定不要在发布环境下启用严格模式！严格模式会深度监测状态树，以此来检测不合规的状态变更——需要确保在发布环境下关闭严格模式，以避免性能损失。可以让构建工具来处理这种情况，代码如下：

```
const store = new Vuex.Store({
  ...
  strict: process.env.NODE_ENV !== 'production'
})
```

13.3.3 表单处理

当在严格模式下使用 Vuex 时，在属于 Vuex 的 state 上使用 v-model 会比较棘手，代码如下：

```
<input v-model="obj.message">
```

假设这里的 obj 是在计算属性中返回的一个属于 Vuex store 的对象，在用户输入时，v-model 会试图直接修改 obj. message。在严格模式中，由于这个修改不是在 mutation 函数中执行的，所以这里会抛出一个错误。

用"Vuex 的思维"去解决这个问题的方法是：在<input>中绑定 value，然后侦听 input 或者 change 事件，在事件回调中调用一种方法，代码如下：

```
<input :value="message" @input="updateMessage">
...
computed: {
  ...mapState({
    message: state => state.obj.message
  })
},
methods: {
  updateMessage (e) {
```

```
    this. $store.commit('updateMessage', e.target.value)
  }
}
```

对应的 mutation 函数代码如下：

```
...
mutations: {
  updateMessage (state, message) {
    state.obj.message = message
  }
}
```

必须承认，这样做比简单地使用“v-model＋局部状态”要烦琐得多，并且也损失了一些 v-model 中很有用的特性。另一种方法是使用带有 setter 的双向绑定计算属性，代码如下：

```
< input v - model = "message">
...
computed: {
  message: {
    get () {
      return this. $store.state.obj.message
    },
    set (value) {
      this. $store.commit('updateMessage', value)
    }
  }
}
```

13.3.4 热重载

Vuex 支持在开发过程中热重载 mutation、module、action 和 getter。对于 mutation 和模块，需要使用 store.hotUpdate()方法，代码如下：

```
//store.js
import Vue from 'vue'
import Vuex from 'vuex'
import mutations from './mutations'
import moduleA from './modules/a'

Vue.use(Vuex)

const state = { ... }
```

```
const store = new Vuex.Store({
  state,
  mutations,
  modules: {
    a: moduleA
  }
})

if (module.hot) {
  //使 action 和 mutation 成为可热重载模块
  module.hot.accept(['./mutations', './modules/a'], ) => {
    //获取更新后的模块
    //因为 Babel 6 的模块编译格式问题,这里需要加上.default
    const newMutations = require('./mutations').default
    const newModuleA = require('./modules/a').default
    //加载新模块
    store.hotUpdate({
      mutations: newMutations,
      modules: {
        a: newModuleA
      }
    })
  }
}
```

如果仅使用模块,则可以使用 require.context 来动态地加载或热重载所有的模块,代码如下:

```
//store.js
import Vue from 'vue'
import Vuex from 'vuex'

//加载所有模块
function loadModules() {
  const context = require.context("./modules", false, /([a-z_]+)\.js$/i)

  const modules = context
    .keys()
    .map(key) => ({ key, name: key.match(/([a-z_]+)\.js$/i)[1] })
    .reduce(
      (modules, { key, name }) => ({
        ...modules,
        [name]: context(key).default
      }),
      {}
```

```
    )

    return { context, modules }
  }

  const { context, modules } = loadModules()

  Vue.use(Vuex)

  const store = new Vuex.Store({
    modules
  })

  if (module.hot) {
    //在任何模块发生改变时进行热重载
    module.hot.accept(context.id, ) => {
      const { modules } = loadModules()

      store.hotUpdate({
        modules
      })
    })
  }
```

13.4 安装初始化案例

为了方便理解，读者可以先将随书提供的代码初始化后，运行起来，看一看效果。以下将逐个介绍随书提供的代码文档和初始化运行过程。

13.4.1 案例代码介绍

本书提供的案例代码文档一共有4个目录和一个文件，分别是：

（1）＋vue-element-admin-master-old：包含权限代码的前端工程。

（2）＋vue-element-admin-master-resource：vue-element-admin前端模板工程（后面详细介绍）。

（3）＋VueElement_Authority：后端Java工程。

（4）＋redis：Redis数据库。

（5）authsys_db.sql：MySQL数据库脚本。

13.4.2 初始化数据库

打开cmd窗口，通过cd命令进入redis目录，执行redis-server./redis.windows.conf

命令，启动 Redis 数据库，如图 13.6 所示。

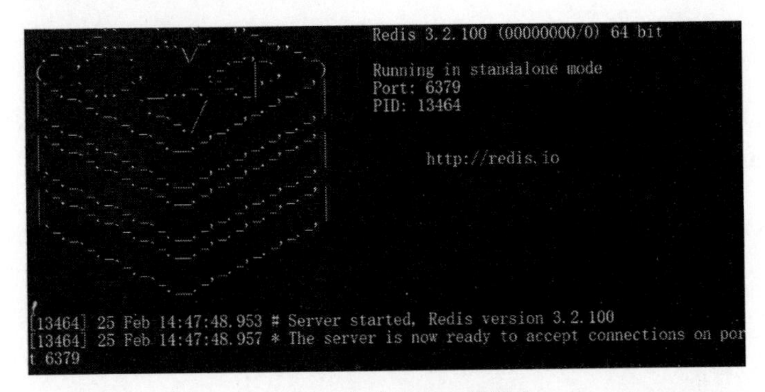

图 13.6　启动 Redis 数据库

创建 authsys_db 数据库，基于 authsys_db. sql 脚本初始化数据库。后端项目的数据库用户名和密码分别是 root 和 123456，读者可以自己设置不同的用户名和密码，只是需要同步后端项目 application. yml 文件里面的 username 和 password。

13.4.3　用 IDEA 打开后端工程

用 IDEA 打开 VueElement_Authority 目录中的后端工程，如图 13.7 所示。

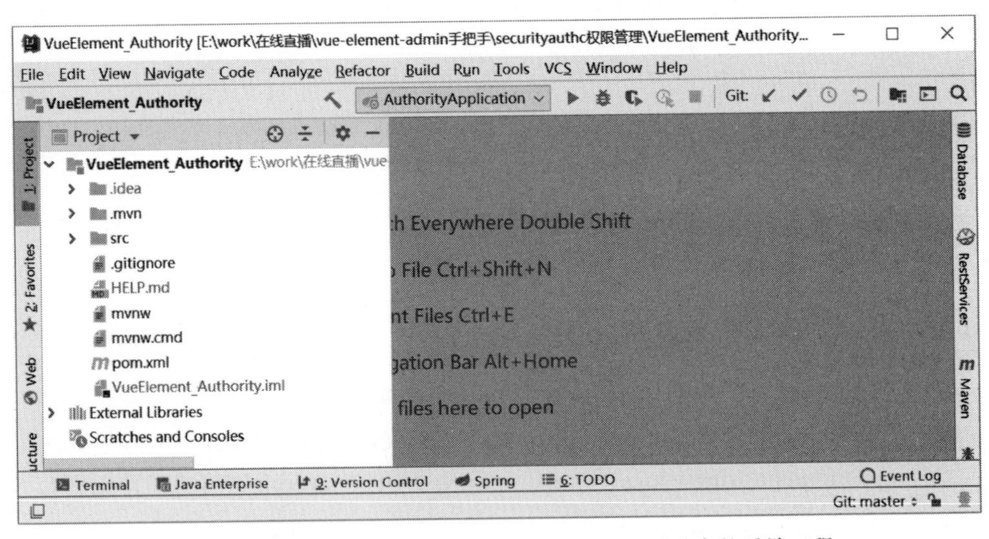

图 13.7　用 IDEA 打开 VueElement_Authority 目录中的后端工程

注意，作者当前使用的是 JDK 1.8，建议读者也使用 JDK1.8 或以上版本。同时，请在 IDEA 中安装 lombok 插件，否则 ResponseResult 里面没有成员变量的 setter 和 getter 方法，因此会有部分代码无法通过编译。当然，读者也可以直接在 cn. com. authority. commons. ResponseResult. java 文件中，添加所有成员变量的 setter 和 getter 方法，来解决部分类无法通过编译的问题。

13.4.4 用 VS Code 打开前端代码

在使用 VS Code 打开前端代码前，需要先安装好 Node.js 和 git，本书案例使用的 Node.js 的版本是 14.16.0，git 的版本是 2.21.0。接着使用 VS Code 打开 vue-element-admin-master-old 目录下的项目，如图 13.8 所示。

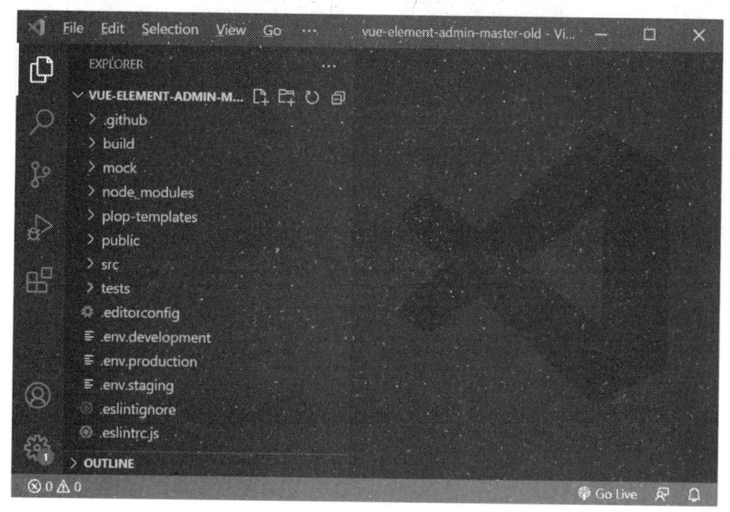

图 13.8　打开 vue-element-admin-master-old 目录下的项目

打开 Terminal 窗口，执行 npm install 命令，安装依赖。依赖需要在线下载，下载速度可能比较慢。如果有部分依赖下载时报错，则需重试几次，或将 node_modules 复制到 vue-element-admin-master-old 目录。node_modules 目录是本书案例下载用到的依赖。

13.4.5 启动测试

启动前后端代码，用浏览器打开前端登录页面，如图 13.9 所示。

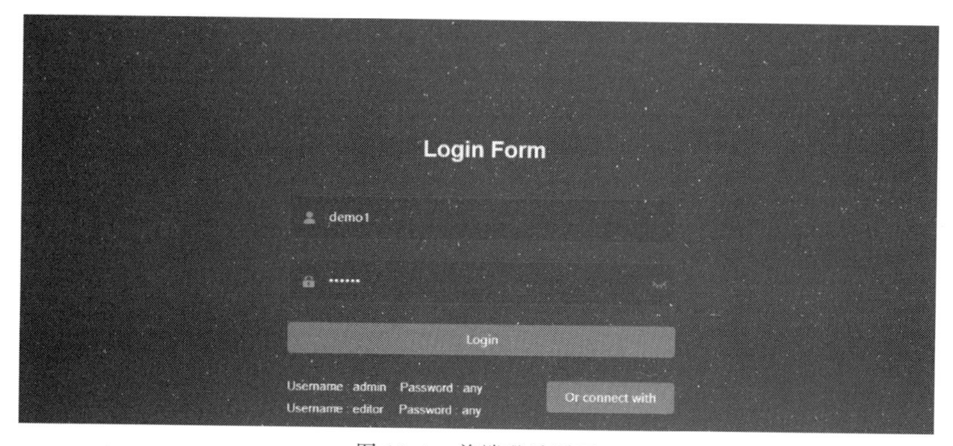

图 13.9　前端登录页面

输入用户名和密码(demo1 和 123456)登录,如图 13.10 所示。

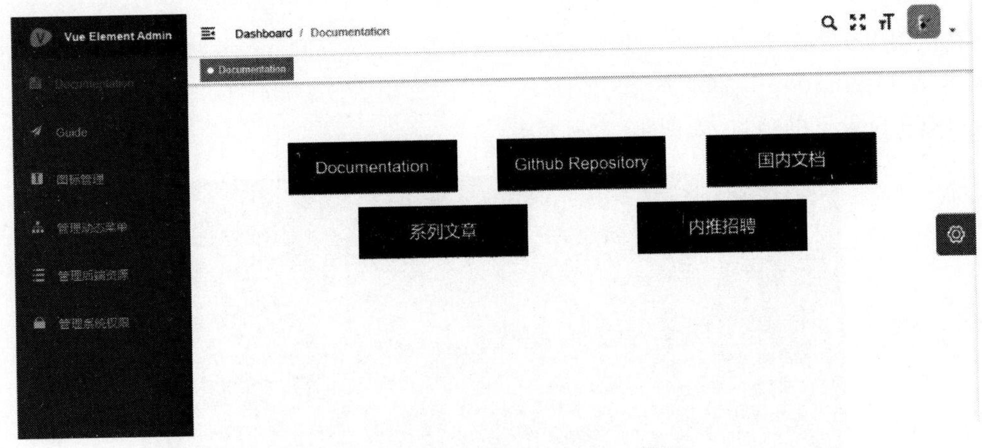

图 13.10 Vue Element Admin 首页

第四篇　Vue.js实战

第 14 章 ShopApp 实战

ShopApp 实战样例给前端开发人员提供了一个开发电商前端的参考。内容包含从软件安装、环境搭建到页面设计和实现思路，以及 Mock 提供的模拟数据，基本涵盖了前端应用开发的整个过程。读者可以从头到尾一步步地按笔者的写作思路通关，也可以根据自己的需求，挑选自己需要的章节阅读。

14.1 准备

14.1.1 安装软件

安装 Node.js(14.16.0 版本)，在安装的过程中，注意选择自动安装相关的依赖工具。

设置淘宝镜像，代码如下：

```
npm config set registry https://registry.npm.taobao.org
npm install -g cnpm --registry=https://registry.npm.taobao.org
```

安装 webpack、webpack-cli 和 Vue-CLI，代码如下：

```
npm install -g webpack webpack-cli
npm install -g @vue/cli@3.0.4
```

检测安装效果，代码如下：

```
webpack -v                //查看 webpack 的版本
vue -V                    //查看 Vue-CLI 的版本
```

14.1.2 创建项目

进入前端工作空间目录，执行 vue create shop-app 命令，选择 Babel、Router、Vuex、CSS Pre-processors 等选项，如图 14.1 所示，完成创建项目。

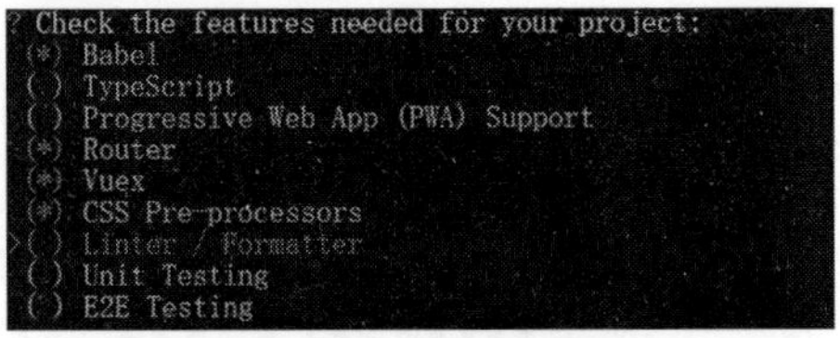

```
? Check the features needed for your project:
 (*) Babel
 ( ) TypeScript
 ( ) Progressive Web App (PWA) Support
 (*) Router
 (*) Vuex
 (*) CSS Pre-processors
>( ) Linter / Formatter
 ( ) Unit Testing
 ( ) E2E Testing
```

图 14.1 选择项目模块

14.1.3 调整项目结构

对使用 vue create 命令创建好的项目目录和文件做如下调整。

添加 vue.config.js 配置文件。在项目根目录下，新建 vue.config.js 文件，代码如下：

```
//目录地址
const path = require('path')

module.exports = {
  //对象和函数都可以,如果要控制开发环境,则可以选择函数
  configureWebpack:{
      resolve:{
        alias:{
          'assets':path.resolve('./src/assets'),
          'common':path.resolve('./src/common'),
          'components':path.resolve('./src/components'),
          'network':path.resolve('./src/network'),
          'views':path.resolve('./src/views')
        }
      }
    },
    devServer: {
      //host: '192.168.5.1',    //将浏览地址的本地服务修改为本机 IP 地址,用于在手机上测试
      //port: 8081,                    //端口号
      disableHostCheck: true,        //设置内网穿透
      proxy: {
        //配置跨域
          '/apis': {
              target: 'http://pv.sohu.com',
              ws: true,
              changOrigin: true,             //允许跨域
              pathRewrite: {
                  '^/apis': ''               //请求的时候使用这个 API 即可
              }
          },
      }
  },
```

```
  //修改打包路径规则
    //publicPath: './',
    //outputDir: 'dist',
    //devServer: {
    //proxy: {
    //  '/api': {
    // target: '',
    // ws: true,
    // changeOrigin: true
    //  }
    //}
    //}
  }
```

调整 store 文件。在 src 下创建 store 目录，将 src 目录下自动生成的 store.js 文件复制到 store 目录下，重命名为 index.js，代码如下：

```
import Vue from 'vue'
import Vuex from 'vuex'

Vue.use(Vuex)

export default new Vuex.Store({
  state: {

  },
  mutations: {

  },
  actions: {

  }
})
```

调整 router 文件。在 src 目录下，创建 router 目录，将 src 目录下自动生成的 router.js 文件复制到 router 目录下，重命名为 index.js，代码如下：

```
import Vue from 'vue'
import Router from 'vue-router'
import Home from '../views/Home.vue'

Vue.use(Router)

const routes = [
]
```

```
export default new Router({
  mode: 'history',
  base: process.env.BASE_URL,
  routes
})
```

在 package.json 文件中添加依赖，代码如下：

```
...
  "dependencies": {
    "babel-plugin-import": "^1.13.3",
    "core-js": "^2.6.5",
    "element-ui": "^2.4.5",
    "vant": "^2.8.0",
    "vue": "^2.6.11",
    "vue-cookies": "^1.7.0",
    "vue-lazyload": "^1.3.3",
    "vue-router": "^3.1.5",
    "vuex": "^3.1.2"
  },
  "devDependencies": {
    "@vue/cli-plugin-babel": "^3.0.4",
    "@vue/cli-service": "^3.0.4",
    "node-sass": "^4.12.0",
    "postcss-px-to-viewport": "^1.1.1",
    "sass-loader": "^8.0.0",
    "vue-template-compiler": "^2.6.10"
  }
...
```

在 assets 目录下，添加必要的 CSS、IMG 和 JS 文件，可从样例工程中复制。

14.1.4 安装项目依赖

执行如下命令：

```
npm install
```

安装完成后应确保没有错误。

14.2 开发前端

14.2.1 调整入口代码

修改 src/App.vue 里面的内容，代码如下：

```
<template>
  <div id = "appHome">
      <!-- exclude = "Detail" -->
      <keep-alive>
          <router-view v-if = " $route.meta.keepAlive"/>
      </keep-alive>
      <router-view v-if = "! $route.meta.keepAlive"/>
      <main-tab-bar></main-tab-bar>
  </div>
</template>

<script>
import MainTabBar from 'components/content/mainTabBar/MainTabBar'

export default {
  name:'app',
  components :{
      MainTabBar
  },
  methods:{

  }

}
</script>

<style lang = "scss">
  @import 'assets/css/base.css';
  #app {
    margin: 0;
  }
</style>
```

14.2.2　实现 TabBar

本小节要实现 ShopApp UI 的底部导航栏，效果如图 14.2 所示。

图 14.2　ShopApp 底部导航栏

通过分析，这部分 UI 可以分解成 MainTabBar、TabBar 和 TabBarItem 共 3 个组件。接下来逐个介绍这 3 个组件的创建和使用。

1. 编写 TabBarItem 组件

创建 src/components/common/tabbar/TabBarItem.vue 文件，代码如下：

```
<template>
  <div class="tab-bar-item" @click="itemClick">
      <!-- item 图标 -->
      <div v-if="!isActive">
          <!-- 普通图标插槽 -->
          <slot name='item-icon'></slot>
      </div>
      <div v-else>
          <!-- 激活图标插槽 -->
          <slot name='item-icon-active'></slot>
      </div>
      <!-- 图标文字 -->
      <div :style='activeStyle'>
          <!-- 图标文本插槽 -->
          <slot name="item-text"></slot>
      </div>
  </div>
</template>

<script>
export default {
  name: 'TabBarItem',
  props: {                      //定义子组件接收父组件传值的属性
      path: String,             //path属性,用来接收单击 TabBarItem 组件后的路由 path
      activeColor: {            //集合颜色,用于定义 TabBarItem 集合状态的颜色
          type: String,
          default: 'red'
      }
  },
  computed: {
      isActive() {              //是否激活 TabBarItem
          return this.$route.path.indexOf(this.path) !== -1
      },
      activeStyle() {           //激活样式
          return this.isActive ? {color:this.activeColor} : {}
      }
  },
  methods: {
      itemClick() {             //TabBarItem 的单击事件
          //this.$router.push(this.path)
      }
  }
}
</script>

<style>
```

```
.tab-bar-item{
  flex: 1;                              /* 长度均等 */
  text-align: center;                   /* 水平居中 */
  height: 49px;
  font-size: 14px;
}
.tab-bar-item img{
  width: 24px;
  height: 24px;
  margin-top: 3px;
  vertical-align: middle;
  margin-bottom: 2px;
}
</style>
```

2. 编写 TabBar 组件

创建 src/components/common/tabbar/TabBar.vue 文件,代码如下:

```
<template>
    <div id="tab-bar">
        <!-- 插槽,用来动态插入 TabBarItem 组件 -->
        <slot></slot>
    </div>
</template>

<script>
export default {
    name: 'TabBar'
}
</script>

<style>
#tab-bar{
  display: flex;                                              //弹性布局容器
  background: #f6f6f6;
  position: fixed;                                            //固定定位
  right: 0;
  left: 0;
  bottom: 0;
  box-shadow: 0px -2px 1px rgba(100, 100, 100, 0.15);  //添加 box 阴影
}
</style>
```

3. 编写 MainTabBar 组件

创建 src/components/content/mainTabBar/MainTabBar.vue 文件,代码如下:

```
<template>
    <tab-bar>
        <!--首页项-->
        <tab-bar-item path='/home' activeColor='red'>
            <img slot="item-icon" src="@/assets/img/tabbar/home.png" alt="">
            <img slot="item-icon-active" src="@/assets/img/tabbar/home1.png" alt="">
            <div slot="item-text">首页</div>
        </tab-bar-item>
        <!--分类-->
        <tab-bar-item path="/category">
            <img slot="item-icon" src="@/assets/img/tabbar/shop.png" alt="">
            <img slot="item-icon-active" src="@/assets/img/tabbar/shop1.png" alt="">
            <div slot="item-text">商城</div>
        </tab-bar-item>
        <!--购物车-->
        <tab-bar-item path="/cart">
            <img slot="item-icon" src="@/assets/img/tabbar/cart.png" alt="">
            <img slot="item-icon-active" src="@/assets/img/tabbar/cart1.png" alt="">
            <div slot="item-text">购物车</div>
        </tab-bar-item>
        <!--个人中心-->
        <tab-bar-item path="/profile">
            <img slot="item-icon" src="@/assets/img/tabbar/my.png" alt="">
            <img slot="item-icon-active" src="@/assets/img/tabbar/my1.png" alt="">
            <div slot="item-text">我的</div>
        </tab-bar-item>
    </tab-bar>
</template>

<script>
    //引入 TabBar 组件
    import TabBar from 'components/common/tabbar/TabBar.vue'
    //引入 TabBarItem 组件
    import TabBarItem from 'components/common/tabbar/TabBarItem.vue'

    export default {
        name: 'MainTabBar',
        components: {
            TabBar, //注册 TabBar
            TabBarItem //注册 TabBarItem
        }
    }
</script>

<style>

</style>
```

4. 编写 AppHome 组件

创建 src/views/AppHome.vue 文件,代码如下:

```
<template>
    <div id = "appHome">
        <! -- exclude = "Detail" -->
        <keep-alive>
            <router-view v-if = "$route.meta.keepAlive"/>
        </keep-alive>
        <router-view v-if = "!$route.meta.keepAlive"/>

        <! -- 显示底部 TabBar/底部菜单 -->
        <main-tab-bar></main-tab-bar>
    </div>
</template>

<script>
//引入 MainTabBar 组件
import MainTabBar from 'components/content/mainTabBar/MainTabBar'

export default {
    name:'app',
    components :{
        MainTabBar //注册 MainTabBar
    }

}
</script>

<style lang = "scss">
  @import 'assets/css/base.css';
  #app {
    margin: 0;
  }
</style>
```

5. 添加路由信息

在 src/router/index.js 文件中添加首页路由,代码如下:

```
...
//动态引入 AppHome 组件
const AppHome = () => import("../views/AppHome.vue");

const routes = [
  { path: "", redirect: "/home" },
```

```
  { path: "/", redirect: "/home" },
  //定义 home 路由
  {
    path: "/home",
    component: AppHome,
  }
]
...
```

6. 测试

运行 npm run serve 命令,结果如图 14.3 所示。

图 14.3 ShopApp 底部导航栏

14.2.3 实现 Home

Home View 用来显示商品列表,里面的功能和组件比较多,在这里分两步进行介绍:第一步,实现 Home View 中的导航和 Search;第二步,实现商品列表。

先介绍如何实现 HomeView 中的导航和 Search。

1. 定义 NavBar 组件

创建 src/components/common/navbar/NavBar.vue 文件,代码如下:

```
<template>
    <div class = "nav - bar">
```

```
        <div class = "left"><slot name = "left"></slot></div>
        <div class = "center"><slot name = "center"></slot></div>
        <div class = "right"><slot name = "right"></slot></div>
    </div>
</template>

<script>
    export default {
        name:'NavBar'
    }
</script>

<style>
.nav-bar{
    display: flex;                          /* 定义成弹性盒子 */
    height: 44px;
    line-height: 44px;
    text-align: center;
    box-shadow: 0px 1px 1px rgba(100, 100, 100, 0.15)
}
.left,.right{
    width: 60px;                            /* 定义左右两边的宽度 */
}
.center{
    flex: 1;                                /* 自动填充中间 */
}
</style>
```

2. 引入 element ui 和 vant 中用到的组件

从样例项目中将 src/plugins 目录复制到当前目录下。在 element.js 文件中引入的是 element ui 的组件，而在 vant.js 文件中引入的是 vant 的组件。

在 src/main.js 文件中引入 element.js 和 vant.js 文件，代码如下：

```
...
//全局按需引入第三方组件库
import "./plugins/element.js";
import "./plugins/vant.js";
...
```

3. 创建 Home 组件

创建 src/views/home/Home.vue 文件，在 Home.vue 文件中使用 NavBar 组件，在首页显示 logo，同时定义 search 框，代码如下：

```
<template>
    <div id = "home">
```

```
        <nav-bar class="home-nav">
          <div slot="center">
            <img src="~assets/img/home/favicon.png" style="width:80%" height="49" alt="" />
          </div>
        </nav-bar>
        <div @click="toSearch">
          <el-input v-model="searchText" placeholder="请输入需要搜索的商品">
            <i slot="prefix" class="el-input__icon el-icon-search"></i>
          </el-input>
        </div>
      </div>
</template>

<script>
    //引入 NavBar 组件
    import NavBar from "components/common/navbar/NavBar.vue";
    export default {
        name: 'Home',
        data(){
            return {
                searchText:''
            }
        },
        components: {
            NavBar
        },
        methods: {
            toSearch(){
                console.log('to search ...')
            }
        }
    }
</script>

<!--定义样式,只在当前 Vue 中使用有效-->
<style scoped>
#home {
  /*padding-top: 44px; */
  height: 100vh;
  position: relative;
}
</style>
```

接下来实现商品列表，首先封装滚动组件，安装 better-scroll 插件，代码如下：

```
npm install better-scroll
```

创建 src/components/common/scroll/Scroll.vue 文件，代码如下：

```
<template>
    <div class = "wrapper" ref = "wrapper">
        <div class = "content">
            <slot></slot>
        </div>
    </div>
</template>

<script>
//引入 better - scroll
import BScroll from "better - scroll";

export default {
    name: "Scroll",
    props: {
        probeType: {
            type: Number,
            default: 0
        },
        pullUpLoad: {
            type: Boolean,
            default: false
        },
        handleToScroll: {
            type: Function,
            default: function() {}
        },
        handleToTouchEnd: {
            type: Function,
            default: function() {}
        }
    },
    data() {
        return {
            scroll: null            //定义 scroll 变量,用于存放 BScroll 对象
        };
    },
    mounted() {
        //DOM 更新后的回调函数
        this.$nextTick(() => {
            this.scroll = new BScroll(this.$refs.wrapper, {
                click: true,
                probeType: this.probeType,
                pullUpLoad: this.pullUpLoad,
```

```
            });
            //监听滚动的位置
            if (this.probeType === 2 || this.probeType === 3) {
                this.scroll.on("scroll", position => {
                    //触发外面的滚动事件,将当前的位置传递到外面
                    this.$emit("scroll", position);
                });
            }
            //监听上拉事件,以便加载更多信息
            if (this.pullUpLoad) {
                this.scroll.on("pullingUp", () => {
                    //触发外面的 pullingUp 事件
                    this.$emit("pullingUp");
                });
            }
            //监听下拉事件,以便刷新
            this.scroll.on("scroll", pos => {
                this.handleToScroll(pos);
            });

            this.scroll.on("touchEnd", pos => {
                this.handleToTouchEnd(pos);
            });
        })
    },
    methods: {
        //返回顶部
        scrollTo(x, y, time = 300) {
            this.scroll && this.scroll.scrollTo && this.scroll.scrollTo(x, y, time);
        },
        finishPullUp() {
            this.scroll.finishPullUp();
        },
        refresh() {
            this.scroll && this.scroll.scrollTo && this.scroll.refresh();
        }
    }
};
</script>

<style scoped>
.wrapper {
}
.content {
}
</style>
```

然后封装 GoodsListItem 组件，创建 src/components/content/goods/GoodsListItem.vue 文件，代码如下：

```
<template>
    <div class="goods-item" @click="itemClick">
        <img v-lazy="showImage" alt/>
        <div class="goods-num">已售{{goodsItem.goodsBuyNum | sellCountFilter}}件</div>
        <div id="goods-info">
            <p>{{goodsItem.goodsName}}</p>
            <span class="price">
                <span>￥</span>
                {{goodsItem.goodsPrice}}
            </span>
            <img class="sc-img" src="~assets/img/common/sc.png" alt />
            <span class="collect">{{goodsItem.goodsFav}}</span>
            <span class="old-price">原价：￥{{goodsItem.goodsOldPrice}}</span>
        </div>
    </div>
</template>

<script>
export default {
    name: 'GoodsListItem',
    props: {
        goodsItem: {
            type: Object,
            default() {
                return {}
            }
        }
    },
    //过滤方法
    filters: {
        sellCountFilter(value) {
            let rselt = value;
            if (value > 10000) {
                rselt = (rselt / 10000).toFixed(1) + "万";
            }
            return rselt;
        },
    },
    computed: {
        showImage() {
            return this.goodsItem.goodsLogo;
        },
    },
```

```
    methods: {
        itemClick() {
            //console.log('单击了 goodsItem')
        },
    },
};
</script>

<style scoped>
    .goods-item {
      position: relative;
      padding-bottom: 52px;
      width: 48%;
      box-shadow: 0 2px 12px 0 rgba(0, 0, 0, 0.1);
      margin-bottom: 5px;
    }
    .goods-item img {
      width: 100%;
      border-radius: 5px;
    }
    #goods-info {
      font-size: 12px;
      position: absolute;
      bottom: 5px;
      left: 0;
      right: 0;
      overflow: hidden;
      text-align: center;
    }
    #goods-info .sc-img {
      position: relative;
      left: -4px;
      top: 2px;
      width: 14px;
      height: 14px;
    }
    #goods-info p {
      overflow: hidden;
      text-overflow: ellipsis;
      white-space: nowrap;
      margin-bottom: 3px;
    }
    #goods-info .price {
      color: var(--color-hight-text);
      margin-right: 20px;
    }
```

```
    #goods-info .collect {
      position: relative;
    }
    #goods-info .old-price {
      color: gainsboro;
      margin-right: 50px;
      text-decoration: line-through;
      font-size: 10px;
      float: left
    }
    .goods {
      height: 10px;
    }
    .goods-num {
      position: absolute;
      top: 170px;
      left: 0;
      right: 0;
      min-width: 30%;
      max-width: 40%;
      height: 1rem;
      line-height: 1rem;
      z-index: 3;
      color: white;
      text-align: left;
      background-size: 100%;
      font-size: 0.3rem;
      text-overflow: ellipsis;
      background: url("~assets/img/category/index.png") 0 no-repeat;
      padding-left: 5px;
    }
</style>
```

再封装 GoodsList 组件，创建 src/components/content/goods/GoodsList.vue 文件，代码如下：

```
<template>
  <div class="goods">
    <goods-list-item v-for="(item, index) in goods" :goods-item="item" :key=
"index"/>
  </div>
</template>

<script>
import GoodsListItem from './GoodsListItem'
```

```
export default {
    name:'GoodsList',
    components:{
        GoodsListItem
    },
    props:{
        goods:{
            type:Array,
            default(){
                return []
            }
        }
    }
}
</script>

<style scoped>
.goods{
    display: flex;
    flex-flow: wrap;
    padding: 3px;
    justify-content: space-around;
}
</style>
```

编写 goods API，创建 src/api/goods.js 文件，定义函数，提供模拟后台数据，代码如下：

```
//获取后台的商品列表
export function getGoodsList() {
    return [{
        "goodsId": 2,
        "goodsName": "小雏菊连衣裙女 2020 新款夏装两件套甜美超仙森系初恋网纱裙子",
        "goodsPrice": 119,
        "goodsFav": 164,
        "goodsOldPrice": "¥171.29",
        "goodsLogo": "/images/goods1.jpg",
        "goodsAre": "浙江省杭州市",
        "goodsBuyNum": 33,
        "goodsIntroduce": null,
        "isDel": 0,
        "topImages": null,
        "num": 0,
        "isFav": false,
        "hot": 75,
        "status": 0
},
...]
}
```

最后在 Home 中添加商品列表，编辑 src/views/home/Home.vue 文件，代码如下：

```
<template>
    <div id="home">
      <nav-bar class="home-nav">
        <div slot="center">
          <img src="~assets/img/home/favicon.png" style="width:80%" height="49" alt="" />
        </div>
      </nav-bar>
      <div @click="toSearch">
        <el-input v-model="searchText" placeholder="请输入需要搜索的商品">
          <i slot="prefix" class="el-input__icon el-icon-search"></i>
        </el-input>
      </div>
      <scroll
        class="content"
        ref="scroll"
        :probe-type="3"
        :pull-up-load="true"
        @pullingUp="loadMore"
        :handleToScroll="handleToScroll">
      <!-- 下拉信息提示 -->
      <ul v-if="isShowLoading">
        <li class="pullDown">
          <i class="el-icon-loading"></i>
          {{ pullDownMsg }}
        </li>
      </ul>
      <goods-list :goods='showGoods'></goods-list>

      </scroll>
    </div>
</template>

<script>
    //公共组件
    import NavBar from "components/common/navbar/NavBar.vue"
    import Scroll from "components/common/scroll/Scroll.vue"
    import GoodsList from "components/content/goods/GoodsList.vue"

    export default {
        name: 'Home',
        data(){
            return {
                //查询文本
                searchText:'',
```

```
                //是否显示下拉信息
                isShowLoading: false,
                //商品列表
                goodsList: [],
                //下拉提示信息
                pullDownMsg: ''
            }
        },
        components: {
            NavBar,
            Scroll,
            GoodsList
        },
        created(){
            //创建组件后,初始化商品列表
            this.getGoodsList()
        },
        computed: {
            //计算属性,监听商品列表的变化并实时显示
            showGoods() {
                return this.goodsList
            }
        },
        methods: {
            getGoodsList(){
                console.log('初始化商品列表')
            },
            toSearch(){
                console.log('to search ...')
            },
            //加载更多,上拉到最底部
            loadMore() {
                //查询更多的商品
                console.log('to search 更多的商品')
            },
            //下拉刷新,控制提示信息
            handleToScroll(pos) {
                // 下拉到顶部
                console.log('下拉到了顶部')
            }
        }
    }
</script>

<!-- 定义样式,只在当前 Vue.js 中使用有效 -->
<style scoped>
```

```
  #home {
    /* padding-top: 44px; */
    height: 100vh;
    position: relative;
  }
  .content {
    /* height: 300px; */
    position: absolute;
    top: 93px;
    bottom: 49px;
    left: 0;
    right: 0;
    /* background-color: #fff; */
    overflow: hidden;
  }
  .pullDown {
    margin: 0;
    padding: 16px;
    text-align: center;
    border: none;
  }
  .pullUp {
    margin: 0;
    text-align: center;
    border: none;
  }
  </style>
```

14.2.4 实现详细信息页面

商品详细信息页面包含路由、导航、图片轮播、商品基本信息、评论和商品详细信息等内容，接下来逐个介绍它们的实现。

1. 实现商品列表到商品详细页面的路由

创建一个空白的商品详细信息页面 Vue(src/views/detail/Detail.vue)，代码如下：

```
<template>
  <div>详细 UI </div>
</template>

<script>
export default {

}
</script>
```

```
<style>

</style>
```

在商品列表上添加路由事件，在 src/components/content/goods/GoodsListItem. vue 文件中，实现 itemClick()方法，代码如下：

```
...
    methods: {
        itemClick() {
            let path = '/detail/' + this.goodsItem.goodsId
            console.log('path:' + path)
            this.$router.push(path)
        },
    },
...
```

在 src/router/index. js 文件中，添加详细页面的路由信息，代码如下：

```
...
export default new Router({
  mode: 'history',
  base: process.env.BASE_URL,
  routes: [
    ...
    {
      path: '/detail/:googsId',
      name: 'detail',
      component: () => import('views/detail/Detail.vue')
    },
  ]
})
```

2. 实现详细组件导航

创建 src/views/detail/child/DetailNavBar. vue 文件，代码如下：

```
<template>
    <div class="detail-nvbar">
      <nav-bar>
        <!-- left 插槽插入返回图标 -->
        <div slot="left" class="back" @click="backClick">
          <img src="~assets/img/common/back.png" alt />
        </div>
        <!-- center 插槽插入详细页面里面的标签头 -->
```

```
        <div slot = "center" class = "title">
          <div
            v - for = "(item, index) in titles"
            class = "title - item"
            :class = "{active: index === currentIndex}"
            @click = "titleClick(index)"
          >{{item}}</div>
        </div>
      </nav - bar>
    </div>
</template>

<script>
import NavBar from "components/common/navbar/NavBar";

export default {
    name: 'DetailNavBar',
    components: {
        NavBar,
    },
    data() {
        return {
            titles: ["商品", "评论", "详情", "推荐"],
            currentIndex: 0,
        }
    },
    methods: {
        titleClick(index) {
            this.currentIndex = index;
            this. $emit("titleClick", index);
        },
        backClick() {
            this. $router.go( - 1);          //返回上一层
        },
        },
}
</script>

<style>
.detail - nvbar {
  position: relative;
  z - index: 8;
  background - color: #ffffff;
}
.title {
  display: flex;
```

```
  font-size: 14px;
}
.title-item {
  flex: 1;
}
.back img {
  width: 20px;
  height: 20px;
  margin-top: 12px;
  margin-left: 12px;
}
.active {
  color: var(--color-hight-text);
}
</style>
```

在 Detail 组件中添加 DetailNavBar，代码如下：

```
<template>
    <div class="detail">
        <!--加入导航栏，绑定 titleClick，单击标签时回传事件-->
        <detail-nav-bar @titleClick="titleClick" ref="detailBar"></detail-nav-
bar>
    </div>
</template>

<script>
    import DetailNavBar from './child/DetailNavBar'

    export default {
        name: 'Detail',
        components: {
            DetailNavBar
        },
        methods: {
            //跳转
            titleClick(index) {
                //单击了 index 标签
            },
        }
    }
</script>

<style scoped>

</style>
```

显示效果如图 14.4 和图 14.5 所示。

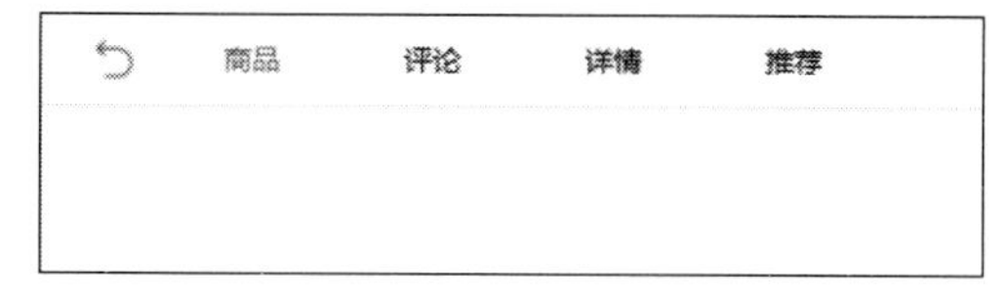

图 14.4 Home 页面顶部导航

图 14.5 Home 页面底部栏

3. 实现图片轮播

1）注册 element ui 的 Carousel 和 CarouselItem 组件

在 src/plugins/element.js 文件中注册 Carousel 和 CarouselItem 组件。

2）定义详细页面上的图片轮播组件

创建 src/views/detail/child/DetailSwiper.vue 文件，代码如下：

```
<template>
  <div>
    <el-carousel height="314px" trigger="click">
      <el-carousel-item v-for="(item, index) in topImages" :key="index">
        <img height="100%" width="100%" v-lazy="item" alt="">
      </el-carousel-item>
    </el-carousel>
  </div>
</template>

<script>
export default {
    name:'DetailSwiper2',
    props:{
      topImages:{
        type:Array,
        default(){
          return []
        }
      }
    },
}
</script>

<style>
.el-carousel__indicator .el-carousel__button {
  background: #faf9f9;
  border-radius: 50%;
  height: 14px;
  width: 14px;
}
```

```
.el-carousel__indicator.is-active .el-carousel__button {
  background: rgb(153, 25, 25);
}
</style>
```

3）在详细组件中添加图片轮播组件

修改 src/views/Detail.vue 文件，代码如下：

```
<template>
    <div class="detail">
        <detail-nav-bar @titleClick="titleClick" ref="detailBar"></detail-nav-bar>
        <!--添加滚动组件-->
        <scroll
            class="content"
            @scroll="contentScroll"
            ref="scroll"
            :probe-type="3"
            :pull-up-load="true"
            >
            <!--添加轮播组件,传入要轮播的图片数组 topImages-->
            <detail-swiper :top-images="topImages"></detail-swiper>

        </scroll>
    </div>
</template>

<script>
    import DetailSwiper from './child/DetailSwiper'
    import DetailNavBar from './child/DetailNavBar'
    import Scroll from "components/common/scroll/Scroll.vue"
    import { Toast } from "vant";
    import { getGoodsInfo } from 'api/goods.js'

    export default {
        name: 'Detail',
        components: {
            DetailSwiper,
            DetailNavBar,
            Scroll
        },
        data() {
            return {
                topImages: [],
                goodsId: 0,
```

```
            goodsInfo: {}
        }
    },
    created() {
        //获取路由参数 goodsId
        this.goodsId = this.$route.params.goodsId;
        //请求详情信息
        this.getGoodsInfo();
    },
    methods: {
        getGoodsInfo() {
            const toast = Toast.loading({
                duration: 0,                  //持续展示 toast
                message: "加载中...",
                forbidClick: true,
                loadingType: "spinner"
            });
            //往后台发请求,以便获取商品详细信息
            const goods = getGoodsInfo(this.goodsId)
            //获取商品详细信息
            this.goodsInfo = goods.info

            //获取商品信息中的轮播图片
            this.topImages = this.goodsInfo.topImages.split(',')
            this.topImages.forEach((img, index) => {
                this.topImages[index] = this.tinyImgPath(img) //'/images/' + img
            })

            Toast.clear();
        },
        //跳转
        titleClick(index) {
            //this.$refs.scroll.scrollTo(0, -this.topY[index], 800);
        },
        //滚动触发事件,控制在轮播图片上的拖动
        contentScroll(position) {
            let i = 0;
            if (-position.y >= 0 && -position.y < this.topY[1]) {
                i = 0;
            } else if (-position.y >= this.topY[1] && -position.y < this.topY[2]) {
                i = 1;
            } else if (-position.y >= this.topY[2] && -position.y < this.topY[3]) {
                i = 2;
            } else if (-position.y >= this.topY[3]) {
                i = 3;
            }
```

```
                this.currentIndex = i;
                this.$refs.detailBar.currentIndex = this.currentIndex;
                //1. 判断 BackTop 是否显示
                this.isShowBackTop = -position.y > 1000;
            },
            tinyImgPath(img){
                return '/images/' + img
            }
        }
    }
</script>

<style scoped>

</style>
```

4. 显示商品的基本信息

1）创建显示基本信息的组件

创建 src/views/detail/child/DetailBassInfo.vue 文件，代码如下：

```
<template>
    <div v-if="Object.keys(goods).length !== 0" class="base-info">
        <!-- 基本信息 -->
        <div class="info-title">{{goods.goodsName}}</div>
        <div class="info-price">
            <span class="n-price">￥{{goods.goodsPrice}}</span>
            <span class="o-price">{{goods.goodsOldPrice}}</span>
            <span v-if="goods.discount" class="discount">{{goods.discount}}</span>
        </div>
        <div class="info-other">
            <span>销量{{goods.goodsBuyNum}}</span>
            <span>收藏{{goods.goodsFav}}</span>
            <span>{{goods.goodsAre}}</span>
        </div>
        <!-- 售后服务 -->
        <div class="info-service">
            <span class="info-service-item" v-for="(item, index) in servesInfo" :key=
"index">
                <img :src="item.icon">
                <span>{{item.name}}</span>
            </span>
        </div>
    </div>
</template>
```

```
<script>
  export default {
    name:'DetailBassInfo',
    props:{
      goods:{
        type:Object,
        default(){
          return {}
        }
      },
      servesInfo:{
        type:Array,
        default(){
          return []
        }
      }
    }
  }
</script>

<style scoped>
    .base - info{
      margin - top: 15px;
      padding: 0 8px;
      color: #999;
      border - bottom: 5px solid #f2f5f8;
    }
    .info - title{
      color: #222;
      overflow: hidden;
      text - overflow:ellipsis;
    }
    .info - price{
      margin - top: 10px;
    }
    .info - price .n - price{
      font - size: 24px;
      color: var( -- color - hight - text);
    }
    .info - price .o - price{
      font - size: 13px;
      margin - left:5px;
      text - decoration: line - through;
    }
    .info - price .discount{
      font - size: 12px;
```

```
        padding: 2px 5px;
        color: #fff;
        border-radius: 8px;
        background-color: var(--color-hight-text);
        margin-left: 5px;
        position: relative;
        top: -12px;
      }
      .info-other{
        margin-top: 15px;
        line-height: 30px;
        display: flex;
        font-size: 13px;
        border-bottom: 1px solid rgba(100, 100, 100, 0.1);
        justify-content: space-between;
      }
      .info-service{
        display: flex;
        justify-content: space-between;
        line-height: 60px;
      }
      .info-service-item img{
        width: 14px;
        height: 14px;
        position: relative;
        top:2px;
      }
      .info-service-item span{
        font-size: 13px;
        color: #333;
      }
</style>
```

2）在 Detail 组件中，使用显示基本信息组件

修改 src/views/Detail.vue 文件，添加 DetailBassInfo 组件，以便显示基本信息，代码如下：

```
<template>
    <div class="detail">
        <detail-nav-bar @titleClick="titleClick" ref="detailBar"></detail-nav-bar>
        <scroll
            class="content"
            @scroll="contentScroll"
            ref="scroll"
            :probe-type="3"
```

```
            :pull-up-load="true"
            >
            <!-- 商品的轮播图片 -->
            <detail-swiper :top-images="topImages"></detail-swiper>
            <!-- 商品的详细信息 -->
            <detail-bass-info :goods="goodsInfo" :servesInfo="serves"></detail-bass-
info>
        </scroll>
    </div>
</template>

<script>
    import DetailSwiper from './child/DetailSwiper'
    import DetailNavBar from './child/DetailNavBar'
    import DetailBassInfo from './child/DetailBassInfo'           //引入 DetailBassInfo 组件
    import Scroll from "components/common/scroll/Scroll.vue"
    import { Toast } from "vant";
    import { getGoodsInfo } from 'api/goods.js'

    export default {
        name: 'Detail',
        components: {
            DetailSwiper,
            DetailNavBar,
            Scroll,
            DetailBassInfo                                          //注册 DetailBassInfo 组件
        },
        data() {
            return {
                topImages: [],
                goodsId: 0,
                goodsInfo: {},                                      //定义商品信息变量
                serves: []                                          //定义服务信息变量
            }
        },
        created() {
            //保存 goodsId
            this.goodsId = this.$route.params.goodsId;
            //请求详情信息
            this.getGoodsInfo();
        },
        methods: {
            getGoodsInfo() {
                const toast = Toast.loading({
                    duration: 0,                                    //持续展示 toast
                    message: "加载中...",
```

```
                forbidClick: true,
                loadingType: "spinner"
            });
            //往后台发请求,以便获取商品详细信息
            const goods = getGoodsInfo(this.goodsId)
            //获取商品详细信息
            this.goodsInfo = goods.info

            //获取商品信息中的轮播图片
            this.topImages = this.goodsInfo.topImages.split(',')
            this.topImages.forEach((img, index) => {
                this.topImages[index] = this.tinyImgPath(img) //'/images/' + img
            })
            //获取商品的服务信息
            this.serves = goods.serves
            this.serves.forEach( (serve, index) => {
                this.serves[index].icon = this.tinyImgPath(serve.icon)
            })

            Toast.clear();
        },
        ...
        tinyImgPath(img){
            return '/images/' + img
        }
    }
  }
</script>

<style scoped>

</style>
```

5. 显示评论

1）创建一个工具类

创建 src/common/utils.js 文件，代码如下：

```
//在时间前面追加 00
function padLeftZero(str) {
  return ("00" + str).substr(str.length);
}

//时间格式化
export function formatDate(date, fmt) {
    if (/(y+)/.test(fmt)) {
```

```
    fmt = fmt.replace(RegExp.$1, (date.getFullYear() + "").substr(4 - RegExp.
$1.length));
  }

  let o = {
    "M+": date.getMonth() + 1,
    "d+": date.getDate(),
    "h+": date.getHours(),
    "m+": date.getMinutes(),
    "s+": date.getSeconds()
  };

  for (let k in o) {
    if (new RegExp(`(${k})`).test(fmt)) {
      let str = o[k] + "";
      fmt = fmt.replace(RegExp.$1, RegExp.$1.length === 1 ? str : padLeftZero(str));
    }
  }

  return fmt;
}
```

2）添加评论组件

创建 src/views/detail/child/DetailCommentInfo.vue 文件，代码如下：

```
<!--评论组件-->
<template>
    <div>
      <!--有评论信息-->
      <div class="comment-info" v-if="Object.keys(commentInfo).length !== 0">
        <!--评论头-->
        <div class="info-header">
          <div class="header-title">买家评价 {{commentInfo.commentSum}}</div>
          <div class="header-more" @click="toMore">
            更多 >
            <i class="arrow-right" />
          </div>
        </div>
        <!-- 展示两条用户评论 -->
        <div v-for="(item,index) in commentInfo.comments" :key="index">
          <div v-if="item !== null">
            <!--发评人-->
            <div class="info-user">
              <img v-lazy="item.commentBody.userImg" alt />
              <span>{{ item.commentBody.userName }}</span>
```

```
            </div>
            <!-- 评论内容 -->
            <div class="info-detail">
              <p>{{ item.commentBody.content }}</p>
              <div class="info-imgs">
                <img :key="index" v-lazy="item" alt v-for="(item, index) in item.
commentImg" />
              </div>
              <div class="info-other">
                <span class="date">{{ item.commentBody.auditTime | showDate }}</span>
                <span>颜色:默认 尺码:默认</span>
              </div>
            </div>
          </div>
        </div>
      </div>
      <!-- 没有评论 -->
      <div class="comment-info" style="color: red;" v-else>暂无用户评论</div>
    </div>
</template>
<script>
//评论信息的数据结构
//const commentInfo = {
// sum: 2,
// comments: [
// {
//commentImg: ['图片'],
//commentBody: {
//     commentId: 0,
//     productId: 1,
//     orderId: 2,
//     userName: '用户名',
//     userImg: '用户头像',
//     imgUrl: '',
//     content: '评论内容',
//     fwValue: 5,
//     wlValue: 6,
//     auditTime: 'audit时间',
//     modifiedTime: '修改时间'
//}
// }
// ]
//}
  import { formatDate } from "@/common/utils";          //引入工具JS中的格式化日期函数
  export default {
    name: "DetailCommentInfo",
```

```
    props: {
      commentInfo: {
        type: Object,
        default() {
          return {};
        },
      },
    },
    filters: {
      showDate: function (value) {
        let date = new Date(value * 1000);
        return formatDate(date, "yyyy - MM - dd hh:mm:ss");
      },
    },
    methods: {
      toMore() {
        console.log(this.commentInfo)
      },
    },
  };
</script>

<style scoped>
  .comment - info {
    padding: 5px 12px;
    color: #333333;
    border - bottom: 5px solid #f2f5f8;
  }
  .info - header {
    line - height: 50px;
    height: 50px;
    border - bottom: 1px solid rgba(0, 0, 0, 0.1);
  }
  .header - title {
    font - size: 12px;
    float: left;
  }
  .header - more {
    font - size: 13px;
    float: right;
    margin - right: 10px;
  }
  .info - user {
    padding: 10px 0 5px;
  }
  .info - user img {
```

```
    width: 42px;
    height: 42px;
    border - radius: 50% ;
  }
  . info - user span {
    font - size: 15px;
    position: relative;
    top:  - 15px;
    margin - left: 10px;
  }
  . info - detail {
    padding: 0 5px 15px;
  }
  . info - detail p {
    font - size: 14px;
    line - height: 1.5;
    color: #777777;
  }
  . info - detail . info - other {
    font - size: 12px;
    margin - top: 10px;
    color: #999999;
  }
  . info - other . date {
    margin - right: 8px;
  }
  . info - imgs {
    margin - top: 10px;
  }
  . info - imgs img {
    width: 70px;
    height: 70px;
    margin - right: 5px;
  }
</style>
```

3）在商品信息中添加评论信息

在 src/api/goods. js 文件的 getGoodsInfo 里面添加商品的评论信息，代码如下：

```
export function getGoodsInfo(id) {
    return {
        ...
        commentInfo: {
            sum: 2,
            comments: [
```

```
                {
                    commentImg: ['/images/comment/img1.jpg'],
                    commentBody: {
                        commentId: 1,
                        productId: 2,
                        orderId: 2,
                        userName: 'zhangsan',
                        userImg: '/images/user/img1.jpg',
                        imgUrl: '/images/goods1.jpg',
                        content: '测试评论 1,不错不错',
                        fwValue: 5,
                        wlValue: 6,
                        auditTime: '1593765746',
                        modifiedTime: '1593765746'
                    }
                },
                {
                    commentImg: ['/images/comment/img2.jpg'],
                    commentBody: {
                        commentId: 1,
                        productId: 2,
                        orderId: 2,
                        userName: 'lisi',
                        userImg: '/images/user/img2.jpg',
                        imgUrl: '/images/goods2.jpg',
                        content: '测试评论 2,不错不错',
                        fwValue: 5,
                        wlValue: 6,
                        auditTime: '1593765746',
                        modifiedTime: '1593765746'
                    }
                }
            ]
        }
    }
}
```

4）在详细页面中添加评论

在 src/views/detail/Detail.vue 文件中添加评论组件，以便显示评论，代码如下：

```
<template>
    <div class = "detail">
        <detail - nav - bar @titleClick = "titleClick" ref = "detailBar" ></detail - nav - bar >
        <scroll
            class = "content"
```

```
            @scroll = "contentScroll"
            ref = "scroll"
            :probe - type = "3"
            :pull - up - load = "true"
            >
            <! -- 商品的轮播图片 -->
            < detail - swiper :top - images = "topImages"></detail - swiper >
            <! -- 商品的详细信息 -->
            < detail - bass - info :goods = "goodsInfo" :servesInfo = "serves"></detail - bass - info >
            <! -- 商品评论 -->
            < detail - comment - info :commentInfo = 'commentInfo'></detail - comment - info >

            <! -- 避免底部内容被 Bar 遮掉 -->
            < div class = "bottom"></div >
        </scroll >
    </div >
</template >

< script >
    ...
    import DetailCommentInfo from './child/DetailCommentInfo'
    ...

    export default {
        name: 'Detail',
        components: {
            ...
            DetailCommentInfo
        },
        data() {
            return {
                ...
                commentInfo: {}
            }
        },
        ...
        methods: {
            getGoodsInfo() {
                ...
                //获取商品评论信息
                this.commentInfo = goods.commentInfo

                Toast.clear();
            },
            ...
        }
```

```
    }
</script>

<style scoped>
  .bottom{
      height: 49px;
  }
</style>
```

6. 显示商品的详细信息

1）创建显示详细信息组件

创建 src/views/detail/child/DetailGoodsInfo.vue 文件，代码如下：

```
<template>
    <div v-html="detailInfo" class="goods-info"></div>
</template>

<script>
  export default {
    name: "DetailGoodsInfo",
    props: ["detailInfo"]
  };
</script>

    <style>
  .goods-info {
    padding: 20px 0;
    border-bottom: 5px solid #f2f5f8;
  }
  .graphic-pic{
    width: 100%;
  }
  .graphic-pic img{
    width: 100%;
  }
</style>
```

2）在 Detail 组件中添加商品详细信息的显示

代码如下：

```
<template>
    <div class="detail">
        <detail-nav-bar @titleClick="titleClick" ref="detailBar"></detail-nav-bar>
        <scroll
            class="content"
```

```
            @scroll = "contentScroll"
            ref = "scroll"
            :probe - type = "3"
            :pull - up - load = "true"
            >
            <! -- 商品的轮播图片 -->
            < detail - swiper :top - images = "topImages"></detail - swiper >
            <! -- 商品的基本信息 -->
            < detail - bass - info :goods = "goodsInfo" :servesInfo = "serves"></detail - bass - info >
            <! -- 商品评论 -->
            < detail - comment - info ref = 'dp' :commentInfo = 'commentInfo'></detail - comment - info >
            <! -- 商品详细信息 -->
            < detail - goods - info ref = "detail" :detail - info = 'detailInfo'></detail - goods - info >

            < div class = "bottom"></div >
        </scroll >
    </div >
</template >

< script >
    ...
    import DetailGoodsInfo from './child/DetailGoodsInfo'
    ...

    export default {
        name: 'Detail',
        components: {
            ...
            DetailGoodsInfo
        },
        data() {
            return {
                ...
                detailInfo: ''
            }
        },
        created() {
            //保存 goodsId
            this.goodsId = this. $route.params.goodsId;
            //请求详细信息
            this.getGoodsInfo();
        },
        methods: {
            getGoodsInfo() {
                ...
                //获取商品的详细信息
```

```
            this.detailInfo = goods.info.goodsIntroduce
            Toast.clear();
        },
        ...
      }
    }
</script>

<style scoped>
    .wrapper {
      height: calc(100vh - 83px);          /* scroll 的高度 */
      /* position: absolute; */
      top: 44px;
      /* bottom: 49px; */
      left: 0;
      right: 0;
      /* background-color: #fff; */
      overflow: hidden;
    }
</style>
```

7. 实现详细信息的滚动

1）计算详细页面中顶部、评论和详细信息的底部 y 值

在 Detail 中添加 topY 属性，用来存放每个部分的 y 值，同时在 created 中添加代码，初始化 topY 属性，代码如下：

```
    export default {
      name: 'Detail',
        components: {
            DetailSwiper,
            DetailNavBar,
            Scroll,
            DetailBassInfo,
            DetailCommentInfo,
            DetailGoodsInfo
        },
        data() {
            return {
                ...
                topY:[]                       //定义 topY,用来存放滚动的 y 值
            }
        },
        created() {
            ...
            //初始化 topY
```

```
            this. $nextTick(() => {
                setTimeout(() => {
                    this.topY = []
                    this.topY.push(0);
                    this.topY.push(this. $refs.dp. $el.offsetTop);
                    this.topY.push(this. $refs.detail. $el.offsetTop);
                },2500)
            })
        },
        ...
}
```

注意：需要在 Detail 组件模板里面的 detail-comment-info 元素和 detail-goods-info 元素上面分别添加属性 ref="dp"和 ref="detail"。

2）刷新 scroll 状态

在 Detail 的 mounted()方法中，刷新 scroll 的状态，以便能及时初始化 scroll 的属性，代码如下：

```
<template>
    <div class="detail">
        <detail-nav-bar @titleClick="titleClick" ref="detailBar"></detail-nav-bar>
        <scroll
            class="content"
            @scroll="contentScroll"
            ref="scroll"
            :probe-type="3"
            :pull-up-load="true"
            >
            <!-- 商品的轮播图片 -->
            <detail-swiper :top-images="topImages"></detail-swiper>
            <!-- 商品的基本信息 -->
            <detail-bass-info :goods="goodsInfo" :servesInfo="serves"></detail-bass-info>
            <!-- 商品评论 -->
            <detail-comment-info ref='dp' :commentInfo='commentInfo'></detail-comment-info>
            <!-- 商品详细信息 -->
            <detail-goods-info ref="detail" :detail-info='detailInfo'></detail-goods-info>

            <div class="bottom"></div>
        </scroll>
    </div>
</template>

<script>
```

```
    ...

    export default {
        ...
        mounted() {
            //刷新 scroll 的状态
            this.refreshScroll();
            //绑定全局总线事件,以便及时刷新 scroll 的状态
            this. $bus. $on('RefreshScrollEvent', this.refreshScroll)
        },
        methods: {
            ...
            refreshScroll() {          //属性 scroll 的状态
                //无抖动刷新,以便 scroll 组件正确地初始化
                const refresh = debounce( this. $refs.scroll.refresh,300)
                refresh()
            }
        }
    }
</script>
...
```

3) 定义样式

初始化 scroll 的显示高度等属性,代码如下:

```
<style scoped>
    .wrapper {
      height: calc(100vh - 83px);
      /* position: absolute; */
      top: 44px;
      /* bottom: 49px; */
      left: 0;
      right: 0;
      /* background-color: #fff; */
      overflow: hidden;
    }
</style>
```

实现 scroll 绑定在@scroll 上面的 contentScroll()函数,代码如下:

```
//根据当前位置判断显示的部分(基本信息、评论和详细信息),显示 detailBar 的对应选项
contentScroll(position) {
        let i = 0;
        if (-position.y >= 0 && -position.y < this.topY[1]) {
                i = 0;
```

```
        } else if ( - position. y >= this. topY[1] && - position. y < this. topY[2]) {
                i = 1;
        } else if ( - position. y >= this. topY[2]) {
                i = 2;
        }
        this. currentIndex = i;
        this. $refs. detailBar. currentIndex = this. currentIndex;
}
```

8. 显示购物栏(底部导航)

1) 编写购物栏组件

在 src/views/detail/child 中创建 DetailBottomBar. vue 文件,代码如下:

```
<template>
  <div class="bottom-bar">
    <div class="bar-item bar-left">
      <div @click="toService">
        <img src="~assets/img/common/kf.png" class="icon service" />
        <span class="text">客服</span>
      </div>
      <div @click="pushCart">
        <el-badge :value="cartLength" :max="99" class="item">
          <img src="~assets/img/common/cart.png" class="icon shop" />
          <span class="text">购物车</span>
        </el-badge>
      </div>
      <div @click="addEnshrine" v-if="!goodsInfo. isFav">
        <img src="~assets/img/common/sc4.png" class="icon select" />
        <span class="text">收藏</span>
      </div>
      <div @click="removeEnshrine" v-else>
        <img src="~assets/img/common/sc5.png" class="icon select" />
        <span class="text">已收藏</span>
      </div>
    </div>
    <div class="bar-item bar-right">
      <div class="cart" @click="addCart">加入购物车</div>
      <div class="buy" @click="buyClick">购买</div>
    </div>
  </div>
</template>

<script>
  export default {
    name: "DetailBottomBar",
```

```
    data() {
      return {
        cartLength: 0
      }
    },
    props:['goodsInfo'],
    created() {
      this.getLength()
    },
    methods: {
      //获取购物车中商品的数量
      getLength(){

      },
      //添加购物车
      addCart() {
        this.$emit("addCart");
      },
      //立即购买
      buyClick() {
        this.$emit("buyClick");
      },
      //添加收藏
      addEnshrine() {
        this.$emit("addEnshrine");
      },
      //联系客服
      toService() {
        this.$emit("toService");
      },
      //跳转到购物车
      pushCart() {

      },
      //取消收藏
      removeEnshrine(){
        this.$emit("removeEnshrine");
      }
    },

  };
</script>

<style scoped>
  .cartSum {
    position: fixed;
```

```
    left: 28% ;
    z - index: 9;
    width: 20px;
    height: 20px;
    text - align: center;
    line - height: 20px;
    background - color: red;
    color: #fff;
    border - radius: 50% ;
  }
  .bottom - bar {
    /* position: relative; */
    position: fixed;
    z - index: 100;
    right: 0;
    bottom: 0;
    left: 0;
    width: 100% ;
    height: 49px;
    display: flex;
    text - align: center;
    background - color: white;
  }
  .bar - item {
    flex: 1;
    display: flex;
  }
  .bar - item > div {
    flex: 1;
  }
  .bar - left .tex {
    font - size: 13px;
  }
  .bar - left .icon {
    display: block;
    width: 22px;
    height: 22px;
    margin: 4px auto 3px;
  }
  .bar - left .service {
    background - position: 0  - 54px;
  }
  .bar - left .shop {
    background - position: 0 98px;
  }
  .bar - right {
```

```
    font-size: 15px;
    color: white;
    line-height: 49px;
  }
  .bar-right .cart {
    background-color: #ffe817;
    color: black;
  }
  .bar-right .buy {
    background-color: #eb4868;
  }
</style>
```

2）在 Detail 中添加购物栏

在 src/views/detail/Detail.vue 文件中添加 DetailBottomBar 组件和对应的事件，代码如下：

```
<template>
    <div class="detail">
        <detail-nav-bar @titleClick="titleClick" ref="detailBar"></detail-nav-bar>
        <scroll
            class="content"
            @scroll="contentScroll"
            ref="scroll"
            :probe-type="3"
            :pull-up-load="true"
            >
            <!-- 商品的轮播图片 -->
            <detail-swiper :top-images="topImages"></detail-swiper>
            <!-- 商品的基本信息 -->
            <detail-bass-info :goods="goodsInfo" :servesInfo="serves"></detail-bass-info>
            <!-- 商品评论 -->
            <detail-comment-info ref='dp' :commentInfo='commentInfo'></detail-comment-info>
            <!-- 商品详细信息 -->
            <detail-goods-info ref="detail" :detail-info='detailInfo'></detail-goods-info>
        </scroll>
        <!-- 显示购物栏 -->
        <detail-bottom-bar
            ref="cart"
            :goodsInfo="goodsInfo"
            @addCart="addCart"
            @buyClick="buyClick"
            @toService="toService"
            @addEnshrine="addEnshrine"
            @removeEnshrine="removeEnshrine">
```

```
        </detail - bottom - bar >
    </div >
</template >

< script >
    ...
    import DetailBottomBar from './child/DetailBottomBar'
    ...

    export default {
        name: 'Detail',
        components: {
            ...
            DetailBottomBar
        },
        ...
        methods: {
            ...
            //添加购物车
            addCart(){

            },
            //单击购买商品,以便唤起 sku
            buyClick(){

            },
            onBuyClicked(sku){

            },

            //收藏商品
            addEnshrine(){

            },
            //联系客服
            toService(){

            },
            //移除收藏
            removeEnshrine(){

            }
        }
    }
</script >
...
```

14.2.5 实现登录

登录页面由两部分组成：登录页面和注册页面，登录页面和注册页面之间可以切换，注册成功后可自动切换到登录页面。

1. 创建 Login 组件

通过 Login 组件能显示登录入口，创建 src/views/login/child/Login.vue 文件，代码如下：

```
<template>
    <div id="login" class="dom">
      <el-form
        ref="loginFormRef"
        :model="loginForm"
        label-width="60px"
        :rules="loginFormRules"
        label-position="top"
        status-icon
      >
        <div class="s1">
          <el-form-item label="账号" prop="loginName">
            <el-input
              v-model="loginForm.loginName"
              name="loginName"
              type="text"
              placeholder="请输入登录账号"
            ></el-input>
          </el-form-item>
        </div>
        <div class="s1">
          <el-form-item label="密码" prop="loginPassword">
            <el-input
              v-model="loginForm.loginPassword"
              name="loginPassword"
              type="password"
              placeholder="请输入登录密码"
            ></el-input>
          </el-form-item>
        </div>
        <div class="s2" @click="remPWD">
          <input type="checkbox" id="remCheck" checked />
          <span>记住密码</span>
        </div>
        <el-button class="btn" @click="onSubmit" type="primary">登 录</el-button>
      </el-form>
```

```
        <div class = "dom - footer">
          <div class = "login - another">
            <a href = "#" @click = "toForgetPWD">找回密码</a>
            <span>|</span>
            <span>还没有注册账号?</span>
            <a href = "#" @click = "toRegister">立即注册</a>
          </div>
        </div>

      </div>
  </template>

  <script>

    export default {
      name: "Login",
      components: {
      },
      created() {
        this.getuserLogin();
      },
      data() {
        return {
          //判断是否记住密码
          isremCheck: true,
          loginForm: {
            loginName: "",
            loginPassword: "",
          },
          isPassing: false,                      //是否验证通过
          show: false,                           //是否展示验证码验证
          //表单验证
          loginFormRules: {
            loginName: [
              { required: true, message: "请输入登录名", trigger: "blur" },
              {
                min: 2,
                max: 10,
                message: "长度在 2～10 个字符",
                trigger: "blur",
              },
            ],
            loginPassword: [
              { required: true, message: "请输入登录密码", trigger: "blur" },
              {
                min: 3,
```

```
          max: 18,
          message: "长度在 3～18 个字符",
          trigger: "blur",
        },
      ],
    },
  };
},
methods: {
  //判断用户是否选择了记住密码,如果是,则获取密码
  getuserLogin() {

  },
  toRegister() {
    this.$emit("toZC");
  },
  remPWD() {
    //修改复选框状态
    this.isremCheck = !this.isremCheck;
    document.getElementById("remCheck").checked = this.isremCheck;
  },
  //前往修改密码
  toForgetPWD() {
    this.$router.push("/forgetPWD");
  },
  //登录表单拦截
  async onSubmit() {
    await this.$refs.loginFormRef.validate(async (result) => {
      if (!result) {
        return false;
      } else {
        this.getUser()
      }
    });
  },
  //调用登录 API
  getUser() {
    //登录显示 loading
    const loading = this.$loading({
      lock: true,
      text: "验证中.....",
      spinner: "el-icon-loading",
      background: "rgba(0, 0, 0, 0.5)",
    });
    this.isPassing = true
    if(this.isPassing) this.loginForm.isPassing = "yes"
```

```
        //关闭登录显示
        loading.close();
      },
    },
  };
</script>

<style>
</style>
```

2. 创建注册组件

创建 src/views/login/child/Register.vue 文件,代码如下:

```
<template>
  <div>
    <el-form
      ref="RegisterFormRef"
      :model="RegisterForm"
      label-width="60px"
      :rules="RegisterFormRules"
      label-position="top"
      status-icon
    >
      <div class="s1">
        <el-form-item label="手机号码" prop="phone">
          <el-input
            v-model="RegisterForm.phone"
            name="phone"
            type="text"
            placeholder="注册手机号为找回密码的唯一凭证"
            ref="input_phone"
          ></el-input>
        </el-form-item>
      </div>
      <div class="s1">
        <el-form-item label="登录名" prop="LoginName">
          <el-input
            v-model="RegisterForm.LoginName"
            name="LoginName"
            type="text"
            placeholder="中、英文均可,不超过 12 个字符"
            ref="input_description"
          ></el-input>
        </el-form-item>
```

```
        </div>
        <div class="s1">
          <el-form-item label="密码" prop="LoginPassword">
            <el-input
              v-model="RegisterForm.LoginPassword"
              name="LoginPassword"
              type="password"
              placeholder="3~18位英文、数字、符号,区分大小写"
            ></el-input>
          </el-form-item>
        </div>

        <el-form-item class="btns">
          <el-button type="primary" @click="onSubmit" id="zc">注册</el-button>
        </el-form-item>
      </el-form>

    </div>
  </template>

  <script>
  import { Toast } from "vant";

  export default {
    name: "Register",
    data() {
      var checkPhone = (rule, value, callback) => {
        if (value) {
          let re = /^(?:(?:\+|00)86)?1(?:(?:3[\d])|(?:4[5-7|9])|(?:5[0-3|5-9])|(?:6
[5-7])|(?:7[0-8])|(?:8[\d])|(?:9[1|8|9]))\d{8}$/;
          if (re.test(value)) {
            return callback();
          } else {
            return callback(new Error("请输入正确的手机号码"));
          }
        }
      };
      return {
        //注册数据
        RegisterForm: {
          LoginName: "",
          LoginPassword: "",
          phone: "",
        },
        //告诉短信组件是哪个组件需要发送短信
        type: "Register",
```

```
        //验证码用于验证是否通过
        codeIsOK: false,
        //登录名是否可用
        nameIsOK: true,
        //该手机号是否已注册,默认为没有被注册
        phoneIsOK: true,
        //注册预验证
        RegisterFormRules: {
          LoginName: [
            { required: true, message: "请输入登录名", trigger: "change" },
            {
              min: 2,
              max: 10,
              message: "长度在 2～10 个字符",
              trigger: "change",
            },
          ],
          LoginPassword: [
            { required: true, message: "请输入登录密码", trigger: "change" },
            {
              min: 3,
              max: 18,
              message: "长度在 3～18 个字符",
              trigger: ["change"],
            },
          ],
          code: [{ required: true, message: "请输入验证码", trigger: "change" }],
          phone: [
            { required: true, message: "请输入注册手机号", trigger: "change" },
            { validator: checkPhone, trigger: "change" },
          ],
        },
      };
    },
    methods: {

      //进行注册
      onSubmit() {

      },
      //手机号格式验证
      checkMobile(str) {
        let re = /^(?:(?:\+|00)86)?1(?:(?:3[\d])|(?:4[5-7|9])|(?:5[0-3|5-9])|(?:6[5-
  7])|(?:7[0-8])|(?:8[\d])|(?:9[1|8|9]))\d{8}$/;
        if (re.test(str)) {
          return true;
```

```
        } else {
          this.$message.error("请输入正确的手机号");
          return false;
        }
      },
    },
  };
  </script>

  <style>
  #zc {
    width: 100%;
  }
  </style>
```

3. 创建登录页面

创建 src/views/login/LoginPage.vue 文件,代码如下:

```
<template>
    <div id="content">
      <nav-bar class="home-nav">
        <div slot="center">
          <img src="~assets/img/home/favicon.png" style="width:80%" alt />
        </div>
      </nav-bar>
      <!--头部-->
      <div class="content-header clearfix">
          <a href="JavaScript:;"
          :class="{'current':isShow}"
          @click="tabClick()"
          >登录</a>
          <a href="JavaScript:;"
          @click="tabRegister()"
          :class="{'current':!isShow}"
          >注册</a>
      </div>
      <!--内容-->
      <div class="content-body">
          <login v-if="isShow" @toZC='tabRegister'></login>
          <register v-else @zcOK='tabClick'></register>
      </div>
  </div>
</template>

<script>
```

```
    import NavBar from "components/common/navbar/NavBar.vue";
    import Login from 'views/login/child/Login'
    import Register from 'views/login/child/Register'
    export default {
      name:'LoginPage',
      components:{
        Login,
        NavBar,
        Register
      },
      data() {
        return {
          //确定是显示登录还是注册,true 为登录状态
          isShow:true,
        }
      },
      methods: {
        tabClick(){
          this.isShow = true
        },
        tabRegister(){
          this.isShow = false
        },
      },
    }
  </script>

<style scoped>
    @import url('~assets/css/LoginPage.css');
    .text-style{
      color: red;
    }
    .nav-bar{
      height: 52px;
    }
</style>
```

4. 添加登录路由信息

在 src/router/index.js 文件中添加登录路由信息,代码如下:

```
...
const Login = () = XCJ2.EPS,JZ;P] import("../views/login/LoginPage")

const routes = [
  ...
```

```
    {
      path: '/login',
      component: Login
    }
]
...
```

在 src/views/detail/Detail.vue 文件里面的 addCart()函数里面添加路由，代码如下：

```
...
//添加购物车
addCart(){
    //路由到
    this.$router.push('/login')
},
...
```

14.2.6 实现购物车

实现购物车，这里按 3 个步骤逐个介绍：第一步，编写购物车框架；第二步，实现购物车商品列表；第三步，实现购物车操作。

1. 编写购物车框架

创建 src/views/car/Car.vue 文件，使用 NavBar 组件显示导航，代码如下：

```
<template>
  <div>
      <nav-bar class="cart-nav">
        <div slot="center">购物车({{shopCount}})</div>
        <div slot="right" @click="edit" v-show="!isedit">管理</div>
        <div slot="right" @click="edit" v-show="isedit">完成</div>
      </nav-bar>
  </div>
</template>

<script>
import NavBar from '../../components/common/navbar/NavBar.vue'
export default {
    name: 'Car',
    components: { NavBar },
    data(){
        return {
            shopCount: 0,
            isedit: false
        }
```

```
    },
    methods:{
        edit(){

        }
    }
}
</script>

<style>

</style>
```

修改 TabBarItem 组件中的 itemClick()方法,实现单击导航逻辑。修改 src/components/common/tabbar/TabBarItem. vue 文件,代码如下:

```
<template>
  ...
</template>

<script>
export default {
  ...
  methods: {
      itemClick() {
          this. $router.push(this.path)
      }
  }
}
</script>

<style>
...
</style>
```

在 src/router/index. js 文件中添加购物车路由信息,代码如下:

```
...
const Car = () => import("../views/car/Car")

const routes = [
  { path: "", redirect: "/home" },
  { path: "/", redirect: "/home" },
  {
    path: "/home",
```

```
        component: AppHome,
        children: [
          ...
          {
            path: "/car",
            component: Car,
            meta:{
              keepAlive:true
            }
          },
          ...
        ]
      },
      ...
    ]
    ...
```

2. 实现购物车商品列表

创建 CheckButton 组件，以便于选中/取消购物车中的商品。创建 src/components/content/checkButton/index.vue 文件，代码如下：

```
<template>
  <div class="check-button">
    <img src="~assets/img/common/xz.png" alt="" :class="{check: isCheck}">
  </div>
</template>

<script>
export default {
    name: "CheckButton",
    props:{
      isCheck:{
        type:Boolean,
        default:false
      }
    }
}
</script>

<style scoped>
    .check-button img{
      -webkit-border-radius: 50%;
      -moz-border-radius: 50%;
      border-radius: 50%;
      width: 28px;
```

```
    height: 28px;
    vertical-align: middle;
    margin-bottom: 6px;
    margin-right: 6px;

  }
  .check {
    border-radius: 50%;
    background-color: red;
  }
</style>
```

创建购物车商品列表项组件，用来显示购物车中的每项商品。创建 src/views/car/child/GoodsListItem.vue 文件，代码如下：

```
<template>
  <div class="list-item">
    <!-- 每项的选择框 -->
    <div class="item-selector">
      <check-button :is-check="product.checked" @click.native="checkClick1" v-show=
"!isedit"></check-button>
      <check-button :is-check="product.checked" @click.native="checkClick2" v-show=
"isedit"></check-button>
    </div>
    <div class="item-img" @click="toGoodsInfo">
      <img v-lazy="product.image" alt="商品图片" @load="imgLoad" />
    </div>
    <div class="item-info">
      <!-- 商品名称 -->
      <div class="item-title" @click="toGoodsInfo">{{product.title}}</div>
      <div class="item-desc"></div>
      <div class="info-bottom">
        <!-- 商品价格 -->
        <div class="item-price">¥{{product.price}}</div>
        <!-- 编辑时显示商品数量 -->
        <div class="input-text" v-show="isedit">
          <van-stepper
            v-model="product.amount"
            theme="round"
            @change="handleChange"
            :min="1"
            :max="10"
          />
        </div>
        <!-- 结算显示商品数量 -->
```

```
          <div class = "item - count" v - show = "!isedit"> x{{product.amount}}</div>
        </div>
      </div>
    </div>
  </template>

  <script>
  import CheckButton from "components/content/checkButton";
  import _ from "lodash";
  import { Toast } from "vant";

  export default {
      name: "GoodsListItem",
    components: {
      CheckButton
    },
    props: {
      product: {
        type: Object,
        default() {
          return {};
        }
      },
      isedit: {
        type: Boolean,
        default() {
          return false;
        }
      }
    },
    data() {
      return {
        cartInfo: {},            //用来确保在商品数量修改失败时还能进行页面的正确显示
        isOk: false              //上一个网络请求是否完成
      };
    },
    created() {
      //深复制一份商品的数据
      this.cartInfo = _.cloneDeep(this.product);
    },
    methods: {
      imgLoad() {
        this.$bus.$emit("itemImageLoad");
      },
      //修改商品的选择状态
      checkClick1() {
```

```
        this.product.checked = !this.product.checked
      },
      //转到商品详情
      toGoodsInfo() {
        this.$router.push("/detail/" + this.product.id);
      },
      //商品数量发生改变时
      handleChange(value) {

      },
      //修改商品的选择状态,仅本地选择,不修改服务器的数据
      //用于确认删除
      checkClick2() {
        return (this.product.checked = !this.product.checked);
      }
    }
  }
</script>

<style scoped>
    .list-item {
      width: 100%;
      display: flex;
      font-size: 0;
      padding: 5px;
      border-bottom: 1px solid #ccc;
    }
    .item-selector {
      width: 20px;
      display: flex;
      justify-content: center;
      align-items: center;
    }
    .item-title,
    .item-desc {
      overflow: hidden;
      white-space: nowrap;
      -ms-text-overflow: ellipsis;
      text-overflow: ellipsis;
    }
    .item-title {
      font-size: 18px;
      margin-bottom: 4px;
    }
    .item-img {
      padding: 5px;
```

```
    }
    .item-img img {
      width: 80px;
      height: 100px;
      display: block;
      -webkit-border-radius: 5px;
      -moz-border-radius: 5px;
      border-radius: 5px;
    }
    .item-info {
      font-size: 17px;
      color: #333;
      padding: 5px 10px;
      position: relative;
      overflow: hidden;
    }
    .item-info,
    .item-desc {
      color: #666;
      font-size: 14px;
    }
    .item-desc {
      margin-top: 15px;
    }
    .info-bottom {
      display: flex;
      justify-content: space-between;
      margin-top: 16px;
    }
    .item-price {
      color: #ffb805;
      font-size: 20px;
    }
    .item-count {
      font-size: 20px;
    }
    </style>
```

创建购物车商品列表，创建 src/views/car/child/GoodsList.vue 文件，代码如下：

```
<template>
  <div id="mmmmm" class="cart-list">
    <scroll
      class="content"
      ref="scroll"
```

```
        :pull-up-load="true"
        @pullingUp="loadMore"
        :handleToScroll="handleToScroll"
        :handleToTouchEnd="handleToTouchEnd"
      >
        <ul v-if="isShowLoading">
          <li class="pullDown">
            <i class="el-icon-loading"></i>
            {{ pullDownMsg }}
          </li>
        </ul>
        <goods-list-item v-for="(item, index) in cartList" :key="index" :product="item" :
isedit="isedit"></goods-list-item>
        <ul v-if="isShowUpLoading">
          <li class="pullUp">
            <i class="el-icon-loading"></i>
            {{ pullUpMsg }}
          </li>
        </ul>
      </scroll>
    </div>
</template>

<script>
import GoodsListItem from "./GoodsListItem";

import Scroll from "components/common/scroll/Scroll";
//功能组件
import { debounce } from "common/utils";          //防抖函数
export default {
  name: "CartGoodsList",
  props: ["cartList",'isedit'],
  data() {
    return {
      isShowLoading: false,
      isShowUpLoading: false,
      isShowBackTop: false,
      pullDownMsg: "",
      pullUpMsg: "加载更多....."
    };
  },
  components: {
    Scroll,
    GoodsListItem
  },
  activated() {
```

```
      this.$refs.scroll.refresh();
    },
    mounted() {
      const refresh = debounce(this.$refs.scroll.refresh, 200);
      //监听 itme 图片是否加载完成
      this.$bus.$on("itemImageLoad", () => {
        //this.refresh();
      });

    },
    methods: {
      //上拉加载更多,分页数据
      loadMore() {
        this.isShowUpLoading = true;
        //this.pullDownMsg = "正在更新中...";
        //调用接口
        return this.$emit("handleToNew");
      },
      handleToScroll(pos) {
        if (pos.y > 60) {
          this.isShowLoading = true;
          this.pullDownMsg = "正在更新中...";
        }
      },
      //下拉刷新
      handleToTouchEnd(pos) {
        if (pos.y > 60) {
          setTimeout(() =>{
            this.isShowLoading = false
          },1500)
          this.$bus.$emit('LoadAgain')
        }
      },
      //返回顶部
      backClick() {
        this.$refs.scroll.scrollTo(0, 0, 500);
      },
      noAny(){
        this.pullUpMsg = '已经到底了'
        setTimeout(() =>{
          this.isShowUpLoading = false
        },1500)
      },
      finish() {
        this.$refs.scroll.finishPullUp();
      },
```

```
    }
};
</script>

<style scoped>
.content {
    /* margin-top: 44px; */
    height: calc(100vh - 49px - 84px);
    overflow: hidden;
}
.pullDown {
    margin: 0;
    padding: 16px;
    text-align: center;
    border: none;
}
.pullUp {
    margin: 0;
    text-align: center;
    border: none;
}
</style>
```

在 Car 组件中添加商品列表，代码如下：

```
<template>
    <div>
        ...
        <goods-list
            ref="myGoodsList"
            @handleToNew="handleToNewP"
            :cartList="cartList"
            :isedit="isedit">
        </goods-list>
    </div>
</template>

<script>
import NavBar from '../../components/common/navbar/NavBar.vue'
import GoodsList from './child/GoodsList'

export default {
    name: 'Car',
    components: { NavBar, GoodsList },
    data(){
        return {
```

```
            shopCount: 0,
            isedit: false,
            cartList: [],
            page: 1,
            pageSize: 5,
            product_id: ""
        }
    },
    activated() {
        //防止出现单击商品到详情页后,返回到当前页面时,出现商品加载错误的情况
        this.page = 1;
        this.cartList = [];
        this.getShopList();
        this.getShopCount();

        this.$bus.$on("LoadAgain", () => {
            this.getShopList();
        })
    },
    methods:{
        //编辑购物车商品
        edit(){

        },
        //获取购物车商品
        getShopList(){

        },
        //获取购物车商品数量
        getShopCount(){

        },
        //获取下一页数据
        handleToNewP() {
            this.page += 1;
            this.getShopList();
        },
        //删除成功,重新请求数据
        deleteOkP() {
            this.cartList = [];
            this.page = 1;
            this.getShopList();
            this.getShopCount();
        }
    }
}
```

```
</script>

<style>

</style>
```

3. 实现购物车操作

自定义消息提示框，创建一个消息提示框组件。创建 src/components/common/toast/Toast.vue 文件，代码如下：

```
<template>
    <div class="toast" v-show="isShow">
      <div>{{message}}</div>
    </div>
</template>

<script>
  export default {
    name:'Toast',
    data() {
      return {
        message: '这是测试',
        isShow: false
      }
    },
    methods: {
      show(messages,duration){
        this.isShow = true;
        this.message = messages;

        setTimeout(() =>{
          this.isShow = false;
          this.message = '';

        },duration)
      }
    },
  }
</script>

<style scoped>
    .toast{
      position: fixed;
      top:50%;
      left: 50%;
```

```
        transform: translate( - 50% , - 50% );
        padding: 8px 10px;
        color: #ffffff;
        background - color: rgba(0, 0, 0, 0.65);
        z - index: 999;
    }
</style>、
```

注册成 Vue.js 全局组件，创建 src/components/common/toast/index.js 文件，代码如下：

```
import Toast from './Toast.vue'

const obj = {}

obj.install = function(Vue){
  //创建组件构造器
  const toastContrustor = Vue.extend(Toast)
  //以 new 的方式,根据组件构造器可以创建出一个组件对象
  const toast = new toastContrustor()
  //将组件对象手动挂载到某一个元素上
  toast.$mount(document.createElement('div'))
  //toast.$el 对应的就是 div
  document.body.appendChild(toast.$el)
  //注册成原型属性
  Vue.prototype.$toast = toast
}

export default obj
```

在 src/main.js 文件中完成 Toast 对象的注册，代码如下：

```
...
//自定义组件,消息框提醒
import toast from "./components/common/toast";
//安装 toast
...
```

定义 CarBottomBar 组件，创建 src/views/car/child/CarBottomBar.vue 文件，代码如下：

```
<template>
    <div>
        <div class = "cart - bottom - bar" v - show = "!isedit">
            <div class = "total - check">
                <check - button :is - check = "isSelectAll" @click.native = "selectAllClick">
</check - button>
```

```
            <span>全选</span>
          </div>
          <div class="total-price">
            <span>合计:￥{{totalPrice}}</span>
          </div>
          <div class="calc" @click="account">去结算({{calcCount}})</div>
        </div>
        <!-- 显示编辑的状态,以及删除的页面 -->
        <div class="cart-bottom-bar" v-show="isedit">
          <div class="total-check">
            <!--      <img src="~assets/img/cart/tick.svg" alt=""> -->
            <check-button :is-check="isSelectAll" @click.native="selectAllClick">
</check-button>
            <span>全选</span>
          </div>
          <div class="total-price">
          </div>
          <div class="calc" @click="deletecart">删除({{calcCount}})</div>
        </div>
        <!-- 删除确认 -->
        <el-dialog
      title="删除确认"
      :visible.sync="centerDialogVisible"
      width="80%"
      center>
      <span>该操作将 <span class="delete">删除({{calcCount}})条 </span>购物车记录,是否确
定?</span>
      <span slot="footer" class="dialog-footer">
        <el-button @click="centerDialogVisible = false">取 消</el-button>
        <el-button type="primary" @click="Confirm">确 定</el-button>
      </span>
    </el-dialog>
      </div>
</template>

<script>
    import CheckButton from "components/content/checkButton";

    export default {
      name: "CartBottomBar",
      props: ["cartList", "isedit"],
      data() {
        return {
          centerDialogVisible: false,
          cartid: '',                        //记录需要删除的购物车 id
          goodsId:'',                        //记录需要购买的商品 id
```

```
    }
  },
  components: {
    CheckButton
  },
  computed: {
    totalPrice() {
      //汇总,先过滤
      return this.cartList
        .filter(item => {
          return item.checked === true;
        })
        .reduce((preValue, item) => {
          return preValue + item.price * item.amount;
        }, 0)
        .toFixed(2); //两位小数,四舍五入
    },
    calcCount() {
      return this.cartList.filter(item => {
        return item.checked === true;
      }).length;
    },
    isSelectAll() {
      if (this.cartList.length === 0) {
        return false;
      }
      return !this.cartList.filter(item => !item.checked).length;
    }
  },
  methods: {
    selectAllClick() {
      if (this.isSelectAll) {
          this.cartList.forEach(item => (item.checked = false));
      } else {
          this.cartList.forEach(item => (item.checked = true));
      }
    },
    //进行付款购买
    account() {
      if (this.calcCount === 0) {
          //console.log(this.$toast)
          return this.$toast.show("你还没选择商品", 2000);
      } else {
          //清空
          this.goodsId = ''
          this.cartList.forEach(item => {
```

```
                if(item.ischecked){
                this.goodsId += item.product_id + ''
                }
            })
            //跳转到确认订单界面,并把商品的 id 传过去

        }
    },
    //删除购物车中的商品
    deletecart(){
      if (this.calcCount === 0) {
          return this.$toast.show("请选择需要删除的商品", 2000);
      }
      //清空
      this.cartid = ''
      this.cartList.forEach(item => {
          if(item.ischecked){
          this.cartid += item.cart_id + ','
          }
      })
      this.centerDialogVisible = true
    },
    //确认删除
    Confirm(){
      this.centerDialogVisible = false
    }
  }
};
</script>

<style scoped>
  .cart-bottom-bar {
    width: 100%;
    height: 40px;
    position: fixed;
    bottom: 49px;
    background: #eeeeee;
    display: flex;
    line-height: 40px;
    text-align: center;
  }
  .total-check {
    width: 25%;
    display: flex;
    justify-content: center;
  }
```

```
    .total - price {
      width: 50% ;
    }
    .calc {
      width: 25% ;
      background - color: #ff5028;
      color: #fff;
    }
    .delete{
      color: red;
    }
  </style>
```

在 src/views/car/Car.vue 文件中添加 CarBottomBar 组件，代码如下：

```
  <template>
    <div>
      ...
      <car - bottom - bar :cartList = "cartList" @deleteOk = "deleteOkP" :isedit = "isedit">
      </car - bottom - bar>
    </div>
  </template>

  <script>
  ...
  import CarBottomBar from './child/CarBottomBar'
  ...
  export default {
      name: 'Car',
      components: { ... ...CarBottomBar },
      ...
  }
  </script>

  <style>

  </style>
```

14.2.7 实现个人中心

个人中心，这里主要介绍5方面：①个人中心的框架和头部显示；②用户积分和余额显示；③订单概要信息显示；④广告显示；⑤其他常用功能。接下来逐个介绍它们的实现。

1. 实现个人中心的框架和头部显示

创建头部组件，创建 src/views/profile/child/ProfileHead.vue 文件，代码如下：

```
<template>
  <div class="profileHead">
    <div class="jq22-take-head">
      <div class="jq22-flex jq22-flex-one">
        <div class="jq22-take-user" @click="detailPage">
          <img :src="userData.image" alt />
        </div>
        <div class="jq22-flex-box" @click="detailPage">
          <h2>{{userData.userName}}</h2>
          <h2>
            <span>{{userData.Name}}</span>
            <span v-if="userData.gender - 0 === 0 ">先生</span>
            <span v-else-if="userData.gender - 0 ===1 ">女士</span>
          </h2>
          <span>
            <i class="icon icon-phone"></i>
            {{userData.phone}}
          </span>
        </div>
        <div class="jq22-take-button" @click="getPoint()">
          <button>
            <span v-if="userData.isSignInPoint">签到领积分</span>
            <span v-else>已签到</span>
          </button>
        </div>
      </div>
      <div class="jq22-flex jq22-flex-two" @click="addPoint()">
        <div class="jq22-flex-box">
          <h3>每天领红包,年卡仅 0.12 元/天</h3>
        </div>
        <div class="jq22-go-button">
          <button>立即开通</button>
        </div>
      </div>

      <img src="~assets/img/profile/images/head.png" alt />
    </div>
  </div>
</template>

<script>
  import { Toast } from "vant";
  export default {
    name: "profileHead",
    props: ["userData"],
    methods: {
```

```
        //跳转到个人详情页
        detailPage() {
          //路由到用户详细信息页面
        },
        //领取积分
        getPoint() {
          if (!this.userData.isSignInPoint) {
            return Toast.success("今日已签到,明天再来吧!");
          }
          const toast = Toast.loading({
            duration: 0,          //持续展示 toast
            message: "签到中...",
            forbidClick: true,
            loadingType: "spinner"
          });
          Toast.clear();
          //签到成功
          return this.$emit("SignInPointIsOK");
        },
        //获取积分
        addPoint() {
          this.$emit("addPoint");
        }
      }
    };
</script>

<style scoped>
  .jq22-take-button button {
    display: inline-block;
    line-height: 30px;
  }
  .jq22-take-button span,
  .jq22-take-button i {
    line-height: 30px;
  }
</style>
```

定义个人中心组件,创建 src/views/profile/Profile.vue 文件,代码如下:

```
<template>
    <div class="profile">
      <scroll
        class="content"
        ref="scroll"
        :probe-type="3"
```

```
        :pull-up-load="true"
        :handleToScroll="handleToScroll"
        :handleToTouchEnd="handleToTouchEnd"
      >
        <ul v-if="isShowLoading">
          <li class="pullDown">
            <i class="el-icon-loading"></i>
            {{ pullDownMsg }}
          </li>
        </ul>
      <section class="jq22-scrollView">
      <!-- 头部 -->
       <profile-head :userData="userData" @addPoint="under" @SignInPointIsOK=
"SignInPointIsOK"/>

      </section>
      </scroll>
    </div>
</template>

<script>
  import ProfileHead from './child/ProfileHead'
  import Scroll from "components/common/scroll/Scroll";

  import { Toast } from "vant";

  export default {
    name:'Profile',
    components:{
      ProfileHead,
      Scroll
    },
    created() {
      this.getUserData()
    },
    data() {
      return {
        //用户详细的数据
        userData: {},
        //显示 loading 文字
        pullDownMsg: "",
        //是否显示 loading
        isShowLoading: false
      };
    },
    methods: {
```

```
        getUserData() {

        },
        handleToScroll(pos) {
          if (pos.y > 50) {
            this.isShowLoading = true;
            this.pullDownMsg = "正在更新中...";
          }
        },
        handleToTouchEnd(pos) {
          if (pos.y > 50) {
            setTimeout(() => {
              this.pullDownMsg = "更新成功";
              this.$message.success('更新成功')
              this.isShowLoading = false;
            }, 2000);
          }
        },
        SignInPointIsOK(){
          this.userData.isSignInPoint = false
          this.userData.point += 1
        },
        under(){

        }
      }
    }
</script>

<style scoped>
  /* @import "assets/css/profile.css"; */
  @import url('../../assets/css/profile.css');
  .content {
    height: (100vh - 49px);
    position: absolute;
    top: 0;
    bottom: 49px;
    left: 0;
    right: 0;
    /* background-color: #fff; */
    overflow: hidden;
  }
  .pullDown {
    margin: 0;
    padding: 16px;
    text-align: center;
```

```
    border: none;
  }
  .profile{
    height:100vh;
    width: 100%;
    background: #efefef;
      color: #666;
  }
</style>
```

整合个人中心页面，在路由表中添加个人中心的路由信息，在 src/router/index.js 文件中添加的代码如下：

```
...
const Profile = () => import("../views/profile/Profile")

const routes = [
  ...
  {
    path: "/home",
    component: AppHome,
    children: [
      ...
      {
        path: "/profile",
        component: Profile,
        meta:{
          keepAlive:true                         //表示需要对页面进行缓存
        }
      },
        ...
    ]
  },
  ...
]
...
```

2. 实现用户积分和余额显示

创建显示余额和积分的组件，创建 src/views/profile/child/ProfilePoint.vue 文件，代码如下：

```
<template>
    <div class = "ProfilePoint">
      <div class = "jq22 - flex jq22 - flex - three">
```

```
            <div class = "jq22 - flex - box jq22 - flex - box - info" @click = "toBalancelLog()">
                <h4 >账号余额 <em >{{userData.money | sellCountFilter}}</em >元</h4 >
            </div >
            <div class = "jq22 - flex - box" @click = "toPointLog()">
                <h4 >积分 <em >{{userData.point | sellCountFilter}}</em >分</h4 >
            </div >
        </div >
    </div >
</template >

<script >
  export default {
    name:'ProfilePoint',
    props:['userData'],
    filters:{
        sellCountFilter(value){
            let rselt = value;
            if(value > 10000){
                rselt = (rselt / 10000).toFixed(1) + '万'
            }
            return rselt
        }
    },
    methods: {
        toBalancelLog(){
        },
        toPointLog(){
        }
    },
  }
</script >

<style >

</style >
```

在个人中心中整合用户余额和积分，在 src/views/profile/Profile.vue 文件中添加 ProfilePoint 组件，代码如下：

```
<template >
    <div class = "profile">
      <scroll
        class = "content"
        ref = "scroll"
        :probe - type = "3"
        :pull - up - load = "true"
```

```
            :handleToScroll = "handleToScroll"
            :handleToTouchEnd = "handleToTouchEnd"
        >
            ...
        <!-- 积分 -->
        <profile-point :userData = "userData"></profile-point>
        </section>
            </scroll>
    </div>
</template>

<script>
    ...
    import ProfilePoint from './child/ProfilePoint.vue';

    export default {
      name:'Profile',
      components:{
        ...
        ProfilePoint
      },
      ...
    }
</script>

<style scoped>
    ...
</style>
```

3. 实现订单概要信息显示

创建显示订单概要信息的组件，创建 src/views/profile/child/ProfileOrder.vue 文件，代码如下：

```
<template>
    <div class = "ProfileOrder">
        <div class = "jq22-take-item">
        <div class = "jq22-flex" style = "padding-bottom:0">
            <div class = "jq22-flex-box">
            <h1>我的订单</h1>
            </div>
            <div class = "jq22-arrow" @click = "toOrder(0)">
            <span>查看全部订单</span>
            </div>
        </div>
```

```
<div class="jq22-palace">
    <a href="JavaScript:;" class="jq22-palace-grid"  @click="toOrder(0)">
    <div class="jq22-palace-grid-icon">
        <el-badge :value="OrderData.all" :max="10" class="item">
        <img src="~assets/img/profile/images/nav-013.png" alt />
        </el-badge>
    </div>
    <div class="jq22-palace-grid-text">
        <h2>全部</h2>
    </div>
    </a>
    <a href="JavaScript:;" class="jq22-palace-grid"  @click="toOrder(1)">
    <div class="jq22-palace-grid-icon">
        <el-badge :value="OrderData.unPaid" :max="10" class="item">
        <img src="~assets/img/profile/images/nav-001.png" alt />
        </el-badge>
    </div>
    <div class="jq22-palace-grid-text">
        <h2>待支付</h2>
    </div>
    </a>
    <a href="JavaScript:;" class="jq22-palace-grid"  @click="toOrder(2)">
    <div class="jq22-palace-grid-icon">
        <el-badge :value="OrderData.isPaid" :max="10" class="item">
        <img src="~assets/img/profile/images/nav-002.png" alt />
        </el-badge>
    </div>
    <div class="jq22-palace-grid-text">
        <h2>待收货</h2>
    </div>
    </a>
    <a href="JavaScript:;" class="jq22-palace-grid"  @click="toOrder(3)">
    <div class="jq22-palace-grid-icon">
        <el-badge :value="OrderData.unEvaluated" :max="10" class="item">
        <img src="~assets/img/profile/images/nav-003.png" alt />
        </el-badge>
    </div>
    <div class="jq22-palace-grid-text">
        <h2>待评价</h2>
    </div>
    </a>
    <!-- <a href="JavaScript:;" class="jq22-palace-grid" @click="toOrder(4)">
    <div class="jq22-palace-grid-icon">
        <el-badge :value="OrderData.afterSale" :max="10" class="item">
        <img src="~assets/img/profile/images/nav-004.png" alt />
        </el-badge>
```

```
                </div>
                <div class="jq22-palace-grid-text">
                    <h2>售后</h2>
                </div>
                </a> -->
            </div>
            </div>
        </div>
</template>

<script>
export default {
    name: "ProfileOrder",
    props:['OrderData'],
    methods: {
        toOrder(index){
            //this.$router.push('/order/' + index)
        }
    },
};
</script>

<style>
</style>
```

在个人中心中整合订单概要信息组件,在 src/views/profile/Profile.vue 文件中使用 ProfileOrder 组件,代码如下:

```
<template>
    <div class="profile">
      <scroll
        class="content"
        ref="scroll"
        :probe-type="3"
        :pull-up-load="true"
        :handleToScroll="handleToScroll"
        :handleToTouchEnd="handleToTouchEnd"
      >
      ...
      <!-- 订单 -->
      <profile-order :OrderData="userData"></profile-order>
      </section>
      </scroll>
    </div>
</template>
```

```
<script>
  …
  import ProfileOrder from './child/ProfileOrder.vue';
  …
  export default {
    name:'Profile',
    components:{
      …
      ProfileOrder
    },
    …
  }
</script>
```

4. 实现广告显示

定义广告组件，创建 src/views/profile/child/ProfileAdvs.vue 文件，代码如下：

```
<template>
    <!-- 广告 -->
    <div class="ProfileAdvs" @click="toAdv()">
    <div class="jq22-flex jq22-flex-fore">
            <div class="jq22-flex-box">
                <img src="~assets/img/profile/images/ad-001.png" alt="">
            </div>
        </div>
    </div>
</template>

<script>
export default {
    name:'ProfileAdvs',
    methods: {
        toAdv(){
            this.$emit('toAdv')
        }
    },
}
</script>

<style>

</style>
```

在个人中心整合广告组件，代码如下：

```
<template>
    <div class="profile">
      <scroll
        class="content"
        ref="scroll"
        :probe-type="3"
        :pull-up-load="true"
        :handleToScroll="handleToScroll"
        :handleToTouchEnd="handleToTouchEnd"
      >
      ...
      <!-- 广告 -->
      <profile-advs :OrderData="userData"></profile-advs>
      </section>
      </scroll>
    </div>
</template>

<script>
  ...
  import ProfileOrder from './child/ProfileAdvs.vue';
  ...
  export default {
    name:'Profile',
    components:{
      ...
      ProfileAdvs
    },
    ...
  }
</script>
...
```

5. 实现其他常用功能

创建 src/views/profile/ProfileFun.vue 文件,代码如下:

```
<template>
  <div class="ProfileFun">
    <div class="jq22-take-item">
                <div class="jq22-flex">
                        <div class="jq22-flex-box">
                                <h1>常用功能</h1>
                        </div>

              </div>
```

```
                <div class = "jq22 - palace">
                        <a href = "JavaScript:;" class = "jq22 - palace - grid" @click =
"toShip">
                            <div class = "jq22 - palace - grid - icon">
                                <img src = "~assets/img/profile/images/nav -
006.png" alt = "">
                            </div>
                                <div class = "jq22 - palace - grid - text">
                            <h2>收货地址</h2>
                            </div>
                        </a>
                        <a href = "JavaScript:;" class = "jq22 - palace - grid" @click =
"toFavorite">
                            <div class = "jq22 - palace - grid - icon">
                                <img src = "~assets/img/common/sc.png" alt = "">
                            </div>
                            <div class = "jq22 - palace - grid - text">
                                <h2>我的收藏</h2>
                            </div>
                        </a><a href = "JavaScript:;" class = "jq22 - palace - grid"
@click = "footprint">
                            <div class = "jq22 - palace - grid - icon">
                                    <img src = "~assets/img/profile/images/
foot.png" alt = "">
                            </div>
                            <div class = "jq22 - palace - grid - text">
                                <h2>浏览记录</h2>
                            </div>
                        </a>
                        <a href = "JavaScript:;" class = "jq22 - palace - grid" @click =
"toSetting">
                            <div class = "jq22 - palace - grid - icon">
                                <img src = "~assets/img/profile/images/nav -
012.png" alt = "">
                            </div>
                            <div class = "jq22 - palace - grid - text">
                                <h2>设置</h2>
                            </div>
                        </a>

                </div>
            </div>
  </div>
</template>

<script>
```

```
import { Toast } from 'vant';
export default {
  name:'ProfileFun',
  methods: {
          toShip(){
                  // this. $router.push('/ship')
          },
          toFavorite(){
                  //this. $router.push('/favorite')
          },
          toSetting(){
                  //this. $router.push('/setting')
          },
          footprint(){
                  //this. $router.push('/footprint')
          }
    },
}
</script>

<style>

</style>
```

将常用功能组件整合到个人中心，代码如下：

```
<template>
    <div class = "profile">
      <scroll
        class = "content"
        ref = "scroll"
        :probe - type = "3"
        :pull - up - load = "true"
        :handleToScroll = "handleToScroll"
        :handleToTouchEnd = "handleToTouchEnd"
      >
      ...
      <! -- 广告 -->
      <profile - fun />
      </section>
        </scroll>
    </div>
</template>

<script>
```

```
    ...
    import ProfileFun from './child/ProfileFun.vue';
    ...
    export default {
      name:'Profile',
      components:{
        ...
        ProfileFun
      },
      ...
    }
  </script>
  ...
```

14.2.8 实现商品分类

1. 实现分类主组件

创建 src/views/category/Category.vue 文件,代码如下:

```
<template>
    <div class="category">
        <nav-bar>
            <div slot="center">商品分类</div>
        </nav-bar>
        <!-- 搜索框 -->
        <div @click="toSearch">
        <el-input  placeholder="请输入需要搜索的商品">
            <i slot="prefix" class="el-input__icon el-icon-search"></i>
        </el-input>
        </div>
    </div>
</template>

<script>
import NavBar from "components/common/navbar/NavBar.vue";
import { Sidebar, SidebarItem,Toast } from 'vant';
export default {
name: "CateGory",
components: {
    NavBar,
},
created() {
},
methods: {
    toSearch(){
```

```
      }
    }
  };
  </script>

  <style scoped>
  </style>
```

2. 创建 TabControl

创建 src/components/content/tabControl/TabControl.vue 文件,代码如下:

```
<template>
  <div class="tab-control">
    <div v-for="(item,index) in titles" class="tab-control-item"
    :class="{active:index === currentIndex}" @click="itemClick(index)">
      <span>{{item}}</span>
    </div>
  </div>
</template>

<script>
export default {
  name:'TabControl',
  props:{
    titles:{
      type:Array,
      default(){
        return []
      }
    }
  },
  data(){
    return {
      currentIndex:0
    }
  },
  methods:{
    itemClick(index){
      this.currentIndex = index;
      this.$emit('tbClick',index)
    }
  }
}
</script>
```

```
<style>
  .tab-control{
    display: flex;
    text-align: center;
    font-size: 16px;
    text-align: 40px;
    line-height: 40px;
  }
  .tab-control-item{
    flex:1;
  }
  .tab-control-item span{
    padding: 5px;
  }
  .active{
    color: var(--color-hight-text);
  }
  .active span{
    border-bottom: 2px solid var(--color-hight-text);
  }
</style>
```

3. 创建 SideBar

创建 src/views/category/child/SideBar.vue 文件，代码如下：

```
<template>
  <div class="SideBar">
    <van-row>
      <van-col span="5">
        <!-- 侧边导航栏 -->
        <van-sidebar v-model="activeKey" @change="onChange">
          <van-sidebar-item
            v-for="(item,index) in sideBarList"
            :key="index"
            :title="item.cateName"
          />
        </van-sidebar>
      </van-col>
      <van-col span="19">
        <!-- 用于商品展示 -->
        <tab-control :titles="['综合','销量','新品']" ref="tabControl" @tbClick=
"tbClick"></tab-control>
        <scroll
          class="content"
          ref="scroll"
```

```
          :probe-type="3"
          :pull-up-load="true"
          @scroll="scroll"
          @pullingUp="loadMore"
          :handleToScroll="handleToScroll"
          :handleToTouchEnd="handleToTouchEnd"
        >
          <ul v-if="isShowLoading">
            <li class="pullDown">
              <i class="el-icon-loading"></i>
              {{ pullDownMsg }}
            </li>
          </ul>
          <goods-list :goods="showGoods"></goods-list>
          <ul v-if="isShowUpLoading">
            <li class="pullUp">
              <i class="el-icon-loading"></i>
              {{ pullUpMsg }}
            </li>
          </ul>
        </scroll>
      </van-col>
    </van-row>
  </div>
</template>

<script>
import GoodsList from "components/content/goods/GoodsList";
import TabControl from "components/content/tabControl/TabControl";

import Scroll from "components/common/scroll/Scroll";
import { debounce } from "common/utils";              //防抖函数
import { Toast } from "vant";
export default {
  name: "SideBar",
  props: ["sideBarList"],
  data() {
    return {
      activeKey: 0,                                    //激活的侧边栏
      goods: {
        syn: { page: 1, list: [] },
        sales: { page: 1, list: [] },
        new: { page: 1, list: [] },
        price: { page: 1, list: [] }
      },                                               //商品的展示信息
      pageSize:10,
```

```
      sortType: "syn",                          //选择的排序
      sideType: "",                             //激活的侧边栏的分类 id
      isShowLoading: false,
      pullUpMsg: "加载更多....",
      isShowUpLoading: false,
      //是否显示,回到顶部按钮
      isShowBackTop: false
    };
  },
  components: {
    TabControl,
    GoodsList,
    Scroll,
  },
  //监听打开事件
  activated() {
    this.$refs.scroll.refresh();
    this.$refs.scroll.scrollTo(0, this.saveY, 0);
  },
  //监听离开事件
  deactivated() {
    //保存 Y 值
    this.saveY = this.$refs.scroll.scroll.y;
  },
  computed: {
    showGoods() {
      return this.goods[this.sortType].list;
    }
  },
  activated() {
    this.$refs.scroll.refresh();
  },
  mounted() {
    const refresh = debounce(this.$refs.scroll.refresh, 200);
    //监听 item 图片加载完成
    this.$bus.$on("itemImageLoad", () => {
      refresh();
    });
  },
  computed: {
    showGoods() {
      return this.goods[this.sortType].list;
    }
  },
  methods: {
    //切换分类
```

```
onChange(index) {
  this.listClear();
  this.getCateGoods(index);
},
//页面创建时进行数据获取
createdData() {
  this.getCateGoods(0);
},
//切换排序
//切换显示的样式：价格、销量
tbClick(index) {
  this.isPrice = false;
  switch (index) {
    case 0:
      //综合
      this.sortType = "syn";
      break;
    case 1:
      //销量
      this.sortType = "sales";
      break;
    case 2:
      //价格
      this.sortType = "new";
      break;
    case 3:
      this.sortType = "price";
      //将价格显示器显示出来
      this.isPrice = true;
      break;
  }
  this.listClear()
  this.getCateGoods(this.activeKey);
},
listClear(){
//接口前处理
  this.goods.syn.page = 1;
  this.goods.sales.page = 1;
  this.goods.new.page = 1;
  this.goods.price.page = 1;

  this.goods.syn.list = [];
  this.goods.sales.list = [];
  this.goods.new.list = [];
  this.goods.price.list = [];
},
```

```
//获取展示的数据
getCateGoods(cateId) {
  //cateId, pageNo, pageSize, sortType
  const params = {
    cateId,
    pageNo: this.goods[this.sortType].page,
    pageSize: this.pageSize,
    sortType: this.sortType
  }
  getCategoryGoodsList(params).then(res => {
    Toast.clear();
    this.goods[this.sortType].list.push(...res.data);
    this.goods[this.sortType].page += 1;

    this.ok()
    return this.finish();
  }).catch( err => {
    Toast.clear();
    Toast.fail("加载失败");
    console.log(err)
  })
},

ok() {
  this.pullDownMsg = "更新成功";
  this.tabIsShow = true;
  this.isShowLoading = false;
  this.isShowUpLoading = false;
},
finish() {
  this.$refs.scroll.finishPullUp();
},
//上拉加载更多,分页数据
loadMore() {
  this.isShowUpLoading = true;
  this.pullDownMsg = "正在更新中...";
  this.getCateGoods(this.activeKey);
},
handleToScroll(pos) {
  if (pos.y > 60) {
    this.isShowLoading = true;
    this.pullDownMsg = "正在更新中...";
  }
},
//下拉刷新
handleToTouchEnd(pos) {
```

```
        if (pos.y > 60) {
          //this.page = 1;
          this.goods[this.sortType].page = 1;
          this.getCateGoods(this.activeKey);
        }
      },
      //确认显示还是隐藏,回到顶部
      scroll(position) {
        //判断 BackTop 是否显示
        this.isShowBackTop = -position.y > 1000;
      },
    }
};
</script>

<style scoped>
.wrapper {
  height: calc(100vh - 173px);
  /* position: absolute; */
  /* top: 66px; */
  bottom: 49px;
  left: 0;
  right: 0;
  /* background-color: #fff; */
  overflow: hidden;
}
.pullDown {
  margin: 0;
  padding: 16px;
  text-align: center;
  border: none;
}
.pullUp {
  margin: 0;
  text-align: center;
  border: none;
}
.back-top {
  border: 1px solid red;
}
</style>
```

14.3 提供 Mock 模拟数据

Mock.js 是一款模拟数据生成器，能根据数据模板生成模拟数据，能模拟 Ajax 请求，生成并返回模拟数据，能根据基于 HTML 的模板生成模拟数据，接下来介绍怎样基于 Mock 给前面的功能提供模拟数据。

14.3.1 搭建 Mock 框架

安装 Mock.js，代码如下：

```
npm install mockjs --save-dev                //--save
```

定义环境变量，在项目根路径下，创建.env.dev 文件，代码如下：

```
//just a flag
ENV = 'dev'
NODE_ENV = 'dev'                              //固定环境变量名称

//base api  自定义环境变量名称,需要以 VUE_APP_为前缀,并且全部大写
VUE_APP_BASE_API = '/dev-api'
```

添加 dev 的启动方式，打开根目录下的 package.json 文件，在 scripts 中添加 dev 的启动项，代码如下：

```
{
  "name": "mock",
  "version": "0.1.0",
  "private": true,
  "scripts": {
    "serve": "Vue-CLI-service serve",
    "dev": "Vue-CLI-service serve --mode dev",         //加载.env.dev 配置文件
    "build": "Vue-CLI-service build",
    "test:unit": "Vue-CLI-service test:unit"
  },
  ...
}
```

创建请求响应，在根目录下创建 mock/requests/user.js 文件，代码如下：

```
module.exports = [
    {
        url: '/user/test',
        type: 'get',
```

```
            response: config => {
                const query = config.query
                console.log(query)

                return {
                    code: 20000,
                    data: {
                        name: '张三',
                        age: 18
                    }
                }
            }
        }
]
```

合并 Mock 请求，在 mock 目录下创建 index.js 文件，代码如下：

```
const user = require('./requests/user')

const mocks = [
    ...user
]

module.exports = {
    mocks
}
```

创建 mock-server，在 mock 目录下创建 mock-sever.js 文件，代码如下：

```
const chokidar = require('chokidar')
const bodyParser = require('body-parser')
const chalk = require('chalk')
const path = require('path')
const Mock = require('mockjs')

const mockDir = path.join(process.cwd(), 'mock')

function registerRoutes(app) {
  let mockLastIndex
  const { mocks } = require('./index.js')
  const mocksForServer = mocks.map(route => {
    return responseFake(route.url, route.type, route.response)
  })
  for (const mock of mocksForServer) {
    app[mock.type](mock.url, mock.response)
    mockLastIndex = app._router.stack.length
```

```
    }
    const mockRoutesLength = Object.keys(mocksForServer).length
    return {
      mockRoutesLength: mockRoutesLength,
      mockStartIndex: mockLastIndex - mockRoutesLength
    }
  }

  function unregisterRoutes() {
    Object.keys(require.cache).forEach(i => {
      if (i.includes(mockDir)) {
        delete require.cache[require.resolve(i)]
      }
    })
  }

  //for mock server
  const responseFake = (url, type, respond) => {
    return {
      url: new RegExp(`${process.env.VUE_APP_BASE_API}${url}`),
      type: type || 'get',
      response(req, res) {
        console.log('request invoke:' + req.path)
        res.json(Mock.mock(respond instanceof Function ? respond(req, res) : respond))
      }
    }
  }

  module.exports = app => {
    //parse app.body
    //https://expressjs.com/en/4x/api.html#req.body
    app.use(bodyParser.json())
    app.use(bodyParser.urlencoded({
      extended: true
    }))

    const mockRoutes = registerRoutes(app)
    var mockRoutesLength = mockRoutes.mockRoutesLength
    var mockStartIndex = mockRoutes.mockStartIndex

    //watch files, hot reload mock server
    chokidar.watch(mockDir, {
      ignored: /mock-server/,
      ignoreInitial: true
    }).on('all', (event, path) => {
      if (event === 'change' || event === 'add') {
```

```
        try {
          //remove mock routes stack
          app._router.stack.splice(mockStartIndex, mockRoutesLength)

          //clear routes cache
          unregisterRoutes()

          const mockRoutes = registerRoutes(app)
          mockRoutesLength = mockRoutes.mockRoutesLength
          mockStartIndex = mockRoutes.mockStartIndex

           console.log(chalk.magentaBright(`\n > Mock Server hot reload success! changed
${path}`))
        } catch (error) {
          console.log(chalk.redBright(error))
        }
      }
    })
}
```

添加 Mock 拦截，在根目录下创建 vue.config.js 文件，代码如下：

```
module.exports = {
    devServer: {
      before: require('./mock/mock-server')
    }
}
```

启动应用，用浏览器请求服务器就可以访问了。

14.3.2 搭建 axios 请求框架

安装 axios，代码如下：

```
npm install axios --save
```

封装 axios 的请求和响应，分支 axios 实例对象，并且定义拦截器拦截请求和响应。具有拦截器的 axios 的请求响应的流程如图 14.6 所示。

创建 src/utils/request.js 文件，内容如下：

```
import axios from 'axios'
import { MessageBox, Message } from 'element-ui'
//创建 axios 实例
const service = axios.create({
    baseURL: process.env.VUE_APP_BASE_API, //url = base url + request url
```

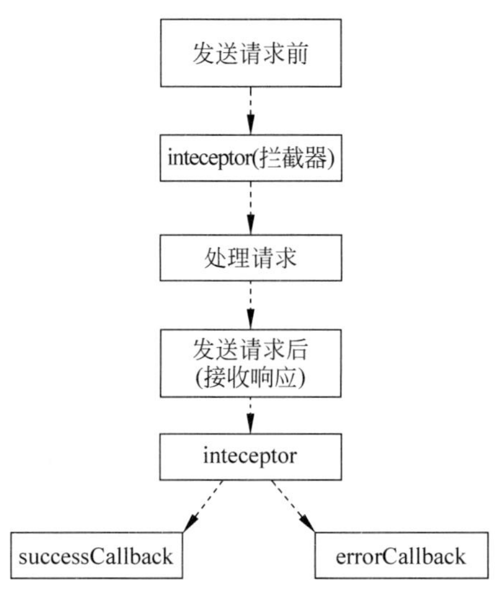

图 14.6 axios 请求响应流程

```
    withCredentials: true,                    //跨域的时候,支持 Cookie
    timeout: 5000 ,                           //超时
    crossDomain: true                         //支持跨域
})

//设置 post、put、delete 默认 Content - Type
service.defaults.headers.post['Content - Type'] = 'application/json'
service.defaults.headers.put['Content - Type'] = 'application/json'
service.defaults.headers.delete['Content - Type'] = 'application/json'

//请求拦截器
service.interceptors.request.use(
    config => {
        if (config.method === 'post' ||
            config.method === 'put' ||
            config.method === 'delete') {
            //post、put 提交时,将对象转换为 string, 为处理 Java 后台解析问题
            config.data = JSON.stringify(config.data)
        }
        //请求发送前进行处理
        return config
    },
    error => {
      //错误处理
      console.log(error) //for debug
      return Promise.reject(error)
    }
```

```
)

//响应拦截器
service.interceptors.response.use(
  /**
   * If you want to get http information such as headers or status
   * Please return  response => response
  */

  /**
   * Determine the request status by custom code
   * Here is just an example
   * You can also judge the status by HTTP Status Code
   */
  response => {
    const res = response.data

    //if the custom code is not 20000, it is judged as an error.
    if (res.code !== 20000) {
      Message({
        message: res.data || 'Error',
        type: 'error',
        duration: 5 * 1000
      })

      //50008: Illegal token; 50012: Other clients logged in; 50014: Token expired;
      if (res.code === 50008 || res.code === 50012 || res.code === 50014) {
        //to re-login
        MessageBox.confirm('登录失效', '确定重新登录?', {
          confirmButtonText: '重新登录',
          cancelButtonText: '取消',
          type: 'warning'
        }).then(() => {
            Message({
              message: '需要重新登录',
              type: 'info',
              duration: 5 * 1000
            })
        })
      }
      return Promise.reject(new Error(res.data || 'Error'))
    } else {
      return res
    }
  },
  error => {
```

```
        console.log('err' + error) //for debug
        Message({
          message: error.message,
          type: 'error',
          duration: 5 * 1000
        })
        return Promise.reject(error)
      }
)

export default service
```

14.3.3 改造 Home

创建商品的请求模块，包含查询商品列表的方法。创建 src/api/goods.js 文件，定义 getGoodsList()函数，代码如下：

```
import request from 'utils/request'

export function getGoodsList(query){
   return request({
      url: '/goods/list',
      method: 'get',
      params: query
   })
}
```

因为引入了 src/utils/request.js 文件，而代码里面写的是'utils/request'，所以需要在 vue.config.js 文件中添加 src/utils 的别名 utils，代码如下：

```
module.exports = {
  //对象和函数都可以,如果要控制开发环境,则可以选择函数
  configureWebpack:{
      resolve:{
        alias:{                           //定义路径的别名
           'assets':path.resolve('./src/assets'),
           'common':path.resolve('./src/common'),
           'components':path.resolve('./src/components'),
           'network':path.resolve('./src/network'),
           'views':path.resolve('./src/views'),
           'api':path.resolve('./src/api'),
           'utils':path.resolve('./src/utils'),
           '@':path.resolve('./src')
        }
```

```
            }
        },
        ...
    }
    ...
}
```

使用 Mock 模拟生成商品列表数据，创建 mock/responses/goods. js 文件，生成商品列表数据，并且导出商品列表，代码如下：

```
const Mock = require('mockjs')

const shopStore = Mock.mock({
    "goodsList|20 - 20":[
        {
            "goodsId| + 1": 1,
            "goodsName|1": ["红色复古连衣裙女夏装2020新款小香风收腰显瘦气质法式长裙子",
                           "格子西装连衣裙女夏装2020新款可甜高腰a字甜美法式初恋裙子",
                           "韩版新款短袖时尚纯棉宽松中袖黑白条纹T恤女简约圆领韩国百搭潮",
                           "时尚套装春韩版打底修身包臀吊带连衣裙+拉链长袖开衫外套两件套"],
            "goodsPrice|100 - 200": 110,
            "goodsFav|10 - 50": 10,
            "goodsOldPrice|100 - 300": 110,
            "goodsLogo|1": ["/images/goods1.jpg","/images/goods2.jpg",
                           "/images/goods3.jpg","/images/goods4.jpg"],
            "goodsAre|1": ["浙江省杭州市","广东省广州市","湖南省长沙市","湖北省武汉市"],
            "goodsBuyNum|1 - 100": 1,
            "goodsIntroduce": "<div class = \"panel - title\"><h1>图文详情</h1></div><!--
描述 -->        <div class = \"graphic - pic\"><div class = \"pic - box\" style2 = \"padding -
bottom:127.97468354430379 %;\"><img class = \"lazy\" style2 = \"left: - 350px;\" src = \"/
images/detail_img1.jpg\" src = \"/images/detail_img2.gif\"></div></div>        <div class = \
"graphic - pic\"><div class = \"pic - box\" style2 = \"padding - bottom:101.7721518987342 %;\">
<img class = \"lazy\" style2 = \"left: - 350px;\" src = \"/images/detail_img3.jpg\" src = \
"/images/detail_img2.gif\"></div></div>        <div class = \"graphic - pic\"><div class = \
"pic - box\" style2 = \"padding - bottom:157.0886075949367 %;\"><img class = \"lazy\" style2 = \
"left: - 350px;\" src = \"/images/detail_img6.jpg\" src = \"/images/detail_img2.gif\"></div>
</div>        <div class = \"graphic - pic\"><div class = \"pic - box\" style2 = \"padding -
bottom:138.35443037974684 %;\"><img class = \"lazy\" style2 = \"left: - 350px;\" src = \"/
images/detail_img7.jpg\" src = \"/images/detail_img2.gif\"></div></div>        <div class = \
"graphic - pic\"><div class = \"pic - box\" style2 = \"padding - bottom:19.746835443037973 %;\">
<img class = \"lazy\" style2 = \"left: - 350px;\" src = \"/images/detail_img8.jpg\" src = \"/
images/detail_img2.gif\"></div></div>        <div class = \"graphic - pic\"><div class = \
"pic - box\" style2 = \"padding - bottom:126.45569620253166 %;\"><img class = \"lazy\" style2 = \
"left: - 350px;\" src = \"/images/detail_img1.jpg\" src = \"/images/detail_img2.gif\"></div>
</div>        <div class = \"graphic - pic\"><div class = \"pic - box\" style2 = \"padding -
```

```
bottom:144.30379746835442 % ;\"> < img class = \"lazy\" style2 = \"left: - 350px;\" src = \"/images/detail_img10. jpg\" src = \"/images/detail_img2. gif\"> </div> </div>    < div class = \"graphic - pic\"> < div class = \"pic - box\" style2 = \"padding - bottom:93. 29113924050633 % ;\"> < img class = \"lazy\" style2 = \"left: - 350px;\" src = \"/images/detail_img11. jpg\" src = \"/images/detail_img2. gif\"> </div > </div >    < div class = \"graphic - pic\"> < div class = \"pic - box\" style2 = \"padding - bottom: 140 % ; \"> < img class = \"lazy\" style2 = \"left: - 350px;\" src = \"/image/detail_img12. jpg\" src = \"/images/detail_img2. gif\"> </div > </div >    < div class = \"graphic - pic\"> < div class = \"pic - box\" style2 = \"padding - bottom:87. 46835443037975 % ;\"> < img class = \"lazy\" style2 = \"left: - 350px; \" src = \"/images/detail_img13. jpg\" src = \"/images/detail_img2. gif\"> </div> </div>",
            "isDel": 0,
            "topImages": "swiperImage1. jpg, swiperImage2. jpg, swiperImage3. jpg, swiperImage4. jpg",
            "num|100 - 300": 100,
            "isFav|1": false,
            "hot|1 - 30": 4,
            "status": 0
        }
    ]
})

function getGoodsList() {
    return shopStore. goodsList;
}
//暴露响应
module. exports = [
    {
        url: '/goods/list',
        type: 'get',
        response: config = > {
            return {
                code: 20000,
                data: getGoodsList()
            }
        }
    }
]
```

注意：新增加了 goods. js 文件，定义了响应请求和生成模拟数据的方法，需要注册到 Mock-Server 中去。

在 mock/index. js 文件中，添加注册 goods. js 文件，代码如下：

```
...
const goods = require('. /responses/goods') //引入

const mocks = [
```

```
    ...demo,
    ...goods  //注册
]
...
```

添加 axios 请求方法，请求模拟数据。创建 src/api/goods.js 文件，添加 getGoodsList() 函数，向 MockServer 发送 axios 请求，以便获得商品列表数据，代码如下：

```
import request from 'utils/request'

export function getGoodsList(query){
    return request({
        url: '/goods/list',
        method: 'get',
        params: query
    })
}
```

完善 Home.vue 文件，支持模拟数据显示，按以下方式调整 src/views/home/Home.vue 文件的代码。

(1) 添加 import { debounce } from "@/common/utils" 代码，以便于加载数据后无抖动刷新。

(2) 在 Vue.js 对象中，添加 refreshScroll()方法，实现无抖动刷新代码。

(3) 修改 Vue.js 对象中的 getGoodsList()方法，改成以 Promise 的方式调用。

最后的样例代码如下：

```
<template>
    ...
</template>

<script>
    ...
    import { debounce } from "@/common/utils"
    import { getGoodsList } from 'api/goods.js'

    export default {
        name: 'Home',
        data(){
            ...
        },
        components: {
            ...
        },
```

```
        created(){
            this.getGoodsList()
        },
        computed: {
            showGoods() {
                return this.goodsList
            }
        },
        methods: {
            getGoodsList(){
                getGoodsList().then(res =>{
                    this.goodsList = res.data
                    this.refreshScroll()
                })
            },
            toSearch(){
                console.log('to search ...')
            },
            //加载更多
            loadMore() {
                //查询更多的商品
                console.log('to search 更多的商品')
            },
            //下拉刷新,控制提示信息
            handleToScroll(pos) {
                if (pos.y > 50) {
                    this.isShowLoading = true;
                    this.pullDownMsg = "正在更新中...";
                    this.$router.go(0);
                }
            },

            refreshScroll() {
                //无抖动刷新,以便 scroll 组件正确地初始化
                const refresh = debounce( this.$refs.scroll.refresh,300)
                refresh()
            },
        }
    }
</script>

<!-- 定义的样式,只在当前 Vue.js 中有效 -->
<style scoped>
...
</style>
```

14.3.4 改造显示详细信息页面

在 Mock 中生成商品详细信息数据，打开 mock/responses/goods.js 文件，在里面添加商品信息生成代码，同时暴露 getGoodsInfo 请求，代码如下：

```
const Mock = require('mockjs')

const shopStore = Mock.mock({
    "goodsList|20-20":[...],
    "goodsInfos|20-20":[
        {
            serves: [{                  //服务
                "icon": "ok.png",
                "name": "延误必赔"
            }, {
                "icon": "ok.png",
                "name": "退货补运费"
            }, {
                "icon": "ok.png",
                "name": "全国包邮"
            }, {
                "icon": "ok.png",
                "name": "七天无理由退货"
            }],
            commentInfo: {              //评论回复
                sum: 2,
                comments: [
                    {
                        commentImg: ['/images/comment/img1.jpg'],
                        commentBody: {
                            commentId: 1,
                            productId: 2,
                            orderId: 2,
                            userName: 'zhangsan',
                            userImg: '/images/user/img1.jpg',
                            imgUrl: '/images/goods1.jpg',
                            content: '测试评论 1,不错不错',
                            fwValue: 5,
                            wlValue: 6,
                            auditTime: '1593765746',
                            modifiedTime: '1593765746'
                        }
                    },
                    {
                        commentImg: ['/images/comment/img2.jpg'],
```

```
                        commentBody: {
                            commentId: 1,
                            productId: 2,
                            orderId: 2,
                            userName: 'lisi',
                            userImg: '/images/user/img2.jpg',
                            imgUrl: '/images/goods2.jpg',
                            content: '测试评论 2,不错不错',
                            fwValue: 5,
                            wlValue: 6,
                            auditTime: '1593765746',
                            modifiedTime: '1593765746'
                        }
                    }
                ]
            }
        }
    ]
})

function getGoodsInfo(id) {
    return {
        info: shopStore.goodsList[id],
        serves: shopStore.goodsInfos[id].serves,
        commentInfo: shopStore.goodsInfos[id].commentInfo
    }
}
...

module.exports = [
    ...
    {                                               //获取商品信息请求
        url:'/goods/info',
        type:'get',
        response: config => {
            const { id } = config.query             //获取参数 id 的值
            return {
                code: 20000,
                data: getGoodsInfo(id)
            }
        }
    }
]
```

改造 Mock 的 XmlHttpRequest，创建 mock/utils.js 文件，定义 param2Obj()和 deepClone()两个函数，代码如下：

```
/**
 * @param {string} url
 * @returns {Object}
 */
function param2Obj(url) {
  const search = decodeURIComponent(url.split('?')[1]).replace(/\+/g, ' ')
  if (!search) {
    return {}
  }
  const obj = {}
  const searchArr = search.split('&')
  searchArr.forEach(v => {
    const index = v.indexOf('=')
    if (index !== -1) {
      const name = v.substring(0, index)
      const val = v.substring(index + 1, v.length)
      obj[name] = val
    }
  })
  return obj
}

/**
 * This is just a simple version of deep copy
 * Has a lot of edge cases bug
 * If you want to use a perfect deep copy, use lodash's _.cloneDeep
 * @param {Object} source
 * @returns {Object}
 */
function deepClone(source) {
  if (!source && typeof source !== 'object') {
    throw new Error('error arguments', 'deepClone')
  }
  const targetObj = source.constructor === Array ? [] : {}
  Object.keys(source).forEach(keys => {
    if (source[keys] && typeof source[keys] === 'object') {
      targetObj[keys] = deepClone(source[keys])
    } else {
      targetObj[keys] = source[keys]
    }
  })
  return targetObj
}

module.exports = {
  param2Obj,
  deepClone
}
```

修改 mock/index.js 文件，添加改造 Mock 的 XmlHttpRequest，代码如下：

```
const { param2Obj } = require('./utils')

...

//for front mock
//please use it cautiously, it will redefine XMLHttpRequest
//which will cause many of your third - party libraries to be invalidated(like progress event).
function mockXHR() {
    //mock patch
    //https://github.com/nuysoft/Mock/issues/300
    Mock.XHR.prototype.proxy_send = Mock.XHR.prototype.send
    Mock.XHR.prototype.send = function() {
      if (this.custom.xhr) {
        this.custom.xhr.withCredentials = this.withCredentials || false

        if (this.responseType) {
          this.custom.xhr.responseType = this.responseType
        }
      }
      this.proxy_send(...arguments)
    }

    function XHR2ExpressReqWrap(respond) {
      return function(options) {
        let result = null
        if (respond instanceof Function) {
          const { body, type, url } = options
          //https://expressjs.com/en/4x/api.html#req
          result = respond({
            method: type,
            body: JSON.parse(body),
            query: param2Obj(url)
          })
        } else {
          result = respond
        }
        return Mock.mock(result)
      }
    }

    for (const i of mocks) {
      Mock.mock(new RegExp(i.url),
                i.type || 'get',
                XHR2ExpressReqWrap(i.response))
```

```
    }
  }

  module.exports = {
      mocks,
      mockXHR           //添加 mockXHR
  }
```

添加获取商品信息的 axios 请求，编辑 src/api/goods.js 文件，添加 getGoodsInfo()函数，代码如下：

```
...
export function getGoodsInfo(id){
    return request({
        url: '/goods/info',
        method: 'get',
        params: {id}
    })
}
```

完善显示详细信息的 Detail 组件，编辑 src/views/detail/Detail.vue 文件的 getGoodsInfo()方法，完善里面的 getGoodsInfo 请求方法的调用，代码如下：

```
<template>
    ...
</template>

<script>
    ...

    export default {
        ...
        methods: {
            getGoodsInfo() {
                ...
                //往后台发送请求获取商品详细(修改成发送 axios 请求的调用方式)
                getGoodsInfo(this.goodsId).then(res => {
                    const goods = res.data
                    //获取商品详细信息
                    this.goodsInfo = goods.info

                    //获取商品信息中的轮播图
                    this.topImages = this.goodsInfo.topImages.split(',')
                    this.topImages.forEach((img, index) => {
                        //'/images/' + img
```

```
                this.topImages[index] = this.tinyImgPath(img)
            })
            //获取商品的基本信息
            this.serves = goods.serves
            this.serves.forEach( (serve, index) => {
                this.serves[index].icon = this.tinyImgPath(serve.icon)
            })
            //获取商品评论信息
            this.commentInfo = goods.commentInfo
            //获取商品的详细信息
            this.detailInfo = goods.info.goodsIntroduce
            Toast.clear();

            this.refreshScroll()
        })
    },
    ...
  }
}
</script>

<style scoped>
...
</style>
```

最后重启,进行测试。

14.3.5 改造登录功能

在 Mock Server 中添加用户数据模拟器,创建 mock/responses/users.js 文件,模拟生成 3 个用户数据,并且定义/user/login 请求响应,代码如下:

```
const Mock = require('mockjs')
//模拟生成 3 个用户名和密码
const userStore = Mock.mock({
  "userList|3-3":[
    {
      "id|+1":1,
      "userName|+1":['admin','zhangsan','lisi'],
      "password":'123456'
    }
  ]
})
/**
 * 根据用户名,查询用户对象并返回
```

```
 * @param {string} userName  用户
 * @returns 用户对象
 */
function getUserByUserName(userName){
    let user;
    userStore.userList.forEach(element => {
        if( element.userName === userName){
            user = element;
            return
        }
    });
    return user;
}
/**
 * 验证用户名和密码
 * @param {string} userName 用户名
 * @param {string} password 密码
 * @returns boolean
 */
function verifyPassword(userName, password){
    let result = false;
    let user = getUserByUserName(userName);
    if(user){
        result = user.password === password;
    }
    return result;
}

module.exports = [
    {
        url: '/user/login',
        type: 'post',
        response: config => {
            const { userName, password } = config.body;
            let result = verifyPassword(userName, password);
            let response = {code:20000, data: 'x-token-value'};
            if(!result){
                response.code = 50003;
                response.data = '用户名或密码错误';
            }
            return response;
        }
    }
]
```

在 mock/index.js 文件中导入并注册用户模拟器，代码如下：

```
...
const users = require('./responses/users')

const mocks = [
    ...
    ...users
]
...
```

定义用户请求接口，创建 src/api/users.js 文件，定义登录 API，代码如下：

```
import request from 'utils/request'

export function login(data){
    return request({
        url: '/user/login',
        method: 'post',
        data
    })
}
```

实现 store 的 user 模块，在 store 中定义一个 user 模块，专门完成用户登录和登出的业务处理。创建 src/utils/auth.js 文件，定义对 Cookie 中 token 的操作方法。注意，需要先使用命令 npm install js-cookie 安装好 js-cookie 插件，代码如下：

```
import Cookies from 'js-cookie'

const TokenKey = 'X-Token'

export function getToken() {
  return Cookies.get(TokenKey)
}

export function setToken(token) {
  return Cookies.set(TokenKey, token)
}

export function removeToken() {
  return Cookies.remove(TokenKey)
}
```

创建 src/store/modules/user.js 文件，里面定义与登录和登出相关的数据和业务，代码如下：

```
import { login, logout } from 'api/users'
import { getToken, setToken, removeToken } from 'utils/auth'

const state = {
    token: getToken(),
    userName: ''
}

const getters = {
    getToken(state){
        return state.token;
    },
    getName(state){
        return state.userName;
    }
}

const mutations = {
    SET_TOKEN: (state, token) => {
        state.token = token ;
    },
    SET_NAME: (state, userName) => {
        state.userName = userName;
    },
    SET_TOKEN_NAME: (state, payload) => {
        state.token = payload.token
        state.userName = payload.userName
    }
}

const actions = {
    login({commit}, loginInfo){
        const {loginName, loginPassword} = loginInfo;
        return new Promise((resolve, reject) => {
            login({userName:loginName.trim(), password:loginPassword.trim()})
            .then( res => {
                const { code, data } = res;
                if(code === 20000 && data){
                    commit('SET_TOKEN_NAME',{
                        token: data,
                        userName: loginName.trim()
                    })
                    setToken(data)
                    resolve(res)
                }else{
                    reject(data)
```

```
                    }
                })
                .catch( err => {
                    reject(err)
                })
            })
        },
        logout({commit, state, dispatch}){
            commit('SET_TOKEN','')
            commit('SET_NAME','')
            removeToken()
        }
    }

    export default {
        namespaced: true,
        state,
        mutations,
        actions,
        getters
    }
```

将 user 模块添加到 Vuex 中，创建 src/store/index.js 文件，完成 Vuex 对象的定义和初始化，注册 user 模块，代码如下：

```
import Vue from 'vue'
import Vuex from 'vuex'
import user from './modules/user'

Vue.use(Vuex)

export default new Vuex.Store({
 modules: {
  user
 },
 state: {

 },
 mutations: {

 },
 actions: {

 }
})
```

注意，需要在 main.js 文件中，完成 Vuex 的全局注册。

实现登录功能，调整 src/views/login/child/Login.vue 文件中的 onSubmit()方法实现，代码如下：

```
<template>
  ...
</template>

<script>

 export default {
  ...
  methods: {
   ...
   //登录 表单拦截
   onSubmit() {
    this.$refs.loginFormRef.validate((result) => {
     if(result){
      this.$store.dispatch('user/login', this.loginForm)
        .then( res => {
         this.$router.go( - 1)
        })
        .catch(err =>{

        })
     }
    });
   },
   ...
  },
 };
 </script>

 <style>
 </style>
```

14.3.6 改造添加购物车功能

封装 JS 的 Map 类，实现数据 key-value 的管理，包括 put、remove、clear 和 values 方法。创建 mock/map.js 文件，代码如下：

```
function Map(){
  /** 存放键的数组(遍历时会用到) */
  this.keys = new Array();
  /** 存放数据 */
```

```
    this.data = new Object();

    /**
     * 放入一个键值对
     * @param {String} key
     * @param {Object} value
     */
    this.put = function(key, value) {
        if(this.data[key] == null){
            this.keys.push(key);
        }
        this.data[key] = value;
    };

    /**
     * 获取某键对应的值
     * @param {String} key
     * @return {Object} value
     */
    this.get = function(key) {
        return this.data[key];
    };

    /**
     * 删除一个键值对
     * @param {String} key
     */
    this.remove = function(key) {
        this.keys.remove(key);
        this.data[key] = null;
    };

    /**
     * 清除 map 的所有值
     */
    this.clear = function(){
        this.keys.splice(0);
        this.data = new Object();
    }

    /**
     * 遍历 Map,执行处理函数
     *
     * @param {Function} 回调函数 function(key,value,index){..}
     */
    this.each = function(fn){
```

```
        if(typeof fn != 'function'){
            return;
        }
        var len = this.keys.length;
        for(var i = 0;i < len;i++){
            var k = this.keys[i];
            fn(k,this.data[k],i);
        }
    };

    /**
     * 获取键值数组(类似 Java 的 entrySet()方法)
     * @return 键值对象{key,value}的数组
     */
    this.entrys = function() {
        var len = this.keys.length;
        var entrys = new Array(len);
        for (var i = 0; i < len; i++) {
            entrys[i] = {
                key : this.keys[i],
                value : this.data[i]
            };
        }
        return entrys;
    };

    /**
     * 判断 Map 是否为空
     */
    this.isEmpty = function() {
        return this.keys.length == 0;
    };

    /**
     * 获取键值对的数量
     */
    this.size = function(){
        return this.keys.length;
    };

    /**
     * 获取所有的 value
     * @return map 里面的所有 value 值数组
     */
    this.values = function(){
        let valueArray = new Array(this.keys.length);
```

```
        this.keys.forEach((key,index) => {
            valueArray[index] = this.data[key];
        })
        return valueArray;
    }

    /**
     * 重写 toString
     */
    this.toString = function(){
        var s = "{";
        for(var i = 0;i < this.keys.length;i++,s += ','){
            var k = this.keys[i];
            s += k + " = " + this.data[k];
        }
        s += "}";
        return s;
    };
}

module.exports = [
    Map
]
```

添加购物车模拟器代码，支持/shopCar/add 和/shopCar/count 两个请求。创建 mock/responses/shopCar.js 文件，代码如下：

```
const Map = require('../map')
shopCarMap = new Map[0]()

function addShop(userName, shop){
    let itemsMap = shopCarMap.get(userName);
    if(!itemsMap){
        itemsMap = new Map[0]()
        itemsMap.put(shop.goodsId, {shop, count:0})
        shopCarMap.put(userName, itemsMap)
    }
    let item = itemsMap.get(shop.goodsId);
    if(!item){
        item = {shop, count:0}
        itemsMap.put(shop.goodsId, item)
    }
    item.count = item.count + 1
}
```

```
function getShopCount(userName){
    let count = 0;
    let itemsMap = shopCarMap.get(userName);
    if(itemsMap){
        let items = itemsMap.values();
        items.forEach(element => {
            count += element.count;
        });
    }
    return count;
}

module.exports = [
    {
        url: '/shopCar/add',
        type: 'post',
        response: config => {
            const { userName, shop } = config.body;
            addShop(userName, shop) ;
            return {
                code: 20000,
                data: 'success'
            }
        }
    },
    {
        url: '/shopCar/count',
        type: 'get',
        response: config => {
            const { userName } = config.query;
            let count = getShopCount(userName);
            return {
                code: 20000,
                data: count
            }
        }
    }
]
```

并且在 mock/index.js 文件中注册 shopCar 模拟器，代码如下：

```
...
const shopCar = require('./responses/shopCar')

const mocks = [
    ...
    ...shopCar
]
```

定义 shopCar API 请求接口，创建 src/api/shopCar. js 文件，包含 addShop 和 getShopCount 两个接口，代码如下：

```
import request from 'utils/request'

export function addShop(data){
    return new request({
        url: '/shopCar/add',
        method: 'post',
        data
    })
}

export function getShopCount(userName){
    return new request({
        url: '/shopCar/count',
        method: 'get',
        params: {userName}
    })
}
```

调整详情页面的代码，实现添加购物车和显示购物车商品数量的功能。编辑 src/views/detail/Detail. vue 文件里面的 addCart()方法，代码如下：

```
<template>
    ...
</template>

<script>
    ...
    import { addShop, getShopCount } from 'api/shopCar'
    ...

    export default {
        .. ...
        methods: {
            ...
            //添加购物车
            addCart(){
                let userName = this. $store.getters['user/getName']
                let token = this. $store.getters['user/getToken']
                if(userName && token){
                    addShop({userName, shop:this.goodsInfo})
                        .then( res => {
                            this.getShopCount(userName)
```

```
                    Message({
                        message: '成功添加到购物车',
                        type: 'info',
                        duration: 5 * 1000
                    })
                })
                .catch( err => {
                Message({
                        message: '添加失败' + err,
                        type: 'error',
                        duration: 5 * 1000
                    })
                })
            }else{
                this.$router.push('/login')
            }
        },
        //获取商品数量
        getShopCount(){
            ...
        },
        ...
    }
  }
</script>

<style scoped>
...
</style>
```

编辑 src/views/detail/Detail.vue 文件里面的 getShopCount()方法，实现购物车商品数量功能，代码如下：

```
<template>
    <div class="detail">
        ...
        <detail-bottom-bar
            ref="cart"
            :shopCount="shopCount"
            :goodsInfo="goodsInfo"
            @addCart="addCart"
            @buyClick="buyClick"
            @toService="toService"
            @addEnshrine="addEnshrine"
            @removeEnshrine="removeEnshrine">
```

```
        </detail-bottom-bar>
    </div>
</template>

<script>
    ...
    export default {
        ...
        data() {
            return {
                ...
                shopCount:0
            }
        },
        created() {
            ...
            let userName = this.$store.getters['user/getName']
            this.getShopCount(userName)
        },
        ...
        methods: {
            ...
            //获取商品数量
            getShopCount(){
                let userName = this.$store.getters['user/getName']
                if(userName){
                    getShopCount(userName).then( res => {
                        this.shopCount = res.data
                    })
                }
            },
            ...
        }
    }
</script>

<style scoped>
...
</style>
```

在 detail-bottom-bar 元素中，新增了一个组件 prop shopCount，用来将购物车商品数量传递到 detail-bottom-bar 的购物车上显示，所以需要修改 src/views/detail/child/DetailBottomBar.vue 文件中的组件，添加 shopCount props 属性和显示，代码如下：

```
<template>
    <div class="bottom-bar">
```

```
        <div class = "bar - item bar - left">
          ...
          <div @click = "pushCart">
            <el - badge :value = "shopCount" :max = "99" class = "item">
              ...
            </el - badge>
          </div>
          ...
        </div>
        ...
      </div>
</template>

<script>
  export default {
    ...
    props:[
      'goodsInfo',
      'shopCount'
    ],
    ...
  };
</script>

<style scoped>
  ...
</style>
```

14.3.7 改造添加购物车列表功能

模拟购物车列表的显示数据。在购物车 Mock 数据生成器中,添加获取购物车商品信息的方法和请求响应。在 mock/responses/shopCar.js 文件中添加的代码如下：

```
...
//获取指定用户购物车中的所有商品列表
function getShopList(userName){
   let result = []
   let itemsMap = shopCarMap.get(userName);
   if(itemsMap){
      result = itemsMap.values()
   }
   return result;
}
//将购物车中的商品转化成产品,以便于在购物车中显示
function convert2ProductList(goodsList){
```

```
    let productList = [];
    const productTemplate = {id:0,
            title:'',
            image:'',
            checked:false,
            amount:0,
            price:0
        };
    goodsList.forEach( (goodsItem, index) => {
        let product = Object.assign({}, productTemplate);
        product.id = goodsItem.shop.goodsId;
        product.image = goodsItem.shop.goodsLogo;
        product.title = goodsItem.shop.goodsName;
        product.amount = goodsItem.count;
        product.checked = goodsItem.checked;
        product.price = goodsItem.shop.goodsPrice;
        productList.push(product);
    })
    return productList;
}
module.exports = [
    ...
    {
        //获取购物车商品列表
        url: '/shopCar/list',
        type: 'get',
        response: config => {
            //页码、页面大小和用户名
            const { pageNo, pageSize, userName } = config.query;
            return {
                code: 20000,
                data: convert2ProductList(getShopList(userName))
            }
        }
    }
]
```

在 src/api/shopCar.js 文件中添加获取购物车商品列表的 API，代码如下：

```
...
export function getShopList(pageNo, pageSize, userName){
    return new request({
        url: '/shopCar/list',
        method: 'get',
        params:{pageNo, pageSize, userName}
    })
}
```

实现购物车组件中的获取数据的方法，编辑 src/views/car/Car.vue 文件，代码如下：

```
<template>
  ...
</template>

<script>
...
import { Toast } from "vant";
import { getShopList } from 'api/shopCar'
export default {
   ...
   methods:{
      ...
      getShopList(){
          const toast = Toast.loading({
             duration: 0,                    //持续展示 toast
             message: "加载中...",
             forbidClick: true,
             loadingType: "spinner"
          });
          let userName = this.$store.getters['user/getName']
          console.log(userName)
          getShopList(this.page, this.pageSize, userName)
             .then( res => {
                   Toast.clear();
                   this.cartList.push(...res.data)
                   this.$refs.myGoodsList.finish();
             })
             .catch( err => {
                   Toast.clear();
                   Toast.fail('加载失败');
                  this.$message.error("可能网络出了问题" + err);
              });
      },
      getShopCount(){
          getShopCount(this.$store.getters['user/getName'])
            .then(res => {
               this.shopCount = res.data
            })
      },
      ...
   }
}
</script>

<style>

</style>
```

14.3.8　改造购物车商品数量

在购物车 Mock 模拟器中，添加修改购物车商品数量的方法，并提供修改数量请求的响应。编辑 mock/responses/shopCar.js 文件，代码如下：

```
...
function editShop(userName, shopId, count){
    let itemsMap = shopCarMap.get(userName);
    if(itemsMap){
        let item = itemsMap.get(shopId);
        if(item){
            item.count = count;
        }
    }
}
...
module.exports = [
    ...
    {
        //修改商品梳理
        url: '/shopCar/edit',
        type: 'put',
        response: config => {
            const {userName, shopId, count } = config.body
            editShop(userName, shopId, count)
            return {
                code: 20000,
                data: 'success'
            }
        }
    }
]
```

添加修改购物车商品数量请求 API，在 src/api/shopCar.js 文件中添加修改数量 API，代码如下：

```
...
export function editShopCount(userName, shopId, count){
    return new request({
        url: '/shopCar/edit',
        method: 'put',
        data: {
            userName,
            shopId,
            count
        }
    })
}
```

在组件里添加修改数量的事件处理代码，修改 src/views/car/GoodsListItem.vue 文件，代码如下：

```
<template>
  ...
</template>

<script>
...
import { editShopCount } from 'api/shopCar'

export default {
    ...
  methods: {
    ...
    //商品数量发生改变时
    handleChange(value) {
      const toast = Toast.loading({
          duration: 0, //持续展示 toast
          message: "修改中...",
          forbidClick: true,
          loadingType: "spinner"
      });
      editShopCount(this.$store.getters['user/getName'], this.product.id, value)
        .then( res => {
          Toast.clear();
        })
        .catch( err => {
          Toast.clear()
          this.$message.error('fail:' + err)
        })
    },
    ...
  }
}
</script>

<style scoped>
  ...
</style>
```

14.3.9 改造删除购物车商品

在购物车 Mock 模拟器中添加删除购物车的方法，并且提供删除购物车商品的请求支持。在 mock/responses/shopCar.js 文件中，添加删除购物车方法和提供删除购物车商品请求支持，代码如下：

```
...
//删除 userName 用户的购物车中 id 为 shopIds(商品 id 的数组)的商品
function deleteShop(userName, shopIds){
    let itemsMap = shopCarMap.get(userName);
    if(itemsMap){
        shopIds.forEach( id => {
            itemsMap.remove(id)
        })
    }
}
...
module.exports = [
    ...
    {
        //删除购物车中的商品
        url: '/shopCar/delete',
        type: 'delete',
        response: config => {
            const {userName, shopIds} = config.body
            deleteShop(userName, shopIds)
            return {
                code: 20000,
                data: 'success'
            }
        }
    }
]
```

添加删除购物车商品的请求 API，在 src/api/shopCar.js 文件中添加删除购物车商品请求 API，代码如下：

```
...
export function deleteShop(userName, shopIds){
    return new request({
        url: '/shopCar/delete',
        method: 'delete',
        data: {
            userName,
            shopIds
        }
    })
}
```

在组件中，添加删除购物车商品的业务代码。在 src/views/car/child/CarBottomBar.vue

文件的 Confirm()方法中，添加删除购物车商品的业务，代码如下：

```
...
        //确认删除
        Confirm(){
          this.centerDialogVisible = false
          let shopIds = []
          //获取所有要删除的商品 id
          this.cartid.split(',').forEach(id =>{
            if(id != ''){
              try{
                    shopIds.push(parseInt(id))
              }catch(e){}
            }
          })
          //调用请求 API 的删除方法,发送删除购物车商品请求
          deleteShop(this.$store.getters['user/getName'], shopIds)
            .then(res =>{
              //删除成功,触发组件绑定事件,调用 Car.vue 文件中的绑定方法刷新购物车列表
              this.$emit('deleteOk')
            })
            .catch(err =>{
              this.$message.error('fail:' + err)
            })
        }
...
```

14.3.10 改造个人中心头信息

完善 Mock 的用户模拟器，在 mock/responses/users.js 文件中完善 userStore 的数据项，添加 getUserByToken()函数（根据 token 获取用户对象），完善 verifyPassword()函数，添加登录成功后 token 的处理，并完善对登录请求的响应处理，登录成功后返回 token，以及添加获取用户信息请求的响应处理，代码如下：

```
const Mock = require('mockjs')
//模拟生成 3 个用户名和密码
const userStore = Mock.mock({
    "userList|3-3":[
        {
            "id|+1":1,
            "userName|+1":['admin','zhangsan','lisi'],
            "password":'123456',
            "image|+1":['/images/user/img1.jpg','/images/user/img2.jpg','/images/user/
img3.jpg'],
```

```
            "name|+1":["管理员",'张三', '李四'],
            "gender|0-1":0,
            "phone|+1":['13688888888', '13788888888', '13888888888'],
            "isSignInPoint":true
        }
    ],
    userTokens:{}
})

function getUserByToken(token){
    for (const name in userStore.userTokens) {
        if(token === userStore.userTokens[name]){
            return name
        }
    }
}

/**
 * 验证用户名和密码
 * @param {string} userName 用户名
 * @param {string} password 密码
 * @returns boolean
 */
function verifyPassword(userName, password){
    let result = false;
    let user = getUserByUserName(userName);
    if(user){
        if(user.password === password){
            result = "token:" + Mock.mock('@integer')
            //记录当前用户对应的 token
            userStore.userTokens[userName] = result;
        }
    }
    console.log(userStore)
    return result;
}
...
module.exports = [
    {                    //登录
        url: '/user/login',
        type: 'post',
        response: config => {
            const { userName, password } = config.body;
            let token = verifyPassword(userName, password);
            let response = {code:20000};
            if(token){
```

```
                response.data = token;
            }else{
                response.code = 50003;
                response.data = '用户名或密码错误';
            }
            return response;
        }
    },
    {
        //获取用户信息
        url: '/user/info\.*',
        type: 'get',
        response: config => {
            const { token } = config.query;
            const userName = getUserByToken(token);
            const res = {code:20000, data: ''}
            if(userName) {
                const user = getUserByUserName(userName);
                res.data = user;
            }else{
                res.code = 50002
                res.data = "没有对应用户或请登录"
            }
            return res
        }
    }
]
```

编写获取用户信息请求的 API。在 src/api/users.js 文件中，添加发送获取用户信息请求的 API，代码如下：

```
...
//获取用户信息
export function getInfo(token){
    return request({
        url: '/user/info',
        method: 'get',
        params: {token}
    })
}
...
```

改造 store 代码，将 user 模块中的局部属性暴露成全局属性，方便操作。添加 src/store/getters.js 文件的代码，封装 user 中的局部属性，代码如下：

```
const getters = {
    token: state => state.user.token,
    userName: state => state.user.userName,
}
```

将 getters 整合到 src/store/index.js 文件中，代码如下：

```
...
import user from './modules/user'
import getters from './getters'

Vue.use(Vuex)

export default new Vuex.Store({
 modules: {
  user
 },
 ...
 getters
})
```

在个人中心组件中，添加获取登录用户的个人信息代码。在 src/views/profile/Profile.vue 文件中，完善 getUserData()方法，代码如下：

```
<template>
    ...
</template>

<script>
    ...
  import { getInfo } from "api/users"
  import { getToken } from "utils/auth"
  import { Toast } from "vant";

  export default {
    ...
    methods: {
      getUserData() {
        let token = this.$store.getters.token
        if(!token){
          token = getToken()
          this.$store.dispatch('user/setToken')
        }
        if(token){
            getInfo(token).then(res =>{
```

```
                this.userData = res.data
            }).catch(err =>{
                Toast.fail("数据加载失败:" + err)
            })
        }else{
            this.$router.push('/login')
        }
    },
    ...
  }
}
</script>

<style scoped>
  ...
</style>
```

14.3.11 改造签到积分

在 Mock 用户模拟器中添加新增积分的方法和暴露新增积分请求的使用，在 mock/responses/users.js 文件中添加的代码如下：

```
...
/**
 * 给 userName 增加 point 积分
 * @param {string} userName
 * @param {number} point
 */
function addPoint(userName, point){
    console.log(userName)
    let user = getUserByUserName(userName);
    user.point += point
}
...
module.exports = [
  ...
  {
      //新增积分
      url: '/user/addPoint',
      type: 'put',
      response: config => {
          const { userName, point } = config.body
          addPoint(userName, point)
          return {
              code: 20000,
```

```
                data: 'success'
            }
        }
    }
]
```

新增积分请求 API，在 src/api/users.js 文件中添加的代码如下：

```
...
//添加用户积分
export function addPoint(userName, point){
    return request({
        url: '/user/addPoint',
        method: 'put',
        data: {
            userName,
            point
        }
    })
}
...
```

在 src/views/profile/Profile.vue 文件的 SignInPointIsOK()方法中，完善签到领积分逻辑，代码如下：

```
<template>
    ...
</template>

<script>
  ...
  import { getInfo, addPoint } from "api/users"

  export default {
    ...
    methods: {
      ...
      SignInPointIsOK(){
        addPoint(this.$store.getters['user/getName'], 1).then( res => {
          if(res.code === 20000 && 'success' === res.data){
            this.userData.isSignInPoint = false
            this.userData.point += 1
          }else{
            this.$toast.show("操作失败", 2000)
          }
```

```
            }).catch( err => {
                this.$toast.show("操作失败:" + err, 2000)
            })
        },
        ...
    }
  }
</script>

<style scoped>
  ...
</style>
```

14.3.12 改造分类UI和左侧分类

Mock模拟生成左侧分类数据，在mock/responses/goods.js文件中添加获取商品类型列表的函数，并且添加获取商品列表请求的响应接口，代码如下：

```
...
function getCategoryList(){
    return [
        {cateId:1, cateName:'女士春装'},
        {cateId:1, cateName:'女士夏装'},
        {cateId:1, cateName:'女士秋装'},
        {cateId:1, cateName:'女士冬装'},
        {cateId:1, cateName:'女士休闲鞋'},
        {cateId:1, cateName:'女士运动鞋'},
    ]
}
module.exports = [
    ...
    {
        url:'/goods/category',
        type:'get',
        response: config => {
            let result = getCategoryList();
            return {
                code: 20000,
                data: result
            }
        }
    }
]
```

新增获取分类列表的API，在src/api/goods.js文件中，新增加获取分类列表的

getCategoryList 接口,代码如下:

```
...
export function getCategoryList(){
    return request({
        url: '/goods/category',
        method: 'get'
    })
}
```

在商品分类组件中添加分离请求数据,在 src/views/Category. vue 文件中,添加调用 getCategoryList,代码如下:

```
...
<script>
...
import { getCategoryList } from 'api/goods'
export default {
    ...
    methods: {
        ...
        getCategoryList(){
            const toast = Toast.loading({
                duration: 0,                //持续展示 toast
                message: "加载中...",
                forbidClick: true,
                loadingType: "spinner"
            });
            let _sel = this;
            getCategoryList().then((res) =>{
                Toast.clear();
                this.sideBarList = res.data
                //调用子组件的方法,加载展示商品
                setTimeout(() =>{
                    _sel.$refs.sideBar.createdData()
                },200)

            }).catch((err) =>{
                Toast.clear();
                Toast.fail("加载失败" + err);
            })
        }
    }
};
</script>
...
```

14.3.13 改造分类商品

在 Mock 中添加模拟数据，并暴露请求响应接口。在 mock/responses/goods.js 文件中添加模拟数据和请求接口，代码如下：

```
...
function getCategoryGoodsList(cateId, pageNo, pageSize, sortType){
    return getGoodsList()
}
module.exports = [
    ...
    {
        url: '/goods/cateList',
        type: 'get',
        response: config => {
            const { cateId, pageNo, pageSize, sortType } = config.query
            let result = getCategoryGoodsList(cateId, pageNo, pageSize, sortType)
            return {
                code: 20000,
                data: result
            }
        }
    }
]
```

添加请求分类商品的 API，在 src/api/goods.js 文件中添加 getCategoryGoodsList 接口，请求分类商品数据，代码如下：

```
...
export function getCategoryGoodsList(params){
    return request({
        url: '/goods/cateList',
        method: 'get',
        params
    })
}
```

在 SideBar 组件中，调用 getCategoryGoodsList 接口，获取分类商品数据。在 src/view/category/SideBar.vue 文件中，添加代码，调用 getCategoryGoodsList 接口，以便获取分类商品数据，代码如下：

```
<template>
  ...
</template>
```

```
<script>
...
import { getCategoryGoodsList } from 'api/goods.js'

export default {
  ...
  methods: {
    ...
    //页面创建时进行的数据获取
    createdData() {
      this.getCateGoods(0);
    },
    ...
    //获取展示的数据
    getCateGoods(cateId) {
      //cateId, pageNo, pageSize, sortType
      const params = {
        cateId,
        pageNo: this.goods[this.sortType].page,
        pageSize: this.pageSize,
        sortType: this.sortType
      }
      getCategoryGoodsList(params).then(res => {
        Toast.clear();
        this.goods[this.sortType].list.push(...res.data);
        this.goods[this.sortType].page += 1;

        this.ok()
        return this.finish();
      }).catch( err => {
        Toast.clear();
        Toast.fail("加载失败");
        console.log(err)
      })
    },

    ok() {
      this.pullDownMsg = "更新成功";
      this.tabIsShow = true;
      this.isShowLoading = false;
      this.isShowUpLoading = false;
    },
    finish() {
      this.$refs.scroll.finishPullUp();
    },
```

```
    //上拉加载更多,分页数据
    loadMore() {
      this.isShowUpLoading = true;
      this.pullDownMsg = "正在更新中...";
      this.getCateGoods(this.activeKey);
    },
    handleToScroll(pos) {
      if (pos.y > 60) {
        this.isShowLoading = true;
        this.pullDownMsg = "正在更新中...";
      }
    },
    //下拉刷新
    handleToTouchEnd(pos) {
      if (pos.y > 60) {
        //this.page = 1;
        this.goods[this.sortType].page = 1;
        this.getCateGoods(this.activeKey);
      }
    },
    //确认显示还是隐藏,回到顶部
    scroll(position) {
      //判断 BackTop 是否显示
      this.isShowBackTop = -position.y > 1000;
    },
  }
};
</script>
...
```

第15章 权限管理实战

安全认证和授权是每个软件项目都必须具备的功能，本章将基于 vue-element-admin，结合 Spring Boot、SSM、Shiro、Redis 和 JWT 等技术，模拟实现前后台分离的安全模块，功能包括动态路由、前端权限控制、前端动态路由、前端权限菜单管理、后端身份认证、后端授权控制和后端权限管理。

15.1 实现前端安全控制

本书的前端是基于 vue-element-admin-master 框架实现的，大家可以在 GitHub 上找到源码，随书文档样例中，权限管理里面的 vue-element-admin-master-resource 是源码。读者可以先熟悉源码，再理解在这个基础上构建的前端安全控制功能。

15.1.1 vue-element-admin 简介

vue-element-admin(https://panjiachen.gitee.io/vue-element-admin-site/)是一个后台前端解决方案，它基于 vue(https://github.com/vuejs/vue) 和 element-ui (https://github.com/ElemeFE/element)实现。它使用了最新的前端技术栈，内置了 i18 国际化解决方案，动态路由和权限验证提炼了典型的业务模型，提供了丰富的功能组件，它可以帮助开发人员快速搭建企业级后台产品原型。vue-element-admin-master 集成的功能比较多，这点查看官网或直接运行项目就可以明了，这里不做太多说明。vue-element-admin 的基本目录结构如下：

```
├── build                      # 构建相关
├── mock                       # 项目 mock 模拟数据
├── plop-templates             # 基本模板
├── public                     # 静态资源
│   │—— favicon.ico            # favicon 图标
│   └──index.html              # html 模板
├── src                        # 源代码
│   ├── api                    # 所有请求
```

```
|     ├── assets                   # 主题、字体等静态资源
|     ├── components               # 全局公用组件
|     ├── directive                # 全局指令
|     ├── filters                  # 全局 filter
|     ├── icons                    # 项目所有 svg icons
|     ├── lang                     # 国际化 language
|     ├── layout                   # 全局 layout
|     ├── router                   # 路由
|     ├── store                    # 全局 store 管理
|     ├── styles                   # 全局样式
|     ├── utils                    # 全局公用方法
|     ├── vendor                   # 公用 vendor
|     ├── views                    # views 所有页面
|     ├── App.vue                  # 入口页面
|     ├── main.js                  # 入口文件、加载组件、初始化等
|     └──permission.js             # 权限管理
├── tests                          # 测试
├── .env.xxx                       # 环境变量配置
├── .eslintrc.js                   # eslint 配置项
├── .babelrc                       # babel - loader 配置
├── .travis.yml                    # 自动化 CI 配置
├── vue.config.js                  # Vue - CLI 配置
├── postcss.config.js              # postcss 配置
└── package.json                   # package.json
```

补充说明：

(1) 关闭 eslint 校验。在 vue.config.js 文件中，将 lintOnSave 设置成 false，代码如下：

```
module.exports = {
  lintOnSave:false
}
```

(2) 设置访问后端代理。修改 vue.config.js 文件中的 devServer 配置，设置代理并将请求转发给后端，代码如下：

```
//源码
  devServer: {
    port: port,
    open: true,
    overlay: {
      warnings: false,
      errors: true
    },
    before: require('./mock/mock - server.js')
  },
```

```
//修改后的代码
  devServer: {
    port: port,
    open: true,
    overlay: {
      warnings: false,
      errors: true
    },
    hot: true,
    headers: {
      'Access-Control-Allow-Origin': '*',
      'Access-Control-Allow-Credentials': 'true'
    },
    host: '0.0.0.0',
    proxy: {                                          //配置代理
      '/dev-api': {
        target: 'http://localhost:8080',              //目标 URL
        changeOrigin: true,
        ws: true,
        secure: false,
        pathRewrite: {
          '^/dev-api': ''
        }
      }
    },
    //before: require('./mock/mock-server.js')        //注释 mock-server
  },
```

本书案例在 src/api/*.js 文件中为每个请求指定后端的 baseURL，代码如下：

```
...
export function getMenuTree() {
 return request({
  baseURL: 'http://localhost:7080',                  //后端 URL
  ...
 })
}
...
```

同时，在 src/utils/request.js 文件中给创建的 axios 对象设置了 withCredentials 和 crossDomain 属性，实现跨域，代码如下：

```
const service = axios.create({
 baseURL: process.env.VUE_APP_BASE_API, //url = base url + request url
 withCredentials: true, //send Cookies when cross-domain requests
 timeout: 5000 ,//request timeout
```

```
crossDomain: true
})
```

显示左上角的 logo。将 src/setting.js 文件中的 sidebarLogo 属性设置成 true，代码如下：

```
/**
 * @type {boolean} true | false
 * @description Whether show the logo in sidebar
 */
sidebarLogo: true,              //false
```

修改 logo 的图和文字。修改/src/layout/components/Sidebar/Login.vue 文件里面的 title 和 logo 属性，代码如下：

```
...
 data() {
  return {
   title: '后台管理系统',
   logo: 'favicon.ico' //图片放在 public 目录下
  }
 }
...
```

解决 vuecli4 报错：sockjs.js? 9be2：1605 GET http://localhost/sockjs-node/info。可注释掉 node_modules\sockjs-client\dist\sockjs.js 文件里面的第 1609 行，代码如下：

```
try {
  //self.xhr.send(payload);
} catch (e) {
  self.emit('finish', 0, '');
  self._cleanup(false);
}
```

15.1.2 实现有后端支持的登录功能

要实现后端权限控制，必须先登录，而且登录也是安全控制功能的一部分。接下来介绍怎么实现带有后端验证的登录功能。

1. 登录涉及的前端代码

1）src/main.js

在前端应用的入口 JS 里面注册了全局组件，包括 Vue、Cookie、Element、store、router 等组件，并且用 App 组件渲染页面，代码如下：

```
...
Vue.use(Element, {
  size: Cookies.get('size') || 'medium', //set element-ui default size
  locale: enLang              //如果使用中文，则无须设置，可删除
})

//register global utility filters
Object.keys(filters).forEach(key => {
  Vue.filter(key, filters[key])
})

Vue.config.productionTip = false

new Vue({
  el: '#app',
  router,
  store,
  render: h => h(App)
})
```

2）src/App.vue

App.vue 为前端应用的根组件对象，里面定义了一个全局的 router-view，用于路由到默认路径/（根路径），代码如下：

```
<template>
  <div id="app">
    <router-view />
  </div>
</template>

<script>
export default {
  name: 'App'
}
</script>
```

/路由路径是在 src/router/index.js 文件中定义的，代码如下：

```
...
  {
    path: '/',
    component: Layout,
    redirect: '/index',
    children: [
      {
```

```
            path: 'index',
            component: () => import('@/views/documentation/index'),
            name: 'Documentation',
            meta: { title: 'Documentation', icon: 'documentation', affix: false }
        }
      ]
    },
  ...
```

3）src/permission.js

permission.js 文件定义了路由拦截器，前端在路由前都需要执行该拦截器。在该拦截器里面实现了登录用户验证、用户信息获取、根据用户角色动态挂载菜单导航等功能，详细代码如下：

```
...
//路由拦截器
router.beforeEach(async(to, from, next) => {
  //start progress bar
  NProgress.start()

  //set page title
  //设置目标页面的 title(从目标路由的 meta 中获取 title)
  document.title = getPageTitle(to.meta.title)

  //determine whether the user has logged in
  //获取 store 里面的登录令牌,如果有令牌,则表示有登录
  const hasToken = getToken()

  if (hasToken) {
    if (to.path === '/login') {
      //if is logged in, redirect to the home page
      //如果有登录,并且目标路径是/login,则表示路由到首页
      next({ path: '/' })
      NProgress.done()
    } else {
      //determine whether the user has obtained his permission roles through getInfo
      //获取登录用户的角色,如果能获取,则表示有角色信息,否则就调用 getInfo
      //获取当前登录用户的角色信息
      const hasRoles = store.getters.roles && store.getters.roles.length > 0
      if (hasRoles) {
        next()
      } else {
        try {
          //get user info
```

```
          //note: roles must be a object array! such as: ['admin'] or ,['developer','editor']
          //派发 user/getInfo action,获取当前用户的角色信息
          const { roles } = await store.dispatch('user/getInfo')

          //generate accessible routes map based on roles
          //根据用户的角色信息,派发到 permission/generateRoutes action
          //生成动态路由表
          const accessRoutes = await store.dispatch('permission/generateRoutes', roles)

          //dynamically add accessible routes
          //挂载动态路由
          router.addRoutes(accessRoutes)

          //hack method to ensure that addRoutes is complete
          //set the replace: true, so the navigation will not leave a history record
          next({ ...to, replace: true })
        } catch (error) {
          //移除 token,并且跳转到登录页面,以便重新登录
          await store.dispatch('user/resetToken')
          Message.error(error || 'Has Error')
          next(`/login?redirect = ${to.path}`)
          NProgress.done()
        }
      }
    }
  } else {
    /* has no token */

    if (whiteList.indexOf(to.path) !== -1) {
      //in the free login whitelist, go directly
      next()
    } else {
      //other pages that do not have permission to access are redirected to the login page
      next(`/login?redirect = ${to.path}`)
      NProgress.done()
    }
  }
})
...
```

4) src/router/index.js

在 index.js 文件里面定义了所有的路由信息,其中包括登录组件的路由。需要注意的是,要实现前端的动态菜单路由,关键点就是后台返回符合定义在 index.js 文件中的路由格式的数据,代码如下:

```
…
/**
 * constantRoutes
 * a base page that does not have permission requirements
 * all roles can be accessed
 * 基本路由表,所有用户都可以访问的路由表
 */
export const constantRoutes = [
  {
    path: '/redirect',
    component: Layout,
    hidden: true,
    children: [
      {
        path: '/redirect/:path(.*)',
        component: () => import('@/views/redirect/index')
      }
    ]
  },
  {
    path: '/login',
    component: () => import('@/views/login/index'),
    hidden: true
  },
  …
]
…
```

5）src/views/login/index.vue

在 index.vue 文件中定义了登录组件，即定义了登录模板和登录操作功能，代码如下：

```
<template>
  …
</template>

<script>
…
export default {
  name: 'Login',
  components: { SocialSign },
  data() {
    const validateUsername = (rule, value, callback) => {
      if (!validUsername(value)) {
        callback(new Error('Please enter the correct user name'))
      } else {
```

```
        callback()
      }
    }
    const validatePassword = (rule, value, callback) => {
      if (value.length < 6) {
        callback(new Error('The password can not be less than 6 digits'))
      } else {
        callback()
      }
    }
    return {
      loginForm: {
        username: 'demo1',
        password: '123456'
      },
      loginRules: {
        username: [{ required: true, trigger: 'blur', validator: validateUsername }],
        password: [{ required: true, trigger: 'blur', validator: validatePassword }]
      },
      passwordType: 'password',
      capsTooltip: false,
      loading: false,
      showDialog: false,
      redirect: undefined,
      otherQuery: {}
    }
  },
  ...
}
</script>

<style lang="scss">
...
</style>
```

6）src/utils/validate.js

在 validate.js 文件里面定义了若干数据的验证函数，其中 validUsername()函数在用户登录前被用来对用户名进行验证，代码如下：

```
...
/**
 * @param {string} str
 * @returns {Boolean}
 */
export function validUsername(str) {
```

```
  const valid_map = ['admin', 'editor']
  return valid_map.indexOf(str.trim()) >= 0
}
...
```

7) src/store/modules/user.js

作为 Vuex 的一个模块,在 User.js 文件中定义了与登录用户信息相关的全局操作方法,代码如下:

```
...
const state = {
  token: getToken(),
  name: '',
  avatar: '',
  introduction: '',
  roles: []
}

const mutations = {
  SET_TOKEN: (state, token) => {
    state.token = token
  },
  SET_INTRODUCTION: (state, introduction) => {
    state.introduction = introduction
  },
  SET_NAME: (state, name) => {
    state.name = name
  },
  SET_AVATAR: (state, avatar) => {
    state.avatar = avatar
  },
  SET_ROLES: (state, roles) => {
    state.roles = roles
  }
}

const actions = {
  //user login
  login({ commit }, userInfo) {
    ...
  },

  //get user info
  getInfo({ commit, state }) {
    ...
```

```
  },

  //user logout
  logout({ commit, state, dispatch }) {
    …
  },

  //remove token
  resetToken({ commit }) {
    …
  },

  //dynamically modify permissions
  async changeRoles({ commit, dispatch }, role) {
    …
  }
}

export default {
  namespaced: true,
  state,
  mutations,
  actions
}
```

8）src/api/user.js

在 user.js 文件中定义了前端同后端关于 user 进行交互的接口，如登录功能里面的登录和获取用户信息请求都定义在这个 JS 文件中，代码如下：

```
…
export function login(data) {
  return request({
    url: '/vue-element-admin/user/login',
    method: 'post',
    data
  })
}

export function getInfo(token) {
  return request({
    url: '/vue-element-admin/user/info',
    method: 'get',
    params: { token }
  })
}
…
```

9）src/util/request.js

request.js 文件是对前后端交互对象 Axios 的封装，里面还定义了请求和响应拦截器，分别实现请求前和响应后的统一处理，代码如下：

```
...
//create an axios instance
const service = axios.create({
 baseURL: process.env.VUE_APP_BASE_API, //url = base url + request url
 //withCredentials: true, //send Cookies when cross - domain requests
 timeout: 5000 ,//request timeout
})

//request interceptor
service.interceptors.request.use(
 config => {
  ...
 },
 error => {
  ...
 }
)

//response interceptor
service.interceptors.response.use(
  /**
   * If you want to get http information such as headers or status
   * Please return  response => response
  */

  /**
   * Determine the request status by custom code
   * Here is just an example
   * You can also judge the status by HTTP Status Code
   */
 response => {
  ...
 },
 error => {
  ...
 }
)

export default service
```

2. 实现带后台数据的登录功能

1）复制个人头像

如果读者想设置自己的头像图片，则可以将图片文件保存到 public 目录下。本书使用框架自带的 profile.gif 文件，此文件保存在 public 目录下。

2）修改登录用户名的验证规则

原框架的用户在登录前需要基于表单验证规则，对输入的用户名和密码进行验证，如 src/views/login/index.vue 文件的代码所示，在 el-form 元素中，使用 rules 属性绑定了 loginRules 对象，进行表单验证，代码如下：

```
<template>
 <div class="login-container">
  <el-form ... :rules="loginRules">
   ...
  </el-form>
 ...
</template>

<script>
...
export default {
...
 data() {
  ...
  return {
   ...
   loginRules: {
    username: [{ required: true, trigger: 'blur', validator: validateUsername }],
    password: [{ required: true, trigger: 'blur', validator: validatePassword }]
   },
   ...
  }
  ...
 }
}
</script>

<style lang="scss">
...
</style>
```

根据源码，最终调用 src/utils/validate.js 文件中的 validUsername()函数实现了用户名验证，代码如下：

```
/**
 * @param {string} str
 * @returns {Boolean}
```

```
*/
export function validUsername(str) {
 const valid_map = ['admin', 'editor']
 return valid_map.indexOf(str.trim()) >= 0
}
```

在原始代码中，只支持 admin 和 editor 两个用户。对方法进行修改，以便支持任何长度且大于 0 的用户名，代码如下：

```
export function validUsername(str) {
 return str.trim() > 0
}
```

3）修改登录请求 API 和获取用户信息 API

修改 src/api/user.js 文件中的 login()和 getInfo()函数，调用后端 API，实现后端登录验证和获取用户信息，代码如下：

```
...
export function login(data) {
 return request({
  baseURL: 'http://localhost:7080',
  url: '/auth/login', //'/vue-element-admin/user/login',
  method: 'post',
  data
 })
}

export function getInfo(token) {
 return request({
  baseURL: 'http://localhost:7080',
  url: '/auth/info', //'/vue-element-admin/user/info',
  method: 'get',
  params: { token }
 })
}
...
```

4）后端代码

在后端提供用户登录和获取用户信息的接口，代码如下：

```
package cn.com.authority.controller;
...
@RestController
@RequestMapping("/auth")
```

```
public class AuthController {
...
    @PostMapping("/login")
    public ResponseResult login(@RequestBody AuthSysAccount account, HttpSession session){
// @RequestMapping("/login")
// public ResponseResult login(AuthSysAccount account, HttpSession session){
    //System.out.println(session.getId());
    ResponseResult result = new ResponseResult();
    //1. 从 Shiro 框架中,获取一个 Subject 对象,代表当前会话的用户
    Subject subject = SecurityUtils.getSubject();
    //2. 认证 subject 的身份
    //2.1 封装要认证的身份信息
     AuthenticationToken token = new UsernamePasswordToken(account.getUsername(),
EncryptUtil.getMD5Str(account.getPassword()));
    //2.2 认证:认证失败,抛出异常
    try {
        subject.login(token);
        String resToken = JwtUtil.sign(account.getUsername());
        result.setData(result.new Token(resToken));
        //在 session 中保存登录标记
// session.setAttribute(resToken, account.getUsername());
        this.redisUtil.set(resToken, account.getUsername(),30 * 60);
    }catch(Exception e){
        e.printStackTrace();
        result.setData("用户名或密码错误");
    }
    return result;
    }

    @GetMapping("/info")
    public ResponseResult info(@RequestHeader("X-Token") String token, HttpSession
session){
        //System.out.println(session.getId());
        ResponseResult result = new ResponseResult();

        String userName = JwtUtil.verityToken(token, "loginName");
        if(userName == null){
            throw new RuntimeException("没有登录");
        }else{
            AuthSysUser user = this.userService.findByUserName(userName);
            if(user == null){
                throw new RuntimeException("没有登录");
            }else{
                result.setData(user);
            }
        }
```

```
            return result;
        }
    ...
    }
```

5）测试

启动前端如图 15.1 所示，分别输入 demo1 和 123456，登录管理主界面，如图 15.2 所示。

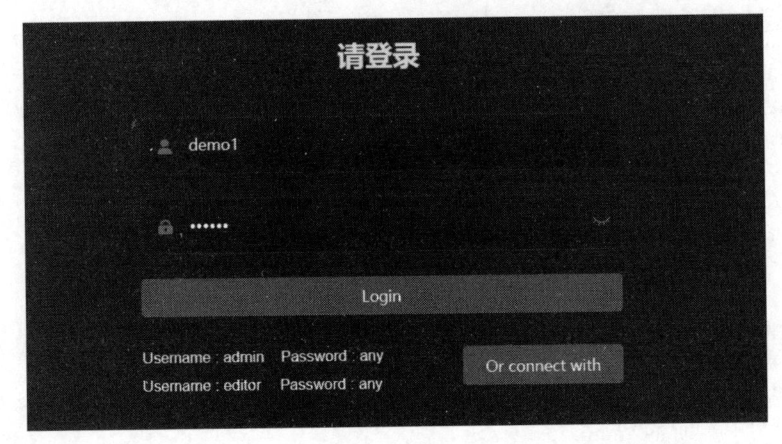

图 15.1　登录页面

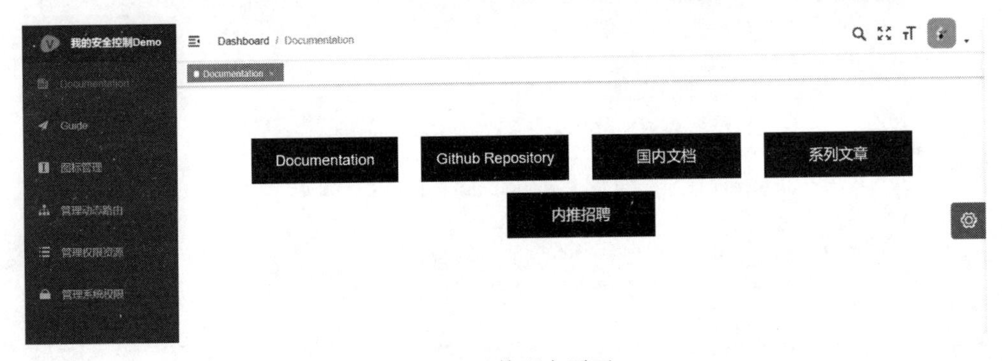

图 15.2　管理主页面

15.1.3　动态显示路由菜单

在 vue-element-admin 原始框架中，在登录后的主页面中，左边菜单内容是通过加载定义在 src/router/index.js 文件中的路由信息结合当前登录用户的角色，综合动态地显示。只需将获取 accessedRoutes 的方式，从 src/router/index.js 文件中导入改成发送请求从后端获取。需要注意的是，从后端返回的数据结构，需要同定义在 src/router/index.js 文件中的 asyncRoutes 一致，代码如下：

```
//src/store/modules/permission.js
...
const actions = {
  generateRoutes({ commit }, roles) {
    return new Promise(resolve => {
      let accessedRoutes
      if (roles.includes('admin')) {
        accessedRoutes = asyncRoutes || []        //获取定义在 router/index.js 的路由信息
      } else {
        accessedRoutes = filterAsyncRoutes(asyncRoutes, roles)
      }
      commit('SET_ROUTES', accessedRoutes)
      resolve(accessedRoutes)
    })
  }
}
...
```

以下是具体实现前端动态路由的步骤。

首先编写后端代码，返回符合要求的路由信息。前端接受的路由信息格式的样例代码如下：

```
...
[
  {
    path: '/permission',
    component: Layout,
    redirect: '/permission/page',
    alwaysShow: true, //will always show the root menu
    name: 'Permission',
    meta: {
      title: 'Permission',
      icon: 'lock',
      roles: ['admin', 'editor'] //you can set roles in root nav
    },
    children: [
      {
        path: 'page',
        component: () => import('@/views/permission/page'),
        name: 'PagePermission',
        meta: {
          title: 'Page Permission',
          roles: ['admin'] //or you can only set roles in sub nav
        }
      },
```

```
        {
          path: 'directive',
          component: () => import('@/views/permission/directive'),
          name: 'DirectivePermission',
          meta: {
            title: 'Directive Permission'
            //if do not set roles, means: this page does not require permission
          }
        },
        {
          path: 'role',
          component: () => import('@/views/permission/role'),
          name: 'RolePermission',
          meta: {
            title: 'Role Permission',
            roles: ['admin']
          }
        }
      ]
    },

    {
      path: '/icon',
      component: Layout,
      children: [
        {
          path: 'index',
          component: () => import('@/views/icons/index'),
          name: 'Icons',
          meta: { title: 'Icons', icon: 'icon', noCache: true }
        }
      ]
    },
    ...
  ]
  ...
```

路由信息是一个JSON对象数组，每个JSON对象是一个菜单，都以children属性的形式包含自己的子菜单。JSON对象的各个属性和意义如下：

```
hidden: true                      //设置菜单是否隐藏(不显示在左边菜单,缺省值为false)
alwaysShow: true                  //设置在一级菜单上,是否总是显示
                                  //true表示总是显示
                                  //false表示一级菜单只会在有多个子菜单情况下才显示
redirect: noRedirect              //如果设置成noRedirect,则面包屑不会重定向
```

```
name:'router-name'                  //路由名称,必须设置,并且唯一
meta : {
  roles: ['admin','editor']         //允许路由的角色列表
  title: 'title'                    //显示在左边导航栏和面包屑上的名称
  icon: 'svg-name'/'el-icon-x'      //图标名称
  noCache: true                     //页面是否不要缓存,默认值为 false
  affix: true                       //标签页面是否总是显示(不能关闭)
  breadcrumb: false                 //默认值为 true,表示是否以面包屑显示页面
  activeMenu: '/example/list'       //在左边菜单上高亮显示
 }
```

注意：如果没有子菜单，则该 JSON 对象不要有 children 属性，否则会影响菜单显示。

后台根据前端的需求，只需返回对应结构的数据，样例代码可以参考后端工程的 AuthController.java 类的 getMenu()方法，代码如下：

```
...
public class AuthController {
   ...
   @GetMapping("/routes")
   public ResponseResult getMenu(){
      ResponseResult result = new ResponseResult();
      List<AuthSysMenu> menuList = this.authService.getMenus();
      result.setData(menuList);
      return result;
   }
   ...
}
```

然后添加获取异步路由信息的 API，在前端工程中，创建 src/api/auth.js 文件，添加获取路由菜单的 getRoutes 接口，代码如下：

```
...
export function getRoutes() {
  return request({
   baseURL: 'http://localhost:7080',
   url: '/auth/routes',
   method: 'get'
  })
}
...
```

在 router/index.js 文件中定义组件名称同组件对象的对应 Map。在后台返回的路由菜单 JSON 对象中有个 component 属性，它的值需要是路由的目标组件对象，后台返回的只能是组件的名称，所以需要在挂载到 router 之前将 component 的名称替换成对应的组件对象。

当然替换成对应对象的工作不在这里实现，不过需要在 router/index.js 文件中定义好组件名称同组件对象对应的映射表，代码如下：

```
...
/**
 * 定义组件名称和组件对象的 map 对象
 */
export const componentMap = {
  'layout': require('@/layout').default,
  'redirect_index': () => import('@/views/redirect/index').then(m => m.default),
  'login_index': () => import('@/views/login/index').then(m => m.default),
  'login_auth_redirect': () => import('@/views/login/auth-redirect').then(m => m.
default),
  'error_page_404': () => import('@/views/error-page/404').then(m => m.default),
  'error_page_401': () => import('@/views/error-page/401').then(m => m.default),
  'dashboard_index': () => import('@/views/dashboard/index').then(m => m.default),
  'documentation_index': () => import('@/views/documentation/index').then(m => m.default),
  'guide_index': () => import('@/views/guide/index').then(m => m.default),
  'profile_index': () => import('@/views/profile/index').then(m => m.default),
  'permission_menu': () => import('@/views/permission/menu').then(m => m.default),
  'permission_resource': () => import('@/views/permission/permResource').then(m => m.
default),
  'permission_role': () => import('@/views/permission/role').then(m => m.default),
  'user_role': () => import('@/views/permission/user').then(m => m.default),
  'icons_index': () => import('@/views/icons/index').then(m => m.default)
}
...
```

最后修改 permission.js 文件，获取异步路由信息，以便实现动态路由。在 src/store/permission.js 文件中，调整 generateRoutes action 方法，以便从后台动态地请求路由菜单信息，并挂载到 router 上动态显示，代码如下：

```
...
//替换 route 对象中的 component
function replaceComponent(comp){
  if(comp.component && typeof(comp.component) == 'string'){
    comp.component = componentMap[comp.component];
  }

  if(comp.children && comp.children.length > 0){
    for(let i = 0;i < comp.children.length;i++){
      comp.children[i] = replaceComponent(comp.children[i]);
    }
  }else{
    delete comp.children
```

```
    }
    return comp
}

const actions = {
    generateRoutes: async function({ commit }, roles) {
        //从后台请求所有的路由信息
        let res = await getRoutes()
        //定义一个变量,用来存放可以访问的路由表
        let dbAsyncRoutes = res.data;

        let myAsyncRoutes = dbAsyncRoutes.filter(curr => {
            if(curr.children == null || curr.children.length == 0){
                delete curr.children
            }
            return replaceComponent(curr);
        })

        let accessedRoutes
        //判断当前的角色列表中是否已包含 admin
        if (roles.includes('admin')) {
            //所有路由都可以被访问,将 asyncRoutes 改造成从数据库中获取
            accessedRoutes = myAsyncRoutes || []
        } else {
            //根据角色,过滤掉不能访问的路由表
            accessedRoutes = filterAsyncRoutes(myAsyncRoutes, roles)
        }
        //commit
        commit('SET_ROUTES', accessedRoutes)
        //成功返回
        //resolve(accessedRoutes)
        return accessedRoutes
    }
}
...
```

15.1.4　动态控制页面内容

15.1.3 节介绍了怎样根据用户角色,控制显示对应角色拥有的有权限操作的路由菜单,如果在某个页面里面有部分内容,例如按钮,也同角色的权限相关,有的角色有权限,需要显示,而有的角色没有权限,不能显示,在 vue-element-admin 框架中,也提供了相关的解决方案。

例如在样例中,demo1 用户是超级管理员,它拥有所有权限,能对角色进行所有管理,角色管理界面显示如图 15.3 所示,而 demo2 用户是普通管理员用户,它在角色管理方面只

有部分权限，角色管理界面显示如图 15.4 所示。

维护角色

新建角色

序号	角色名称	角色描述	Operations
1	admin	超级管理员	编辑角色信息和分配角色路由 分配角色权限 删除角色
2	角色权限分配员	测试，可以访问 管理系统权限 菜单，但是并不能 创建角色	编辑角色信息和分配角色路由 分配角色权限 删除角色
3	后台权限测试人员	能看到管理系统权限 菜单里面所有操作，但是不能发送创建角色请求	编辑角色信息和分配角色路由 分配角色权限 删除角色
4	1	1	编辑角色信息和分配角色路由 分配角色权限 删除角色
5	2	2	编辑角色信息和分配角色路由 分配角色权限 删除角色

图 15.3　demo1 用户的维护角色 UI

维护角色

新建角色

序号	角色名称	角色描述	Operations
1	admin	超级管理员	编辑角色信息和分配角色路由
2	角色权限分配员	测试，可以访问 管理系统权限 菜单，但是并不能 创建角色	编辑角色信息和分配角色路由
3	后台权限测试人员	能看到管理系统权限 菜单里面所有操作，但是不能发送创建角色请求	编辑角色信息和分配角色路由
4	1	1	编辑角色信息和分配角色路由
5	2	2	编辑角色信息和分配角色路由

图 15.4　demo2 用户的维护角色 UI

在 src/utils/permission.js 文件中，定义了一个 checkPermission()函数，代码如下：

```
...
/**
 * @param {Array} value
 * @returns {Boolean}
 * @example see @/views/permission/directive.vue
 */
export default function checkPermission(value) {
  if (value && value instanceof Array && value.length > 0) {
    //获取当前登录用户拥有的角色列表
    const roles = store.getters && store.getters.roles
    const permissionRoles = value
    //判断 value 和 roles 两个集合是否有交集
    const hasPermission = roles.some(role => {
      return permissionRoles.includes(role)
    })
    return hasPermission
  } else {
```

```
      console.error(`need roles! Like v-permission="['admin','editor']"`)
      return false
    }
  }
  ...
```

checkPermission()函数能判断出当前的用户是否拥有给定的角色。传入的 value 是数组，即需要判断的角色，返回值是 boolean。如果当前用户拥有 value 里面的一个或多个角色，则返回值为 true，否则返回值为 false。

前端开发人员可以基于 checkPermission()函数，实现不同角色的用户在同一个页面中不同内容的显示和隐藏。

在 src/views/permission/role.vue 文件中使用 checkPermission 结合 v-if 指令控制“新建角色”按钮的动态显示，即有权限的用户才显示，代码如下：

```
<template>
 <div class="app-container">
  <!-- 第一部分,新建角色 -->
<el-button type="primary" @click="handleAddRole" v-if="checkPermission
(createRoleRoles)">新建角色</el-button>
...
 </div>
</template>
...
```

其中 createRoleRoles 是从后台请求的用户“新建角色”权限的所有角色，如 role.vue 中的代码如下：

```
...
  async getOperateRoles(){
    //新建角色的角色集合
    let res = await getOperateMenuRoleId('createRole')      //获取能够操作"创建角色"
    //功能的所有角色 id
    this.createRoleRoles = res.data
    ...
},
...
```

15.1.5　管理动态路由菜单

15.1.3 节和 15.1.4 节介绍了怎样在前端结合后台，实现路由菜单的动态控制和页面内部内容的动态控制。要实现这两个目标，还需要基于后台，实现这些控制信息的动态维护，也就是对对应控制数据的增、删、改、查。在这里，就不进行详细介绍了，只是给读者列出样例中实现这些功能的对应代码，读者可以根据这些提示，参考对应代码的实现。

1. 维护权限资源

样例中的资源是指样例应用的系统资源，也就是说，通过系统可以做哪些操作，等同于系统的模块。系统的权限限定了对资源的操作，也就是指某个人是否有权限操作某个资源。

作为安全管理模块，首先要管理的就是系统资源。也就是维护系统资源和对系统资源的操作，从而形成系统权限的最小单元。当然，对资源的操作，也是构成动态路由菜单的基本元素。样例应用中有对系统资源的维护，并且做了基本的实现，维护界面如图 15.5～图 15.7 所示。

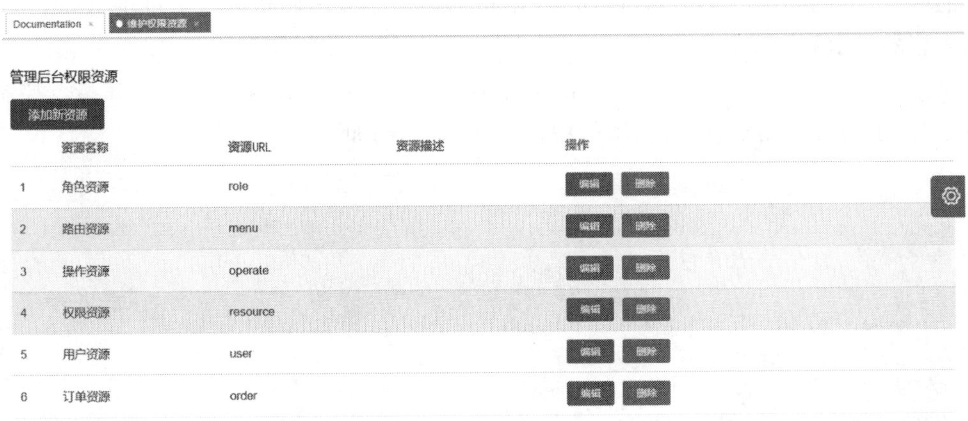

图 15.5 系统资源列表

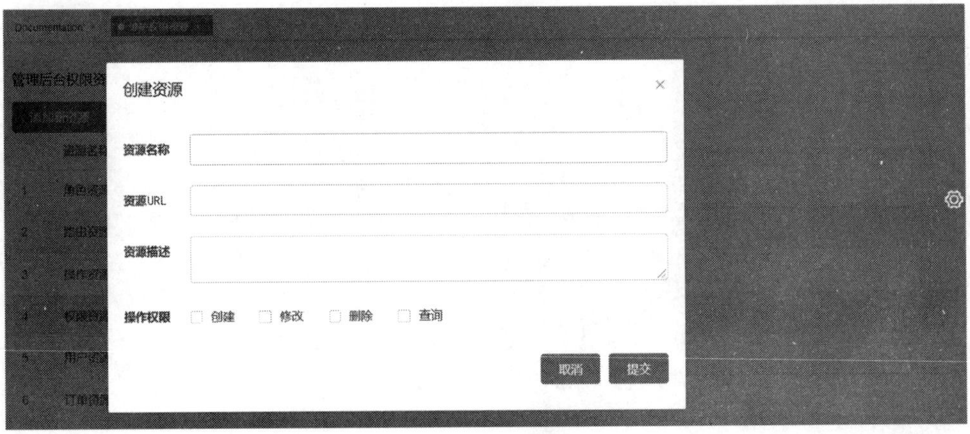

图 15.6 创建资源

维护资源的样例代码和说明如下。

(1) src/views/permission/permResource.vue：前端 Vue 组件。

(2) src/api/resource.js：前端同后端交互的 API 文件。

(3) cn. com. authority. controller. PermResourceController. java：后端资源控制器。

(4) authsys_resources：系统资源表。

(5) authsys_operator：资源操作表。

(6) authsys_permission：权限表。

图 15.7 修改资源

2. 维护路由信息

这里的路由信息是前端显示的，可以是单击的按钮和超链接信息，当用户单击页面上的按钮和超链接时，将会路由到对应的组件并显示。路由信息构成了前端页面能控制权限的最小单元。通过维护路由信息，从而达到动态控制前端路由操作的目的。维护路由信息的界面如图 15.8～图 15.10 所示。

图 15.8 路由信息列表

维护路由信息的代码和说明如下。

(1) src/views/permission/menu. vue：路由 Vue 组件。

(2) src/api/menu. js：维护路由操作同后端交互的 API 文件。

(3) cn. com. authority. controller. MenuController. java：后端路由控制器。

新增<图标管理>的子菜单

路径　路由path

名称　路由名称

组件名称　layout

重定向路径　重定向路径

主题　路由主题

图标　路由图标

高亮菜单　高亮菜单

隐藏　总是显示　显示缓存　affix　面包屑

取消　提交

图 15.9　添加路由信息

修改<图标管理>菜单

路径　/icon

名称　null

组件名称　layout

重定向路径　重定向路径

主题　图标管理

图标　icon

高亮菜单　高亮菜单

隐藏　总是显示　显示缓存　affix　面包屑

取消　提交

图 15.10　修改路由信息

(4) authsys_menu：路由信息表。

3. 维护角色

系统角色是权限管理控制中的一个重要组成部分，通过给角色分配不同的权限，再将角色分配给用户，从而实现权限控制。为了实现系统用户权限的动态控制，需要有对角色动态维护的实现。在样例应用中，维护角色的基本操作如图 15.11～图 15.14 所示。

实现维护角色的样例代码和说明如下。

(1) src/views/role.vue：前端角色 Vue.js 组件。

序号	角色名称	角色描述	Operations
1	admin	超级管理员	编辑角色信息和分配角色路由 分配角色权限 删除角色
2	角色权限分配员	测试，可以访问 管理系统权限 菜单，但是并不能 创建角色	编辑角色信息和分配角色路由 分配角色权限 删除角色
3	后台权限测试人员	能看到管理系统权限 菜单里面所有操作，但是不能发送创建角色请求	编辑角色信息和分配角色路由 分配角色权限 删除角色
4	1	1	编辑角色信息和分配角色路由 分配角色权限 删除角色
5	2	2	编辑角色信息和分配角色路由 分配角色权限 删除角色

图 15.11　角色列表

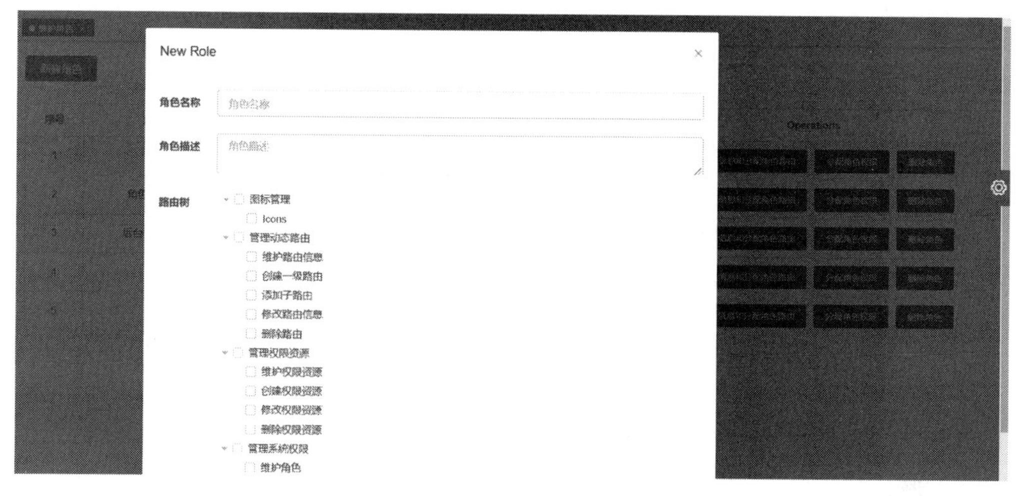

图 15.12　新建角色

Edit Role
角色名称 admin
角色描述 超级管理员
路由树
图标管理
Icons
管理动态路由
维护路由信息
创建一级路由
添加子路由
修改路由信息
删除路由
管理权限资源
维护权限资源
创建权限资源
修改权限资源

图 15.13　修改角色信息和给角色分配路由树

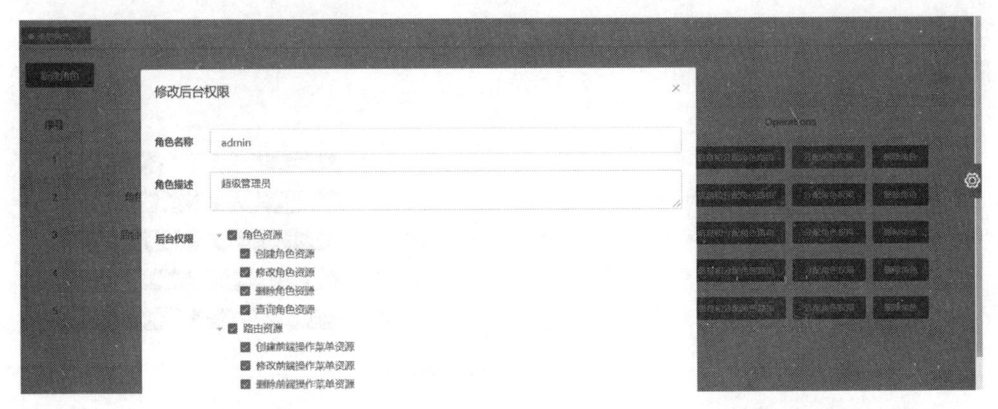

图 15.14　给角色分配权限

(2) src/api/role.js：维护角色操作通过后端交互 API 文件。

(3) cn.com.authority.controller.RoleController.java：后端角色控制器。

(4) authsys_role：角色信息表。

(5) authsys_mn_rl：角色路由关系表。

(6) authsys_rl_prm：角色权限关系表。

4. 维护用户

用户的意义,这里不做过多的赘述。样例代码基本实现了维护用户里面的用户列表和分配用户角色这两个功能,操作界面如图 15.15 和图 15.16 所示。

● 维护用户 ×

序号	用户名	姓名	Operations
1	demo1	测试1	分配用户角色
2	demo2	测试2	分配用户角色
3	demo3	测试3	分配用户角色
4	demo4	测试4	分配用户角色
5	demo5	测试5	分配用户角色
6	demo6	测试6	分配用户角色

图 15.15　用户列表

图 15.16　分配用户角色

样例应用维护用户的实现代码和说明如下。

(1) src/views/user.vue：前用户端 Vue.js 组件。

(2) src/api/user.js：前端操作用户同后端交互 API 文件。

(3) cn.com.authority.controller.UserController.java：后用户端控制器。

(4) authsys_user：用户信息表。

(5) authsys_ur_rl：用户角色关系表。

15.2　实现后端安全控制

如果用户仅仅通过提供的前端界面操作系统，在前端实现权限控制，确实可以限制不同用户，只能在他们自己的权限范围内操作系统，但是如果用户绕过前端操作界面，直接用其他方式向后端发送 URL 请求，这样在安全方面就没有设防了。为了解决这样的安全问题，系统需要在后端，对每个有权限限制的请求，做出对应的权限验证。接下来介绍怎样基于 Shiro 安全框架，在后台完成用户身份认证和授权控制。

15.2.1　Shiro 简介

Apache Shiro 是 Apache 软件基金会(Apache Software Foundation)提供的一个软件开源项目，它为我们实现了一个功能强大且使用方便的 Java 安全框架，该框架实现了权限的认证、验证、加密、会话管理、缓存管理等公共功能，能运用到任何 Java 项目中，从简单的命令行应用到 Web 应用及手机应用，甚至分布式企业级应用。

当然，除了 Apache Shiro 安全框架可实现安全控制功能外，也有其他的安全框架，例如 JavaEE 中有 Java Security 解决方案；Spring 内部也提供了 Spring Security 模块。

Apache Shiro 安全框架相对其他 Java 安全框架有以下优势：

(1) 有一套易于理解的 Java Security API。

(2) 简单的身份认证(登录)，支持多种数据源(LDAP、JDBC、ActiveDirectory、ini 配置等)。

(3) 支持角色基本的和自定义细粒度的访问控制。

(4) 支持缓存，提高系统性能。

(5) 自定义的会话管理，同时支持 Web 和非 Web 应用。

(6) 支持异构的客户端。

(7) 内置加密服务 API。

(8) 不同于任何框架捆绑，可独立运行。

Shiro 安全框架的最主要目标是为基于 Shiro 安全框架开发的程序员，提供 4 个最基本的核心的系统安全功能：身份认证/登录(Authentication)、授权(Authorization)、会话管理(Session Manager)、加密(Cryptography)。

(1) Authentication：身份认证/登录，验证用户的身份是否正确。

（2）Authorization：授权，即验证权限，验证某个已经证明身份的人是否拥有操作某个功能的权限。如判断用户是不是拥有某个角色，或判断用户是否能修改某个数据等。

（3）Session Manager：会话管理。一个用户从登录到退出的所有信息都保存在会话中。Shiro 不仅支持 Web 环境下的会话管理，还支持 J2SE 环境下的会话管理。

（4）Cryptography：加密。使用加密技术保证数据的安全性，例如用户密码都是加密后才保存到数据库中，避免明码的不安全性。

Shiro 安全框架除了提供了上面的 4 个核心功能外，还提供了一些其他的扩展辅助功能，我们在需要的时候，可以自由地将它们集成到系统中。

（1）Web Support：对 Web 应用的支持，能很方便地将 Shiro 集成到 Web 应用中。

（2）Caching：缓存支持。例如用户登录后，该用户的权限信息可以保存在缓存中，避免每次授权的时候都要访问数据库，从而提高系统性能。

（3）Concurrency：支持多线程并发认证。可以在一个线程中开启另外一个线程，自动获得传播权限。

（4）Testing：内置测试模块，方便测试。

（5）Run As：支持一个用户冒充另外一个用户的身份进行访问。

（6）Remember Me：记住用户的登录状态，在一定的时间内，不需要重复登录验证。

Shiro 安全框架的体系结构，如图 15.17 所示。

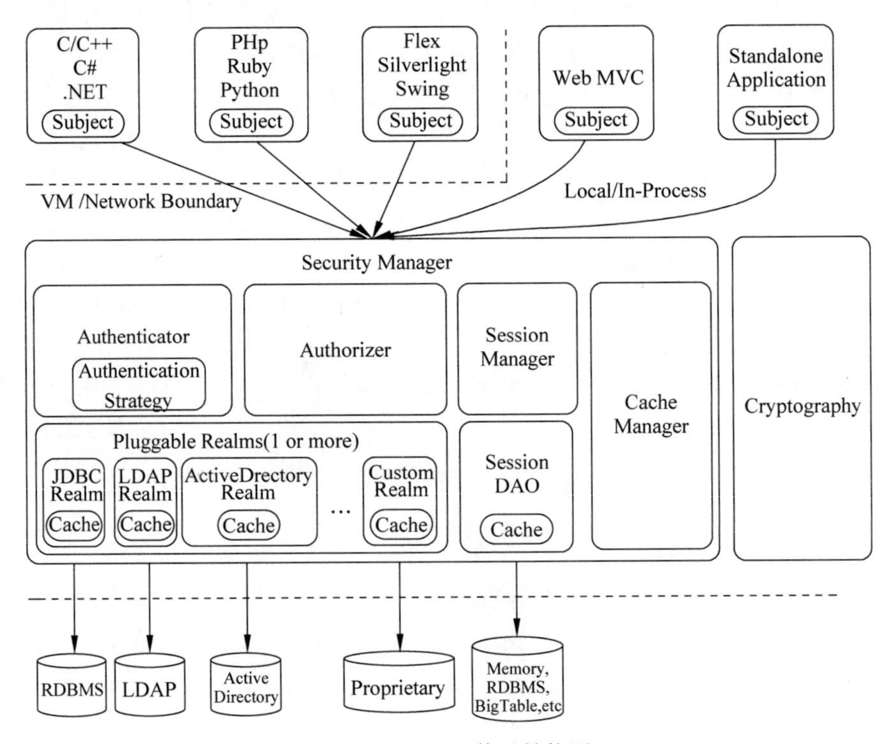

图 15.17　Shiro 体系结构图

(1) Subject(org. apache. shiro. subject. Subject)：代表需要基于 Shiro 实现安全操作的用户(用户、第三方系统或独立任务等)的主题。

(2) SecurityManager(org. apache. shiro. mgt. SecurityManager)：安全管理器。上面提到过，SecurityManager 是 Shiro 的核心，它像一个主事的总管，协调被它管理的所有对象可以顺利地完成所有的安全操作功能。它同时还管理着所有同 Shiro 进行交互的用户，它知道所有用户的安全权限。

(3) Authenticator(org. apache. shiro. authc. Authenticator)：认证器。在用户登录的时候，由 Authenticator 完成对用户身份的认证，验证用户是不是真正的用户，也就是认证用户凭证是否正确。当 Authenticator 认证用户凭证真实性的时候，会自动调用 Realm 对象，从我们的系统中获取对应用户的账号信息(一般是用户名和密码)。

(4) Authentication Strategy(org. apache. shiro. authc. pam. AuthenticationStrategy)：如果在 SecurityManager 中配置了多个 Realm，就由 AuthenticationStrategy 协调按什么方式进行用户凭证的认证，例如只要一个 Realm 认证通过了就算认证通过呢，还是需要所有的 Realm 认证都通过了才算认证通过。

(5) Authorizer(org. apache. shiro. authz. Authorizer)：授权器。Authorizer 负责判断用户是否有权限进行安全操作。也就是说，当前用户能做什么事情，由 Authorizer 说了算。同 Authenticator 一样，Authorizer 知道怎样同后台数据进行交互，从而根据用户拥有的角色和权限进行判断允许用户做什么事情。

(6) SessionManager(org. apache. shiro. session. mgt. SessionManager)：Session 管理器。SessionManager 会根据每个用户创建一个 Session 来维护当前用户的会话周期，并且实现用户会话的过期控制。对用户会话管理是 Shiro 特有的功能，其他安全框架都没有单独提供的。Shiro 自动地对所有用户都进行会话管理，哪怕当前应用没有运行在 Web 容器中或 EJB 容器中。Shiro 会自动地使用已有的容器环境进行 Session 管理，如果整个应用没有运行在容器中，则 Shiro 也会独立实现 Session 管理并且控制会话的过期。同样地，也可以基于 SessionDAO 实现 Session 对象的持久化。

(7) SessionDAO(org. apache. shiro. session. mgt. eis. SessionDAO)：SessionDAO 为 SecurityManager 提供了对 Session 对象的持久化操作(CRUD)。根据用户提供的不同插件，可以实现各种方式的持久化，例如数据库、内存、缓存等都支持。

(8) CacheManager(org. apache. cache. CacheManager)：缓存管理器。CacheManager 是 Shiro 中创建并且管理缓存对象的组件。因为 Shiro 在进行认证、授权和 Session 管理的时候，有很多数据需要交互，如果每次都直接访问后面的数据源，则性能会比较低。有了缓存后，就可以先访问缓存中的数据，如果没有找到合适的数据，再访问后面的数据源，这样就可以提高系统的性能。Shiro 提供了对缓存操作的接口，具体使用哪个缓存组件，用户可以根据自己的需要选择，例如 Ehcache 等。

(9) Cryptography(org. apache. shiro. crypto. *)：加密模块。加密模块是企业级安全框架的一个功能。Shiro 的 crypto 包中定义了很多容易理解，并且可以简单方便地使用基于 Ciphers、Hashes 等不同算法的加密方法。我们可以基于这些实现类，实现数据的加密解密，从而避免明码的不安全性。

(10) Realms(org. apache. shiro. realm. Realm)：域。前面提到了，Realm 等用于一个桥梁或连接器，实现 Shiro 同后台安全数据之间的交互。当 Shiro 要进行认证、授权的时候，自动通过配置的 Realm 同后台数据进行交互，获取必要的认证或授权的依据数据。我们需要在 SecurityManager 中配置至少一个 Realm，也可以配置自定义的 Realm，以便实现 Shiro 同后台数据的交互。

15.2.2 搭建 Shiro 框架

使用 Shiro 实现身份认证和授权控制前，需要先在项目中集成 Shiro，具体步骤如下。

第一步，添加依赖，考虑到同 Spring Boot、Redis 和 Jwt 集成，需要加上相关依赖，代码如下：

```
...
<dependency>
    <groupId>com.auth0</groupId>
    <artifactId>java-jwt</artifactId>
    <version>3.5.0</version>
</dependency>

<!-- Shiro -->
<dependency>
    <groupId>org.apache.shiro</groupId>
    <artifactId>shiro-spring</artifactId>
    <version>1.7.0</version>
</dependency>

<dependency>
    <groupId>org.springframework.boot</groupId>
    <artifactId>spring-boot-starter-data-redis</artifactId>
</dependency>

<!-- https://mvnrepository.com/artifact/redis.clients/jedis -->
<dependency>
    <groupId>redis.clients</groupId>
    <artifactId>jedis</artifactId>
    <version>2.9.0</version>
</dependency>

<!-- https://mvnrepository.com/artifact/org.crazycake/shiro-redis -->
<dependency>
    <groupId>org.crazycake</groupId>
    <artifactId>shiro-redis</artifactId>
```

```
    <version>3.1.0</version>
</dependency>
...
```

第二步，编写 Shiro 配置文件，编写 ShiroConfiguration.java 文件，代码如下：

```
package cn.com.authority.config;

...
@Configuration
@Slf4j
public class ShiroConfiguration {
    @Autowired
    private RedisUtil redisUtil;
    @Autowired
    private IPermissionService permissionService;
    ...
    private static final String CACHE_KEY = "shiro:cache:";
    private static final String SESSION_KEY = "shiro:session:";
    private static final String NAME = "custom.name";
    private static final String VALUE = "/";

    @Bean("shiroFilter")
    public ShiroFilterFactoryBean getShiroFilterFactoryBean(
@Qualifier("securityManager")DefaultWebSecurityManager
     securityManager) {
        ShiroFilterFactoryBean shiroFilterFactoryBean = new ShiroFilterFactoryBean();

        shiroFilterFactoryBean.setSecurityManager(securityManager);
        return shiroFilterFactoryBean;
    }

    @Bean("securityManager")                //固定
    public DefaultWebSecurityManager createSecurityManager(
@Qualifier("shiroRealm") ShiroAuthorizingRealm shiroRealm){
        //创建 SecurityMananger 对象,并且装配 Realm 对象
        DefaultWebSecurityManager manager =
new DefaultWebSecurityManager(shiroRealm);
        //自定义缓存实现,使用 redis
        manager.setCacheManager(cacheManager());
        //自定义 session 管理,使用 redis
        manager.setSessionManager(sessionManager());

        /*
         * 关闭 shiro 自带的 session,详情见文档
```

```
        * http://shiro.apache.org/session-management.html#SessionManagement-
StatelessApplications%28Sessionless%29
        */
       DefaultSubjectDAO subjectDAO = new DefaultSubjectDAO();
       DefaultSessionStorageEvaluator defaultSessionStorageEvaluator =
new DefaultSessionStorageEvaluator();
       defaultSessionStorageEvaluator.setSessionStorageEnabled(false);
subjectDAO.setSessionStorageEvaluator(defaultSessionStorageEvaluator);
       manager.setSubjectDAO(subjectDAO);

       return manager;
   }

   /**
    * Session Manager
    * 使用的是 shiro-redis 开源插件
    */
   @Bean
   public DefaultWebSessionManager sessionManager() {
       DefaultWebSessionManager sessionManager =
new DefaultWebSessionManager();
       sessionManager.setSessionDAO(redisSessionDAO());
       //允许支持基于 Cookie 传递的会话 id
       sessionManager.setSessionIdCookieEnabled(true);
       //支持请求的 RUL 重新(通过 URL 传递会话 id)
       sessionManager.setSessionIdUrlRewritingEnabled(true);

       SimpleCookie simpleCookie = new SimpleCookie();
       simpleCookie.setName(NAME);
       simpleCookie.setValue(VALUE);
       //设置传递会话 id 的 Cookie 的属性
       sessionManager.setSessionIdCookie(simpleCookie);

       return sessionManager;
   }

   //准备 RedisSessionDAO
   /**
    * RedisSessionDAO shiro sessionDao 层的实现,通过 redis
    * 使用的是 shiro-redis 开源插件
    */
   @Bean
   public RedisSessionDAO redisSessionDAO() {
       RedisSessionDAO redisSessionDAO = new RedisSessionDAO();
       redisSessionDAO.setRedisManager(redisManager());
       redisSessionDAO.setExpire(86400);
```

```
        redisSessionDAO.setKeyPrefix(SESSION_KEY);
        return redisSessionDAO;
    }
    /**
     * cacheManager 缓存 redis 实现
     * 使用的是 shiro - redis 开源插件
     *
     * @return
     */
    public RedisCacheManager cacheManager() {
        RedisCacheManager redisCacheManager = new RedisCacheManager();
        //装配 redisManager
        redisCacheManager.setRedisManager(redisManager());
        redisCacheManager.setExpire(86400);
        redisCacheManager.setKeyPrefix(CACHE_KEY);
        return redisCacheManager;
    }
    /**
     * 配置 shiro redisManager
     * 使用的是 shiro - redis 开源插件
     *
     * @return
     */
    public RedisManager redisManager() {
        RedisManager redisManager = new RedisManager();
        redisManager.setHost("127.0.0.1");
        redisManager.setPort(6379);
        redisManager.setTimeout(0);
        //redisManager.setPassword(password);
        return redisManager;
    }
}
```

15.2.3 基于Shiro实现身份认证

1. 编写 ShiroUser.java

ShiroUser 封装了 Shiro 中登录用户的身份凭证信息，包括用户名、密码、id 和 cacheKey。当用户登录时，就以用户唯一的身份凭证的形式将信息保存到 Shiro 上下文中，保存的信息贯穿整个安全控制过程，代码如下：

```
package cn.com.authority.service.shiro;

import lombok.Data;
```

```
import java.io.Serializable;

@Data
public class ShiroUser implements Serializable {
    private String userName;
    private int id;
    private String authCacheKey;
    private String password;

    public String toString(){
        return this.authCacheKey;
    }
}
```

2. 编写 ShiroAuthorizingRealm.java

ShiroAuthorizingRealm 类继承自 AuthorizingRealm 类，用于实现 doGetAuthorizationInfo 和 doGetAuthentictionInfo 接口，给 Shiro 安全框架提供身份凭证信息和拥有的权限信息，用来做身份认证和授权认证，代码如下：

```
package cn.com.authority.service.shiro.realm;

import cn.com.authority.entity.AuthSysUser;
import cn.com.authority.service.IUserService;
import cn.com.authority.service.shiro.ShiroUser;
import cn.com.authority.utils.EncryptUtil;
import org.apache.shiro.authc.*;
import org.apache.shiro.authz.AuthorizationInfo;
import org.apache.shiro.authz.SimpleAuthorizationInfo;
import org.apache.shiro.realm.AuthorizingRealm;
import org.apache.shiro.subject.PrincipalCollection;
import org.springframework.beans.factory.annotation.Autowired;
import org.springframework.stereotype.Component;

import java.util.List;

@Component("shiroRealm")
public class ShiroAuthorizingRealm extends AuthorizingRealm {
    @Autowired
    private IUserService userService;
    @Override
    protected AuthorizationInfo doGetAuthorizationInfo(
PrincipalCollection principalCollection) {
        //已知用户名
        ShiroUser shiroUser = (ShiroUser) principalCollection.getPrimaryPrincipal();
        String username = shiroUser.getUserName();
```

```
        //资源:操作:id      user:create
        List<String> permissionList = this.userService.getUserPermissions(username);

        SimpleAuthorizationInfo simpleAuthorizationInfo = new SimpleAuthorizationInfo();
        simpleAuthorizationInfo.addStringPermissions(permissionList);

        return simpleAuthorizationInfo;
    }

    /**
     * 身份认证信息
     * 重写该方法,从数据库中获取注册用户和对应的密码,封装成 AuthenticationInfo 对象并返回
     **/
    @Override
    protected AuthenticationInfo doGetAuthenticationInfo(
AuthenticationToken authenticationToken) throws AuthenticationException {
        //authenticationToken 当前登录用户的身份标识
        //已知用户名
        String username = authenticationToken.getPrincipal().toString();
        //再根据用户名从数据库中寻找密码
        AuthSysUser user = this.userService.findByUserName(username);
        if(user == null){
            throw new AccountException("account not exist");
        }
        String password = user.getUrPassword();
        //用 AuthenticationInfo 对象封装 username 和 password
        ShiroUser shiroUser = new ShiroUser();
        shiroUser.setId(user.getUrId());
        shiroUser.setUserName(user.getUrUserName());
        shiroUser.setPassword(password);
        shiroUser.setAuthCacheKey(user.getUrUserName() + user.getUrId());
        SimpleAuthenticationInfo info = new SimpleAuthenticationInfo(shiroUser, password,
this.getName());
        return info;
    }
}
```

3. 编写 TokenAuthFilter.java

TokenAuthFilter 继承自 AccessControlFilter,重写了 isAccessAllowed()方法,在过滤请求时,验证当前用户的登录是否为有效登录,代码如下:

```
package cn.com.authority.service.shiro.filter;


import cn.com.authority.commons.ResponseResult;
import cn.com.authority.utils.JwtUtil;
```

```
import cn.com.authority.utils.RedisUtil;
import cn.hutool.json.JSONObject;
import org.apache.shiro.subject.Subject;
import org.apache.shiro.web.filter.AccessControlFilter;
import org.apache.shiro.web.filter.authc.BasicHttpAuthenticationFilter;
import org.apache.shiro.web.util.WebUtils;
import org.springframework.beans.factory.annotation.Autowired;
import org.springframework.stereotype.Component;

import javax.servlet.ServletException;
import javax.servlet.ServletRequest;
import javax.servlet.ServletResponse;
import javax.servlet.http.HttpServletRequest;
import java.io.IOException;
import java.io.PrintWriter;
public class TokenAuthFilter extends AccessControlFilter {
    @Autowired
    private RedisUtil redisUtil;
    @Override
    protected boolean isAccessAllowed(
ServletRequest request, ServletResponse response, Object mappedValue) {

        boolean bool = false;
        String token = ((HttpServletRequest)request).getHeader("X-Token");
        if(token == null || redisUtil.get(token) == null){
            try {
                request.getRequestDispatcher("/auth/noLogin").forward(request,response);
            } catch (ServletException e) {
                e.printStackTrace();
            } catch (IOException e) {
                e.printStackTrace();
            }
        }else{
            bool = true;
        }

        return bool;
    }

    @Override
    protected boolean onAccessDenied(
ServletRequest request, ServletResponse response) throws Exception {

        return false;
    }

    public void setRedisUtil(RedisUtil redisUtil) {
        this.redisUtil = redisUtil;
    }
}
```

4. 在 AuthController 中添加 Shiro 登录支持

在 AuthController 的 login 登录请求处理方法中，基于 Shiro 安全框架的身份验证机制，完成用户的身份凭证的验证，代码如下：

```
package cn.com.authority.controller;

...
@RestController
@RequestMapping("/auth")
public class AuthController {
...

    @PostMapping("/login")
    public ResponseResult login(@RequestBody AuthSysAccount account
, HttpSession session){
// @RequestMapping("/login")
// public ResponseResult login(AuthSysAccount account, HttpSession session){
        //System.out.println(session.getId());
        ResponseResult result = new ResponseResult();
        //1. 从 Shiro 框架中,获取一个 Subject 对象,代表当前会话的用户
        Subject subject = SecurityUtils.getSubject();
        //2. 认证 subject 的身份
        //2.1 封装要认证的身份信息
        AuthenticationToken token =
new UsernamePasswordToken(account.getUsername(), EncryptUtil.getMD5Str(account.getPassword()));
        //2.2 认证:认证失败,抛出异常
        try {
            subject.login(token);
            String resToken = JwtUtil.sign(account.getUsername());
            result.setData(result.new Token(resToken));
            //在 session 中保存登录标记
// session.setAttribute(resToken, account.getUsername());
            this.redisUtil.set(resToken, account.getUsername(),30 * 60);
        }catch(Exception e){
            e.printStackTrace();
            result.setData("用户名或密码错误");
        }
        return result;
    }
    ...
    @PostMapping("/logout")
    public ResponseResult logOut(@RequestHeader("X - Token") String token){
        ResponseResult result = new ResponseResult();
        //session.removeAttribute(token);
        //session.invalidate();
```

```
        SecurityUtils.getSubject().logout();
        //删除 Redis 中的标记
        this.redisUtil.del(token);
        return result;
    }
}
```

5. 注册 TokenAuthFilter

在 ShiroConfiguration 中，注册 TokenAuthFilter 和声明需要登录验证的请求，代码如下：

```
package cn.com.authority.config;
...

@Configuration
@Slf4j
public class ShiroConfiguration {
    ...

    @Bean("shiroFilter")
    public ShiroFilterFactoryBean getShiroFilterFactoryBean(@Qualifier("securityManager")
DefaultWebSecurityManager securityManager) {
        ShiroFilterFactoryBean shiroFilterFactoryBean = new ShiroFilterFactoryBean();

        //创建一个 Map 对象,包含所有要集成到 Shiro 框架中的 filter 对象
        //注册哪些过滤器
        Map<String, Filter> filters = new HashMap<String, Filter>();
        //<filter>
        //<filter-mapping
        this.tokenAuthFilter = new TokenAuthFilter();
        this.tokenAuthFilter.setRedisUtil(this.redisUtil);

        filters.put("authcToken", this.tokenAuthFilter);

        //在 Shiro 中注册过滤器
        shiroFilterFactoryBean.setFilters(filters);

        //设置哪些过滤器,拦截哪些请求
        Map<String,String> permChainMap = new HashMap<>();
        //permChainMap.put("/user/**", "tokenAuthc");
        List<AuthSysPermission> permissionList = this.permissionService.searchPermissions();
        if(permissionList != null){
            for(AuthSysPermission permission:permissionList){///user/**
                permChainMap.put("/" + permission.getResources().getRscurl() + "/**", "authcToken");
```

```
            }
        }

        shiroFilterFactoryBean.setFilterChainDefinitionMap(permChainMap);

        //其他设置
        shiroFilterFactoryBean.setLoginUrl("/auth/login");
        //shiroFilterFactoryBean.setSuccessUrl();
        shiroFilterFactoryBean.setUnauthorizedUrl("/auth/noAuth");

        shiroFilterFactoryBean.setSecurityManager(securityManager);
        return shiroFilterFactoryBean;
    }
...
}
```

15.2.4　基于 Shiro 实现授权

1．编写 PermissionAuthFilter.java

编写 PermissionAuthFilter 类，继承自 AccessControlFilter，校验当前过滤的请求是否有权限访问，代码如下：

```
package cn.com.authority.service.shiro.filter;

...
public class PermissionAuthFilter extends AccessControlFilter {
    @Override
    protected boolean isAccessAllowed(ServletRequest servletRequest,
ServletResponse servletResponse, Object o) throws Exception {
        boolean bool = true;
        //拼凑当前请求 URL 对应的权限字符串    /user/create?xxxxx
        HttpServletRequest httpRequest = (HttpServletRequest)servletRequest;
        String requestURI = httpRequest.getRequestURI();

        String[] permStrs = requestURI.split("/");
        if(permStrs != null && permStrs.length > 2) {
            String perm = permStrs[1] + ":" + permStrs[2];
            //资源名称:操作对象 id
            //user:create,update,delete,search,other: *
            //shop:create,update,delete...: *
            //who what how
            //获取当前用户对应的 subject 对象
            Subject subject = this.getSubject(servletRequest,servletResponse);
            //Object primary = subject.getPrincipals().getPrimaryPrincipal();
```

```
            //调用 subject 对象的方法,判断是否有权限
            //返回判断结果
            bool = subject.isPermitted(perm);
        }
        return bool;
    }

    @Override
    protected boolean onAccessDenied(ServletRequest servletRequest,
ServletResponse servletResponse) throws Exception {
((HttpServletRequest)servletRequest).getRequestDispatcher("/auth/noAuth")
.forward(servletRequest,servletResponse);
        return false;
    }
}
```

2. 在 ShiroConfiguration 中注册使用 PermissionAuthFilter

在 ShiroConfiguration 中注册 PermissionAuthFilter,并声明需要使用 PermissionAuthFilter 进行授权认证的请求,代码如下：

```
package cn.com.authority.config;
...

@Configuration
@Slf4j
public class ShiroConfiguration {
    ...

    @Bean("shiroFilter")
    public ShiroFilterFactoryBean getShiroFilterFactoryBean(@Qualifier("securityManager")
DefaultWebSecurityManager securityManager) {
        ShiroFilterFactoryBean shiroFilterFactoryBean = new ShiroFilterFactoryBean();

        //创建一个 Map 对象,包含所有要集成到 Shiro 框架中的 filter 对象
        //注册哪些过滤器
        Map<String, Filter> filters = new HashMap<String, Filter>();
        //<filter>
        //<filter-mapping
        this.tokenAuthFilter = new TokenAuthFilter();
        this.tokenAuthFilter.setRedisUtil(this.redisUtil);
        this.permissionAuthFilter = new PermissionAuthFilter();

        filters.put("authcToken", this.tokenAuthFilter);
        filters.put("permAuthc", this.permissionAuthFilter);
```

```
        //在 shiro 中注册过滤器
        shiroFilterFactoryBean.setFilters(filters);

        //设置哪些过滤器拦截哪些请求
        Map<String,String> permChainMap = new HashMap<>();
        //permChainMap.put("/user/**", "tokenAuthc");
        List<AuthSysPermission> permissionList = this.permissionService.searchPermissions();
        if(permissionList != null){
            for(AuthSysPermission permission:permissionList){///user/**
                permChainMap.put("/" + permission.getResources().getRscurl() + "/**", "authcToken,
permAuthc");
            }
        }

        shiroFilterFactoryBean.setFilterChainDefinitionMap(permChainMap);
...
        shiroFilterFactoryBean.setSecurityManager(securityManager);
        return shiroFilterFactoryBean;
    }
...
}
```

图书推荐

书 名	作 者
鸿蒙应用程序开发	董昱
HarmonyOS 应用开发实战(JavaScript 版)	徐礼文
鸿蒙操作系统开发入门经典	徐礼文
鸿蒙操作系统应用开发实践	陈美汝、郑森文、武延军、吴敬征
HarmonyOS 移动应用开发	刘安战、余雨萍、李勇军 等
HarmonyOS App 开发从 0 到 1	张诏添、李凯杰
HarmonyOS 从入门到精通 40 例	戈帅
JavaScript 基础语法详解	张旭乾
华为方舟编译器之美——基于开源代码的架构分析与实现	史宁宁
鲲鹏架构入门与实战	张磊
华为 HCIA 路由与交换技术实战	江礼教
Android Runtime 源码解析	史宁宁
深度探索 Go 语言——对象模型与 runtime 的原理、特性及应用	封幼林
Flutter 组件精讲与实战	赵龙
Flutter 组件详解与实战	[加]王浩然(Bradley Wang)
Flutter 实战指南	李楠
Dart 语言实战——基于 Flutter 框架的程序开发(第 2 版)	亢少军
Dart 语言实战——基于 Angular 框架的 Web 开发	刘仕文
IntelliJ IDEA 软件开发与应用	乔国辉
Vue+Spring Boot 前后端分离开发实战	贾志杰
Vue.js 企业开发实战	千锋教育高教产品研发部
Python 从入门到全栈开发	钱超
Python 全栈开发——基础入门	夏正东
Python 全栈开发——高阶编程	夏正东
Python 游戏编程项目开发实战	李志远
Python 人工智能——原理、实践及应用	杨博雄 主编,于营、肖衡、潘玉霞、高华玲、梁志勇 副主编
Python 深度学习	王志立
Python 预测分析与机器学习	王沁晨
Python 异步编程实战——基于 AIO 的全栈开发技术	陈少佳
Python 数据分析实战——从 Excel 轻松入门 Pandas	曾贤志
Python 数据分析从 0 到 1	邓立文、俞心宇、牛瑶
Python Web 数据分析可视化——基于 Django 框架的开发实战	韩伟、赵盼
Python 玩转数学问题——轻松学习 NumPy、SciPy 和 matplotlib	张骞
Pandas 通关实战	黄福星
深入浅出 Power Query M 语言	黄福星
FFmpeg 入门详解——音视频原理及应用	梅会东
云原生开发实践	高尚衡